ALBINISM IN AFRICA

ALBINISM IN AFRICA

HISTORICAL, GEOGRAPHIC, MEDICAL, GENETIC, AND PSYCHOSOCIAL ASPECTS

Edited by

JENNIFER KROMBERG
University of the Witwatersrand and National Health Laboratory Service, Johannesburg, South Africa

PRASHIELA MANGA
New York University School of Medicine, New York, NY, United States; University of the Witwatersrand, Johannesburg, South Africa

Academic Press is an imprint of Elsevier
125 London Wall, London EC2Y 5AS, United Kingdom
525 B Street, Suite 1800, San Diego, CA 92101-4495, United States
50 Hampshire Street, 5th Floor, Cambridge, MA 02139, United States
The Boulevard, Langford Lane, Kidlington, Oxford OX5 1GB, United Kingdom

Notices
Knowledge and best practice in this field are constantly changing. As new research and experience broaden our
understanding, changes in research methods, professional practices, or medical treatment may become necessary.

Practitioners and researchers must always rely on their own experience and knowledge in evaluating and using
any information, methods, compounds, or experiments described herein. In using such information or methods
they should be mindful of their own safety and the safety of others, including parties for whom they have a
professional responsibility.

To the fullest extent of the law, neither the Publisher nor the authors, contributors, or editors, assume any liability
for any injury and/or damage to persons or property as a matter of products liability, negligence or otherwise, or
from any use or operation of any methods, products, instructions, or ideas contained in the material herein.

Library of Congress Cataloging-in-Publication Data
A catalog record for this book is available from the Library of Congress

British Library Cataloguing-in-Publication Data
A catalogue record for this book is available from the British Library

ISBN: 978-0-12-813316-3

For information on all Academic Press publications visit our website at
https://www.elsevier.com/books-and-journals

www.elsevier.com • www.bookaid.org

Publisher: Mica H. Haley
Acquisition Editor: Peter B. Linsley
Editorial Project Manager: Timothy Bennett
Production Project Manager: Mohanapriyan Rajendran
Designer: Mark Rogers

Typeset by TNQ Books and Journals

Cover image. The editors would like to thank adjunct Professor Rosemary Crouch (University of the Witwatersrand,
South Africa) for her original idea for the cover image. The people of Africa and their many different skin colors
(including the pale colors of those with albinism) are represented by the many black, brown, cream and white dots
on the map of Africa.

Dedication

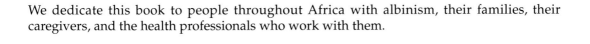

We dedicate this book to people throughout Africa with albinism, their families, their caregivers, and the health professionals who work with them.

No one is born hating another person because of the color of his skin, or his background, or his religion. People must learn to hate, and if they can learn to hate, they can be taught to love, for love comes more naturally to the human heart than its opposite.

Nelson Mandela
Long walk to freedom: the autobiography of Nelson Mandela.
Author: Nelson Mandela
Publisher: Boston: Little, Brown, ©1994.

Contents

14. A Personal Perspective: Living With Albinism

NOMASONTO G. MAZIBUKO AND
JENNIFER G.R. KROMBERG

15. Summary and Conclusion

JENNIFER G.R. KROMBERG AND PRASHIELA MANGA

List of Contributors

Jon Beale   Standing Voice, London, United Kingdom

Sam Clarke   Standing Voice, London, United Kingdom

Sian Hartshorne   Mediderm, Plettenberg Bay, South Africa

Rebecca L. Kammer   Standing Voice, London, United Kingdom; Kammer Consulting, Anaheim, CA, United States

Robyn Kerr   University of the Witwatersrand and National Health Laboratory Service, Johannesburg, South Africa

Jennifer G.R. Kromberg   University of the Witwatersrand and National Health Laboratory Service, Johannesburg, South Africa

Patricia M. Lund   Coventry University, Coventry, United Kingdom

Prashiela Manga   New York University School of Medicine, New York, NY, United States; University of the Witwatersrand, Johannesburg, South Africa

Nomasonto G. Mazibuko   Albinism Society of South Africa, Johannesburg, South Africa

Mark Roberts   African Institute for Mathematical Sciences (AIMS), Bagamoyo, Tanzania; University of Surrey, Guildford, United Kingdom

Susan E.I. Williams   University of the Witwatersrand, Johannesburg, South Africa

Foreword

Albinism is one of the classic genetic conditions. Recognized since antiquity and one of the very first human disorders shown to follow Mendelian inheritance, it has been the subject of detailed genetic studies for more than a century. In recent years, thanks to molecular genetic analysis, it has become possible to separate its different forms definitively and to recognize carriers of the condition reliably.

Albinism occurs worldwide, and it has long been recognized that its frequency varies considerably between countries, with a notably high frequency in several Amerindian populations and many parts of sub-Saharan Africa. The reasons behind these differences remain largely unclear, despite much study and speculation.

In Northern countries albinism is often thought of as a mild condition, with few medical problems apart from its effects on vision. This is far from being the case, though, in tropical countries, notably in Africa, where exposure to the intense sunlight results in much morbidity and mortality from skin cancers, though this can fortunately now be ameliorated by a combination of careful preventive measures and education.

This book is the first to bring together the various strands of our knowledge of albinism in the light of our new understanding of its molecular basis; the authors have themselves not only made valuable original contributions to the genetics and epidemiology of the disorder but also take a wide ranging and holistic approach to the different aspects. The book also covers the various clinical aspects in chapters by expert specialists in these areas.

An even more important aspect that is fully covered is that of the social effects of albinism, in particular, the discrimination, and stigmatization that it produces, as seen from the viewpoint of the affected individual as well as the professional. Albinism must surely be one of the very few genetic disorders where those affected are at risk of murder, not just lesser forms of discrimination. This situation demands active support from us all to end it permanently.

Finally, it is highly appropriate that this book is written from Africa, where the authors live and/or work, and that it is largely based on experience and studies gained from African patients and populations. The knowledge, both practical and theoretical, brought together here, will benefit not only workers across Africa but also others throughout the world trying to help in different ways those many people affected by and involved with albinism; we must be grateful to the authors and their colleagues for their most valuable contribution to this endeavor.

Peter Harper, Cardiff, November 2016

Preface

This book originated out of many years of study on oculocutaneous albinism in South Africa and her neighboring countries, namely Botswana, Lesotho, and Swaziland. These studies culminated in a PhD thesis entitled "A genetic and psychosocial study of albinism in Southern Africa" (editor JK, 1985). Research on the subject started in 1971 when the focus was the large Black population in Soweto, a suburb on the southwestern periphery of Johannesburg with nearly a million inhabitants at the time. The people from this area presented a fertile field for research as very little was known about albinism, let alone about the many other less recognizable genetic disorders occurring in the local Black population. However, people had observed that albinism appeared to be quite "common" in Black ethnic groups and so the first aim was to undertake a prevalence study. Thereafter, the study was extended in many directions, including the clinical, psychosocial, cultural, and later the molecular genetics, and many articles on these and other aspects of the condition were published.

Studies continued into the 1990s but at that stage were mostly laboratory based. The first molecular studies were aimed at mapping the gene for the common southern African form of albinism and then identifying the causative mutations. Once this had been achieved, attention shifted to the rarer forms of albinism, brown and rufous. Both forms were mapped and pathogenic mutations identified as reported in a PhD thesis entitled "Identification and molecular characterization of the genes for brown and rufous oculocutaneous albinism in southern Africa" (editor PM, 1997).

About 45 years after the initiation of the project, we decided that the extensive knowledge of the editors and that accumulated in the Division of Human Genetics at the University of the Witwatersrand, Johannesburg, would form a strong basis from which to write a book on albinism in Africa. Most of the research findings had been published in the form of articles and dissertations or theses for higher degrees, but some of the work remained unpublished, and documenting an overview of the cumulated research was deemed worthwhile. Furthermore, it presented an opportunity to share the extensive experience of the book editors, staff members of the Division, and local medical specialists in ophthalmology and dermatology working in the field. To achieve a wider view of the situation across Africa, various experts working elsewhere in Africa were also approached and agreed to contribute chapters to the book. Then one of the editors (JK) attended the *First International Workshop on Albinism* held in the Cameroon in 2015, where she met delegates from many different Africa countries and learned a great deal from their presentations. Similarly, in 2016, the Centre for Human Rights at the University of Pretoria held a workshop on "Advancing the rights of persons with albinism." JK attended and met lawyers for human rights and a range of concerned experts from all over Africa, as well as the Independent Expert on Albinism at the United Nations, and various Disability Rights representatives, and expanded her perspective.

In this book, we hope to provide a review of not only current knowledge on albinism but also the historical background of this knowledge. To pursue this aim, several old references have been consulted, some now out of print, others published in defunct medical journals, and a few unpublished records of medical personnel working with affected people. Furthermore, because the research in the field in Africa is limited, the chapter authors often had to draw on a small pool of articles, which were generally focused on a few countries where research had been carried out. This was not representative of most of the countries in Africa, all of whom have a significant number of people with albinism. Nevertheless, gathering the information and knowledge from the available sources into a single book would assist in expanding the general knowledge base on albinism, alerting health professionals and others concerned to the challenges, and eventually reducing the barriers to provision of care for affected people.

A concern when addressing a condition that impacts oculocutaneous pigmentation is the broader context of skin color–based discrimination. We have therefore made an active effort, in line with researchers in the biological sciences, to phase out the use of prejudicial terminology. For the purpose of this publication, we have followed the guidelines on ethnicity and race definitions from the National Institutes of Health, United States. Attempts to foster tolerance will be supported by disseminating what is known about skin pigmentation and its role in the health and well-being of every individual. Thus we sought to summarize in this book what is currently known about the biology of human pigmentation.

Because of JK's background in the social sciences, there is an emphasis on social issues that have neither been researched nor covered very thoroughly in the worldwide literature on albinism, to date. Consequently, we hope that the book, which covers many different aspects of albinism, will be of practical use to professionals in the medical and general healthcare fields, including those working in psychosocial areas, those providing community services, educationalists, and lawyers for human rights, as well as to members of albinism societies and their advocates, who work with affected people and would like more information on their condition.

People with albinism in Africa are at risk of rejection by society, discrimination on many levels, and stigmatization from birth onward. Some have lost their lives to superstitious beliefs and have been the victims of harmful practices triggered by widespread myths in specific communities. Ignorance and superstition in these communities contribute significantly to the way in which people with albinism are treated. The editors are hopeful that by publishing a book of this nature, accurate information about albinism may be broadly disseminated. Such widespread education on the condition should lead to benefits for the people with albinism living in Africa, foster better community acceptance, and improve their quality of life while expanding hopes for a better future.

Jennifer G.R. Kromberg
University of the Witwatersrand
and National Health Laboratory Service,
Johannesburg, South Africa

Prashiela Manga
New York University School of Medicine,
New York, NY, United States;
University of the Witwatersrand,
Johannesburg, South Africa

Acknowledgments

Many colleagues have contributed in many ways to this book. The first of them is JK's (Editor, Jennifer Kromberg) long-term research assistant, Sr. Esther Moipone Zwane (born 1953, died 2008); without her assistance, over about 20 years, many of the studies undertaken in Southern Africa would have been impossible. She not only acted as a translator, but also often as a navigator, as we set out into the bush to look for rural families and, due to her warm personality she was able to promote greater than expected involvement from those who participated in our studies. Secondly, the Head of the Department of Human Genetics at the University of the Witwatersrand, Professor Trefor Jenkins, was very supportive of the research and raised all the necessary funding so that the long-term project could develop, expand, and continue over the years.

Thanks are also due to the present head of the division, Professor Amanda Krause, who has not only obtained honorary positions in the division for the editors (JK, now retired, and PM (Editor, Prashiela Manga), now based at New York University School of Medicine) but also provided office space in her department for JK, supported the idea of the writing of the book, and has read through and commented on many of the chapters. The National Health Laboratory Service (made available office space, a computer, telephone, and an essential parking space) and the University of the Witwatersrand (appointed JK as a Retiree Mellon Mentor for 3 years and then provided a 2-year contract as an honorary professor) are also acknowledged. Further thanks are due to the University for allowing JK and PM to publish tables and other material from their PhD theses, in this book. We are grateful to Professor Michele Ramsay for her encouragement and her long-term contributions to the field. She has supervised many molecular biology research projects (including PM's PhD project) and continues to pursue research on albinism and to advocate on behalf of the albinism community in South Africa. Collaboration on this book began when PM was a short-term faculty exchange fellow at the University of the Witwatersrand's Sydney Brenner Institute for Molecular Biology. The visit was funded by a Carnegie African Diaspora Fellowship awarded to PM and hosted by Professor Ramsay.

Other colleagues of JK's, too many to name, made constructive suggestions and helped on this project, mainly by reading and commenting on chapters of the book, raising points they felt were neglected, bringing relevant references to our attention, modifying some of the illustrations, taking some of the photographs (Tasha Wainstein), and providing administrative assistance (Janet Robbie). As a result the content was enriched, topics were broadened, and the whole was improved. We thank PM's colleagues at New York University School of Medicine including Seth J. Orlow, MD, PhD, who is a mentor and supportive Department Chair; Nada Elbuluk, MD, who offered advice and comments; Ms. Genevieve Torres, Ms. Martha Vega, Ms. Elise Kelman and Mr. Sean Hettenbach who provide unfailing technical and administrative help.

Lastly, our families have been very supportive. JK's brother, Chris, has read and commented on most of the manuscript, her daughter, Jacqui, listened patiently and solved computer problems, when necessary, and her son, Jonathan, living in Australia, maintained an active interest and asked thought-provoking questions. PM's family has also been a constant source of encouragement and strength. Special thanks are due to her mother Dai and sister Sharmila whose home is a welcoming base during visits to South Africa.

Finally, the book could not have been written without the participation of many people with albinism, and their families, in our research. Their ready and willing involvement in the numerous different projects we undertook over many years has been remarkable. We dedicate this book to them.

Jennifer G.R. Kromberg
Prashiela Manga
Johannesburg and New York

Introduction and Historical Background

Jennifer G.R. Kromberg
University of the Witwatersrand and National Health Laboratory Service,
Johannesburg, South Africa

I. INTRODUCTION

Oculocutaneous albinism (OCA) is an autosomal recessive, inherited disorder that is associated with significantly reduced or absent pigmentation in the skin, hair, and eyes. It is a condition that is found in every population worldwide and encountered by health professionals and clinicians in many different specialties, fields, and situations. It has been observed and documented for at least two millennia in Africa and has also been reported widely around the world. Presently, research scientists, such as molecular geneticists, medical scientists, biologists, biochemists, and anatomical pathologists, as well as researchers in the humanities, social sciences, and other disciplines, are involved in investigating the condition on many different levels.

Albinism in Africa
http://dx.doi.org/10.1016/B978-0-12-813316-3.00001-5

1

The general aim of this book is to cover as many of the topics and issues relevant to the common forms of OCA found in Africa as possible, from the perspective of clinicians, epidemiologists, dermatologists, ophthalmologists, medical geneticists, genetic counselors, scientists, molecular geneticists, health professionals, psychologists, sociologists, social and community workers, and anthropologists. Each topic will be discussed comprehensively, with a focus on Africa, using the available published literature, as well as the experience and expertise available in South Africa, where the condition is common and 1 in 4000 black individuals is affected (Kromberg and Jenkins, 1982). The rarer forms of albinism, such as ocular albinism (OA), which affects the eyes only, and the types associated with rare syndromes, such as Hermansky–Pudlak, Chediak–Higashi (Witkop et al., 1983), and Griscelli syndrome (Menasche et al., 2003), will be mentioned where necessary but otherwise will not be covered at length.

The book attempts to examine OCA from several different angles so as to be of use to clinicians and researchers, as well as to students of the condition and families affected by it. It also attempts to correct some of the misconceptions that surround the condition, especially the tragic one that has come to the fore recently and led to the murders of affected people for body parts, which are used as powerful medicines by traditional healers, in some countries in Central and Southern Africa.

To place albinism in context, this chapter will outline the relevant writings and research publications that have originated over many centuries and that illustrate how the knowledge on the condition slowly accumulated. In Chapters 2–4, the characteristics associated with the condition, its epidemiology and population genetics, in countries around the world, will be covered. The modern development of molecular biology, the genetics of albinism and the rapidly expanding expertise on these topics will be described in Chapter 5. The dermatological, ophthalmologic, and low-vision rehabilitation aspects will be discussed in detail in Chapters 6–8 respectively. Thereafter, because people with albinism are challenged with many psychosocial and cultural issues (including myths, superstitions, and the resulting stigmatization), several of these issues will be covered in Chapter 9. This discussion will be followed by two chapters, 10 and 11, on genetic counseling for the condition and on genetic testing, prenatal diagnosis, and the ethical and other issues associated with these procedures. Chapter 12 will deal with marginalization, disability and human rights issues, and community work undertaken to advance the rights of people with albinism; whereas in Chapter 13 preventive management, empowerment, advocacy, support services, support groups, and albinism associations will be discussed. The next chapter, Chapter 14, will present the personal perspective and life story of a person with albinism. Lastly, the concluding chapter, 15, will summarize all the key points discussed in the book, make suggestions for research directions, and consider possible future developments in the field, as well as the implications for improving the quality of life for persons with albinism in Africa.

The following sections of this chapter will cover definitions of OCA, clarification of some of the terms used throughout the book, and a broad view of the historical background of the condition, as understood from the available publications and the documented research carried out over past centuries. Because the condition has been observed frequently in Africa, many of the writings and much of the research undertaken on the African continent is presented. However, some research from countries on other continents is also covered (e.g., the extensive work undertaken by Witkop et al. (1983) on African Americans and American Indians and that of Froggatt (1960a) in Northern Ireland). In addition, the informative studies on some isolates, such as the San Blas Indians in Panama (Keeler, 1953), and Hopi Indians in Arizona, United States (Woolf and Grant, 1962), in which high rates of the condition have been found and who were subsequently investigated and reported on in some detail, are discussed.

Some possibly well-known information will be presented in this chapter, while several reports that have been published in rare and/or discontinued journals, as well as some that are little known or have not been published at all, will be included. Various sources have been used, some being ancient, some more modern, some extracted from archival material, and others from unpublished documents available to the writer.

Definitions

In 1971 Witkop defined albinism (somewhat imprecisely) as a "hereditary defect in the metabolism of melanin" resulting in a decrease or absence of this pigment in skin, hair, and eyes. However, in 1983 Witkop et al. recommended that use of the term albinism should be restricted to congenital hypomelanosis that affects only the eyes (OA) or involves both the eye and integument (OCA) and is accompanied by nystagmus, photophobia, and decreased visual acuity. They specified that the terminology used in classifying the different types of albinism should indicate the extent and nature of the pigmentary involvement, wherever possible.

In 1996 Oetting et al. (p. 330) offered a more precise definition: "Albinism represents a group of inherited abnormalities of melanin synthesis. It is characterized by a congenital reduction or absence of melanin pigment in association with specific developmental changes in the optic system resulting from the hypopigmentation."

In the heterogeneous group of disorders called albinism there are two main categories: the defect impacting either the entire melanocyte system, resulting in broad phenotypic effects (e.g., OCA), or more prominently affecting melanocytes at a specific site with localized phenotypic effects (e.g., OA). The former is generally inherited as an autosomal recessive condition, whereas the latter is usually inherited as an X-linked recessive, although there have been some reports of autosomal recessive inheritance. The present book is concerned only with the former (OCA) and more common category (Fig. 1.1 shows an affected individual and her family).

The term albinism includes a number of inherited disorders that affect pigmentation. The condition was originally thought to be caused by different mutations at one genetic locus, but it is now understood that the genetics of albinism is complex and that there is genetic locus heterogeneity. The nomenclature and classification of the condition are difficult and currently being clarified as new molecular genetic information becomes available. A detailed description of the various forms of albinism is provided in Chapter 5. However, the focus of the book will be on the three common types of albinism (see Oetting and King, 1999), which include OCA1 (previously called tyrosinase-negative, based on the lack of tyrosinase enzyme activity), OCA2 (previously called tyrosinase-positive) and its subtype brown OCA (also called brown albinism or OCA2B), and OCA3 (also called rufous or red albinism or ROCA).

The terms "person with albinism" and "people with albinism" (sometimes abbreviated to PWA) will usually be used in the text because these are the more politically correct (and

FIGURE 1.1 An individual with OCA2 and her family. The affected individual has developed a cancerous lesion on her face. *Photograph courtesy of Dr. P. Manga, New York University School of Medicine, USA.*

accepted by those with the condition) and less stigmatized terms, at the time of writing, compared with the other terms used in the past. However, the word "albino" will also be used occasionally, in certain contexts and where the authors of articles under discussion used the term (for a discussion of this issue see the National Organization for Albinism and Hypopigmentation, or NOAH, website). Nomenclature for the indigenous people of Africa is also somewhat problematic. Previously, some authors have used the words Negro (or negro), Negress, or Negroid, but these terms are no longer appropriate nor acceptable in Africa; presently the term Black (or black) is deemed preferable. However, in this book, where necessary, the term used by the author of the article under discussion may be used, but in general the term Black will be applied preferentially. The word White is used preferentially, but Caucasian or Caucasoid are applied where they were used by the original authors, to refer to people of European extraction.

II. HISTORICAL BACKGROUND

According to Sorsby (1958) it is possible that the first written description of a person with albinism appeared in the Pseudepigrapha in the book of Enoch. The birth of Noah is described to the prophet Enoch (Noah's great grandfather) in this extra-Biblical book of Methuselah (Enoch's son and Noah's grandfather). Methuselah, referring to Noah, stated that: "His color is whiter than snow, he is redder than a rose, the hair of his head is whiter than wool, his eyes are the rays of the sun and when he opened them he illuminated the whole house." Noah's father, Lamech, according to the story, questioned the faithfulness of his wife, after seeing the baby. It has been suggested that Lamech and his wife were first cousins, which is possibly why a double dose of a recessive gene was inherited, causing the condition in their child.

Another very early account is that of Pliny who wrote in the first century AD. He quotes Isigoneus of Nicaea, who stated that in Albania there were men whose eyes were of a seagreen color, who had white hair from childhood, and who could see better at night than in the day (Plinius Secundus, 1942).

The term albino is derived from the Latin word "albus" meaning white. It was first used in about 1600 by Portuguese explorer and historian Balthazar Tellez to describe the white Negroes he had seen in Africa (Pearson et al., 1913).

de Maupertuis, a French naturalist who lived in the 18th century studied a few common traits in man, such as extra digits (polydactyly) and absence of skin and hair pigmentation (albinism) (Emery, 1979). From examining pedigrees, he first showed that these two conditions were inherited in different ways.

Albinism was one of the first inherited disorders to be investigated from an epidemiology perspective using population ascertainment (Froggatt, 1960a). This investigation was undertaken by Raseri in 1879 in Italy and he estimated that the condition occurred in 1 in 29,000 Italians (Pearson et al., 1913). Garrod (1908) proposed that albinism was the result of an inborn error of metabolism due to the lack of an "intracellular enzyme," which synthesized melanin. In 1910, Davenport and Davenport, having analyzed many pedigrees of families with an affected member, concluded that the condition was inherited as a simple Mendelian recessive. It appeared in 25% of the offspring from the marriage of carriers but not in the children of people with albinism with normally pigmented spouses.

In their noteworthy "Monograph on Albinism in Man," Pearson et al. (1913) presented the results of a worldwide survey, undertaken between 1911 and 1913 and published in 11 volumes (3 in 1911 and 8 in 1913), into the prevalence and nature of the condition. The monograph also included an extensive historical and bibliographical background and a discussion on the characteristics and heredity of albinism, as well as some of the psychosocial and cultural issues associated with the disorder. The geographical distribution of affected persons in the light-skinned, black-skinned, and other races (terms used by Pearson et al., 1913) was described with the support of research findings and many photographs.

Witkop et al. (1983), working in the United States in the last few decades of the 20th century, contributed much to the understanding of albinism and defined, delineated, and described several subgroups, as well as, initially, clarifying the terminology. Later, the molecular basis of the pigment genes associated with albinism and its different subtypes were investigated (Oetting and King, 1999) and these molecular studies continue at the present time.

Studies on Albinism in Africa

Because people with albinism are so visually different from their peer group in heavily pigmented populations, individuals have been described, commented on, examined, and investigated over the centuries by explorers, missionaries, anthropologists, travelers, and medical practitioners working in many different countries in Africa (see Fig. 1.2).

This review of the literature on albinism in Africa is somewhat selective and focuses on those more comprehensive articles, which have bearing on the topics covered in this book. Reports on biochemical studies, with a few exceptions, have generally been omitted here, as has the discussion of the legend of a white native race, which is frequently mentioned in the oldest writings (Pearson et al., 1913) but the existence of which seems unlikely (Froggatt, 1960b). Each paper is critically reviewed, summarized, and presented in chronological order, arbitrarily divided into five sections: the early observational and descriptive studies including the comprehensive monograph of Pearson et al. (1913), which covers reports from all over the world, sent to him between 1911 and 1913, as well as many from Africa (0–1913);

FIGURE 1.2   Political map of Africa. *By Peter Hermes Furian, from https://stock.adobe.com/. File #:117993060.*

the slightly more detailed and scientific studies starting with the work of Stannus (1913) in Malawi (1913–51); the more systematic data collecting research initiated by the study of Barnicot (1952) in Nigeria (1952–72); the more sophisticated studies of 1973–84; and those of more recent years (1985–date). Some of these papers were found in obscure and/or discontinued journals (such as the Transvaal Medical Journal published in South Africa from 1905 to 1913). Much information in the first section (0–1913) was obtained from Pearson et al. (1913), but some of the references used by them could not be checked in the original sources because, in some cases, they were centuries old and unavailable. The early observations and descriptions of people with albinism, mentioned by them, were generally not based on scientific study but on the possibly biased and exaggerated reports of travelers and explorers and their local informers. Also, the studies reported here, with few exceptions (such as Okoro, 1975), have been conducted by European workers visiting Africa and have generally neither been as extensive nor in such depth as some of the European studies (e.g., the study of Froggatt (1960a) in Northern Ireland). Nevertheless, this review presents a broad background, which is both necessary to obtain an overall view of the subject and relevant to the topics, which will be covered.

0–1913

One of the earliest references to a group of African people who probably had albinism was that of Pliny, writing in the first century AD (Plinius Secundus, 1942). He described the Leucaethiopes (or "white" Ethiopians) living in North Africa. In 1666 Vossius described the ancient state of knowledge concerning people with albinism, and according to a lengthy footnote in Latin in the monograph by Pearson et al. (1913) he stated that not only were affected people or Leucaethiopes found in the lands around the "fountains" (probably the sources) of the Nile but also in the midlands of Africa. He described them as being so pale that one could see them from far away and mistake them for Belgians or Germans, with their blue eyes and blond or red hair. He also observed that they had black parents and that they avoided people, as if their condition was contagious. They disliked the sun and used ointment and fat to cover their skin and maintain good health. These Leucoaethiopes were also called "albini" and, although they were nearly blind, they were said to be very strong.

In the same century Ludof (in 1691, according to Pearson et al., 1913) referred to people with albinism in the court of the King of Loango and attributed the first use of the word "albino" to the Portuguese explorer and historian Balthazar Tellez who traveled through Africa in the early years of that century. In 1688 De La Croix stated that although affected people were attendants at the King's court, they were regarded as monsters. In support of this view, a century later Equiani, an African, is quoted as stating "I remember while in Africa to have seen three Negro children who were tawny and another quite white, who were universally regarded by myself and the natives in general as deformed" (Pearson et al., 1913). Bowdich (in 1819, according to Pearson et al., 1913) found at the court of Ashanti (in present-day Ghana) nearly 100 thin diseased Negroes of different colors from red to copper and white.

In the mid-19th century, David Livingstone, missionary, explorer, and doctor, observed people with the condition during his travels in Africa. He reported, in detail, the case of an affected boy brought to him in present-day Botswana (previously Bechuanaland) and stated that "he was remarkably intelligent for his age. The pupil of his eye was of a pink color and the eye itself unsteady in vision. The hair, or rather wool, was yellow and the features were those common among the Bechuanas" (Livingstone, 1857, p. 493). The mother of this boy clung to him for many years, against the father's orders, and as a result had to live separately with her son and apart from the father. She, eventually tired of this arrangement, murdered the boy and returned to her husband. This appeared to be a common practice in Botswana and Angola, and Livingstone commented that the authorities took no action against the woman. He did not see as many people with albinism further North in Africa but had read, in accounts published by the Portuguese, that they were favored as doctors by certain chiefs. Livingstone also noted that the bodies of affected people he met always "blistered" following sun exposure and that their skin was more "tender" than that of black people. One affected woman wished to be made black, so Livingstone gave her nitrate of silver, which she took internally, but he stated that it did not produce "its usual effect."

In 1910 observations were made during a study on pigmentation and cancer by Watkins-Pitchford, a pathologist in South Africa, who in 1913 became the first Director of the South African Institute for Medical Research in Johannesburg. He commented that a person with albinism should have the greatest risk of all to skin cancer because this susceptibility appeared to increase as pigmentation decreased. He added that rodent ulcers, for example, were rarely seen in the "natives" of Natal and that on diligent enquiry only one such case was reported to him and that was in a person with albinism.

Early in the 20th century, Pearson et al. (1913) attempted to determine the geographical distribution of albinism in Africa. On the basis of published articles and on reports they received from numerous cooperating medical practitioners, travelers, and other correspondents, they compiled a list of types of albinism and suggested that these might be more easily observed in the black people of Africa than in lighter skinned people. Their rough classification was as follows: (1) the xanthous (derived from the Greek word for yellow) negro; (2) the complete albino with possible subclasses; (3) the spotted freckled negro; (4) the yellow-eyed, white-skinned negro; (5) the blue-eyed, white-skinned negro with photophobia and nystagmus but not red pupils; and (6) the partial albino or piebald negro. Pearson et al. (1913) established that people with albinism occurred throughout Africa and that they had been mentioned and described by travelers for centuries. They were said by some to marry and bear normal children (unless they married an affected person, in which case all the offspring were said to be affected too) and by others to have no posterity. In the Sudan, according to Raffenel (Pearson et al., 1913) they lived without having to work, asked for whatever they needed and were never refused. Among the Foulah people of the Sudan everyone gave freely to people with albinism to win favor from heaven. The Foulahs were said to believe that God did not wish affected people to work like other men, so he gave them a white skin color. However, in the Foulah language they were called "danedio," which meant white, not "toulako," which meant White man. In the western Sudan, reports asserted that children born from two people with albinism were black.

In Guinea in 1830, Dr. T. Winterbottom observed that the natives considered albinism a great deformity and a misfortune for the family, but that none of them appeared to be intellectually disabled (Pearson et al., 1913). Reports of confusion between mulattos and people with albinism, and an attempt by affected people to pass for White in Sierra Leone were also received. Albinism was noted to occur in Liberia, the Ivory Coast, Angola, and also in Madagascar. In Ashanti (now in Ghana) the kings, who kept affected women at court, occasionally offered them as wives to passing travelers. Persons with albinism were believed to be under special divine protection. However, sometimes they were sent as exhibits to international exhibitions in London and elsewhere.

Dr. Henry Strachan reported, in a letter from Nigeria, to Pearson et al. (1913) that albinism occurred there with its usual variation and problems. Earliest reports from the Congo stated that affected children, called "dondos," were always presented to the King in Loango, they were then educated as sorcerers, and nobody dared to offend them. They were sometimes described as incapable of coitus and, in other reports, as not being permitted to multiply. They were generally regarded as somewhat sacred and inviolate, and no man might strike them. Further north, affected people in Gabon were regarded as unlucky and frequently killed, unless missionaries intervened. In Angola, however, they appeared to be treated with great respect and fear. From Southern Africa, reports were received on people with albinism in the Transkei (now part of the Eastern Cape province, South Africa), in Basotholand (now

Lesotho), and in Zululand (now part of the Kwazulu/Natal province of South Africa), where red-skinned people were also described; infanticide of affected children was reported to occur quite frequently, and men queried the faithfulness of their wives when the baby was affected. In Basotholand, however, affected women commanded greater bride wealth (known locally as "lobola") than normally pigmented women (Pearson et al., 1913).

According to Pearson et al. (1913) the Black population was said, sometimes, to consider people with albinism as being the product of illicit relations between a Black woman and a gorilla or baboon, and although no authority was given for this legend, they wrote that it appeared repeatedly in writings on the subject. A similar legend has been reported in Malaysia with the orangutan as the non-human parent and in Russia, with the monkey being involved. Furthermore, the word "monkey" ("nkau" in one local dialect) is often used in South Africa with regard to people with albinism (Kromberg and Jenkins, 1984).

Dr. G.A. Turner, medical officer to the Witwatersrand Native Labour Association in South Africa, gave Pearson and his colleague, Dr. Usher, much information in letters and in a report, which was initially sent to the University of Aberdeen, Scotland. This information does not appear to have been published by Dr. Turner, but Pearson et al. (1913) state that Turner reported on 7000 males who he examined for employment on the mines. Among these he found 11 with xanthism, they all had yellowish brown or reddish skin, brown eyes, most had brown hair (although several had black hair and one had light brown reddish hair on the temples), and one had nystagmus. Turner only saw one man with albinism in this series of men and commented that infanticide was still commonly practiced and that affected people were unlikely to seek work on the mines, if they could exist elsewhere. The family histories he collected on the xanthic individuals suggested that there could be a connection between xanthism and albinism as the two conditions were occasionally found together in one family. Turner wrote that he had seen people with albinism in all the Bantu-speaking tribes of South Africa but not in the Bushmen (San), Hottentots (Khoi), or Griqua. Affected people seemed to get on well in their communities and yet young men were afraid to take an affected girl as a wife and fathers would not "sell" their daughters to a man with albinism.

Sir Harry Johnstone, who worked in British Central Africa, reported that albinism was not uncommon there (Pearson et al., 1913). He added that the type where the skin and iris were quite light yellowish brown was much admired especially in women, for there was a tendency to admire a lighter rather than a darker skin, and the wives of chiefs often had this pale yellow brown coloring.

People with albinism were also reported from Kenya and Uganda, where they were apparently looked on as curiosities and kept in establishments of kings and great chiefs. In Somaliland, a report on the ethnographic characteristics of the people included a comment that there was a tendency to admire and accordingly to marry into the lighter shades of skin color (Paulitschke, referred to by Pearson et al., 1913).

From their review of the statements and descriptions of the 213 people with albinism reported to them from Africa, Pearson et al. (1913) concluded that the proportion of "complete" (those with no pigmentation) to "incomplete" (those with some pigmentation) albinos was 3:1 and that where albinism was prevalent, the normally pigmented negro generally had a lighter skin color than other Africans.

1913–51

Later in 1913, the classification of albinism types suggested by Pearson et al. (1913), however, led to some confusion because the interpretation of the color of the eye and the presence of a red reflex appeared to be very subjective, causing a lack of standardization in the reported details. Stannus (1913), who wrote a doctoral thesis on the subject for the University of London, therefore, on the basis of his survey in Nyasaland (now Malawi), attempted to be more precise and suggested an alternative classification:

1. Complete: white skin, straw-colored hair, liquid blue iris with red reflex through pupil, when examined with artificial light;
   a. Spotted: as in (1) but showing spotty pigmentation of the skin;
2. Partial incomplete: as in (1) but the iris light yellow or hazel, photophobia, and nystagmus not always present;
3. Xanthism: reddish brown hair and skin, light brown irides;
4. Incomplete: skin light brown, not red, hair yellow, not brown, irides hazel;
5. Partial albinism: including piebalds and spotlings.

Stannus proceeded to describe his subjects in great detail and to classify them based on his clinical findings. He stated that in all cases the pupil of the eye appeared black in ordinary daylight and that although there was no obvious pigment in many cases, yet on microscopic examination some pigment might be demonstrated in hair and eyes. He found that several people with albinism appeared to be intermediate between two groups. In some families he reported a parent or relative of lighter than average skin color. In another family he reported the custom of killing an affected child at birth. He described the color variations in affected siblings and discussed a family where the older child had xanthism (there appears to be some confusion in the use of this term, which was initially used to describe affected people who had yellowish skin and hair, but appears to be used here by Stannus to describe those with reddish skin) with reddish skin, hazel irides, no nystagmus, and light brown hair, whereas the younger brother had lighter skin, light hazel eyes, nystagmus, and light straw-colored hair. He concluded that albinism was very common in Nyasaland (although he did not attempt to calculate a prevalence rate); it was a "family affliction" and often directly hereditary; there was phenotypic variability in the same family; the degree might vary with age; pigmented spots were common; lanugo body hair was generally associated with the condition; and maldevelopment of the teeth might be found.

McCrackin (1937), who worked in the French Cameroon, reported two cases of twins, one in which both had albinism and the other in which one twin had albinism and the other was normally pigmented. He also commented on local superstitions stating that albinism was said to be caused by maternal impression, or by the mother conceiving the baby by a water spirit, as a result of adultery, or the sins of one or both parents, or stepping over rather than walking around the legs of an affected person while pregnant. Heredity was also recognized and said to be the principal factor held strongly by local tribes. McCrackin observed, what he called, "semialbinos," with brown hair, reddish skin, pale blue eyes, and no photophobia or nystagmus, and added that the apparent mental and physical slowness was most likely a "psychological effect of the albino being regarded as inferior and ridiculous, from the time he was born, and to his difficulty in using his eyes for gaining knowledge." He reported on a case of a father who, as recently as 1924, intended killing his children if they had albinism.

However, he found he could not go through with the murder of his first affected child, so well-meaning friends poisoned the child who subsequently died.

A report on "partial" albinism and nystagmus in two Nigerian families was made in 1944 by Loewenthal. He defined his term "partial albinism" as referring to persons "in whose skin the pigment content is universally reduced though not completely absent." He stated that Turner (the South Africa Medical Practitioner referred to by Pearson et al., 1913) was the originator of the word "xanthism." Loewenthal drew a pedigree of one of Turner's affected families (from apparently unpublished information) showing red-skinned, yellow-skinned, and normally pigmented persons, as well as persons with albinism, in one family. Loewenthal himself had come across a red-skinned person with bright copper-colored skin, lentigines, ginger hair, reddish brown irides, and nystagmus and a yellow-skinned person with clay-colored skin, lentigines, black and gold hairs, hazel irides, and nystagmus. He determined that because of the presence of nystagmus these individuals should be considered as having albinism.

Vallois (1950), an anthropologist, visited Cameroon and on hearing that albinism was quite common there, he cursorily investigated the matter. He found four affected people in the village of Badjoui, which had a population of 10,000 individuals, and in another area he found a further four among 13,000 people. He observed that males appeared to be affected more often than females and that affected people were not allowed to marry. One person with albinism he came across, however, was a sorcerer, who had given himself the right to marry and he had normally pigmented children. Vallois also described and measured skin color in black infants and found that many were very light at birth and only darkened at about 3 days of age.

1952–72

Barnicot (1952) appears to have been the first person to attempt a comprehensive study to estimate the frequency of albinism in Africa. He believed that previous studies had been based mainly on casual observation and, although scientists had claimed that albinism was more common in Negroid than in Caucasoid people, the available data were inadequate to prove the point. In Nigeria he found a frequency of one child with albinism in 2858 school children, which was high in comparison with European rates, which varied, he said, from 1 in 20,000 to 1 in 40,000. In a local Nigerian township, Barnicot estimated a frequency of one person with albinism in 5000 people. He noted: that all the affected people he saw could be classified as incomplete in Pearson's terminology because all had some pigment in the skin, hair, and especially in the irides; that their ages were not markedly divergent from those of the general population; and that there were was an excess of affected males over females (although the difference was not statistically significant).

On detailed examination of a sample of 29 people with albinism, Barnicot (1952) found that 25/29 had pigmented freckles, which were commonest on cheeks and forehead and tended to develop with age. On examination of a biopsy of one of these freckles and adjacent skin, melanoblasts were found to be abundant in both the lightly and heavily pigmented skin, but there was increased pigmentary activity in the heavily pigmented areas. Freckles were not a consistent finding in affected family members. The irides of his subjects were found, on average, to be darker than in European cases, but the darkest was light orange–brown in color, noticeably lighter than the lightest eyes in unaffected Africans. Irides, which were green or blue frequently, had orange-brown pigment at the pupillary margin or brown flecks.

A slightly higher frequency of blue eyes was observed in the youngest age group (1–10 years), suggesting that eyes darken with age, but no sex difference was observed. Nystagmus was obvious in 36/38 cases, the two unaffected subjects being 35 and 75 years old, respectively, and it seemed that nystagmus improved with age. Strabismus and other visual defects were not studied systematically. The hair color ranged from white tinged with yellow to yellowish brown, but there seemed to be little evidence of a correlation either between hair color and age or between hair and eye color.

Barnicot collected hair samples from a group of people with albinism and a group of their normally pigmented relatives for reflectance spectrophotometric studies. The samples from the two groups differed significantly at all wavelengths, the difference being greater at the red end of the spectrum. Barnicot also observed that individuals could have patchy hair of lighter and darker colors and on microscopic examination many (9/13) had alternate light and dark banding on a few of their hairs.

Information on skin cancer and the susceptibility of people with albinism was published by Cohen et al. (1952) who studied malignant disease and skin cancer in South Africa. They found that among "Bantu people" (black people who spoke a Bantu language), one in five individuals with skin cancer also had albinism, which was much higher than the rate of albinism in that population, and that younger age groups and, generally, exposed areas were most often affected. As prophylactic treatment, they suggested that adequate headgear, protective clothing, use of cosmetic ointments, and, perhaps, the growth of beards would afford some protection. They also emphasized that early treatment of solar keratoses would prevent the development of skin cancer.

During a subsequent study of people with albinism in Nigeria, Barnicot (1953) investigated and described 23 xanthous individuals. He observed that skin and hair color were not always closely correlated; for example, one subject had light yellowish red hair with normal skin color, whereas another had brownish-red hair and copper-colored skin. Only three individuals displayed freckling and eye color ranged from medium brown to as "dark as in the general population." Nystagmus was present in only 1 of 23 subjects. In a school survey, which he conducted, he found that red hair occurred in about 1 in 650 children. He commented that the name "xanthism" (a Greek word meaning yellow, the use of which he attributed to Pearson et al., 1913) was not well chosen because in most cases skin and hair were not yellow. He suggested that spectrophotometric analysis of hair and skin color was required.

Further comments were made on the skin of people with albinism by Schrire (1958). He mentioned the anecdotal evidence of a surgeon, in Johannesburg, who had much experience treating affected people and had said that they died early of skin cancer. The "epitheliomas" usually occurred on the shoulders and forearms (areas exposed to the sun and trauma) especially in the young who were often completely unclothed for many years.

Cancer researchers have played an important role in describing albinism and its associated problems in South Africa. Oettle (1963), in the course of his study on skin cancer in Africa, documented cases of albinism to estimate frequency in rural areas of the Transkei (now in the Eastern Cape Province on the East coast of South Africa). He ascertained 197 affected people and estimated a minimum rate of 1 in 3759. The sex ratio in this group was 0.89, with more females being affected; however, males were more likely to migrate to the cities where there were more opportunities for employment. There were also fewer males in the general population. He was unable to explain the high prevalence rate but proposed that if heterozygotes had a paler skin color they would have great selective value as marriage partners, as this was regarded as a mark of beauty and might therefore represent a selective advantage over the remaining population.

Burrell, who was originally engaged in a survey of esophageal cancer in the Northern provinces of South Africa, proceeded to collect further information on people with albinism in many districts but did not publish his findings. Rose (1973) had access to Burrell's register of affected people, compiled in 1964, and she reported that Burrell had found 458 cases in the Transkei. Rose examined 71 of these people in 41 families and on clinical examination alone stated that they all appeared to be of the tyrosinase-positive type (Witkop et al., 1970, had described two types of albinism in the United States, tyrosinase-negative and tyrosinase-positive), and most had "splotchy" freckles and blue to yellowish khaki-colored eyes. People with brown eyes, yellow hair, and red skin were brought to Dr. Rose, while she examined those with albinism, and she found that, when young, their skin was also sensitive to the sun. Some of these red-skinned people, who were called by the same local term as persons with albinism in the Transkei, were reported in Burrell's register as affected people. In a subsequent report, Rose (1974) made an unsubstantiated statement that all the red-skinned people in the sample of families she examined had relatives with albinism and most of those affected had a red-skinned relative (a similar situation in two families had been described by Loewenthal, 1944). She therefore concluded that the red-skinned people had a form of partial albinism. On her retirement in 1980 Dr. Rose handed over Burrell's material to Dr. J. Kromberg (the author), in the Human Genetics Division at the University of the Witwatersrand, Johannesburg.

Burrell's papers proved informative because they not only provided lists of people with albinism in many of the Transkei districts but also included the first seven pages of his preliminary report (which was apparently never completed). In these pages he discussed the ages of the subjects and the nature of the different skin colors and textures in relatives of people with albinism and others, as distinguished by the local Xhosa people. Burrell ascertained 458 affected people, and the population of the area, according to the national Census of 1960, was 1,621,200, which gave a prevalence rate of 1 in 3710 of the population. This figure confirms the later estimate of Oettle (1963) on his smaller sample. From his analysis of the ages of people with albinism, Burrell stated that their population pyramid was similar in pattern to that for the general population, so that there was no indication that affected people died prematurely. He added that only one case of skin cancer was detected, but many people with albinism suffered from solar dermatitis, especially in the winter when they sought the sun for warmth. He compiled pedigrees on 28 affected families and found that in 17 families dusty, light brown skin (which Burrell attributed to Hamitic vestiges) occurred in one or more grandparents, parents, or siblings. The classical signs of photophobia and nystagmus were recorded in less than half the affected people, but because these observations and records were made by poorly trained field assistants, probably without any equipment, this estimate might be unreliable. Eye color was assessed in a similar fashion and 13.2% of his subjects were reported as having pink/red eyes, 40.3% green eyes, 17.6% blue eyes, and 28.9% brown eyes. The people with albinism were often named "wise little person" because they tended to stay in the shadows close to the hut, to overhear conversations and be knowledgeable. They were certainly not "dull witted" but the older ones were considered to be recluses. Teachers, however, found them to be above average scholars. There were 248 females and 210 males in the sample.

On further investigation of his ideas on skin color in families with a member with albinism, Burrell compared the findings on the relatives of affected persons with those for their neighbors in whom there was no history of albinism. He classified subjects as having black, light brown (dusty or smooth), and ordinary skin and found that only 0.3% of relatives of affected people had black skin, whereas 4.6% of their neighbors had black skin. His method of classifying skin color was rather arbitrary, but nevertheless, as an exploratory study and in view of the reports of light skin color in the parents of children with albinism from other studies (Pearson et al., 1913), his findings are worth mentioning here.

Another attempt at estimating prevalence rates of albinism in South African Black population groups was made by Hitzeroth and Hofmeyer (1964). These two research workers collected their data from the passport-type photographs of African females presenting to the Registration Bureau for purposes of registering for identity documents. They found an average prevalence rate of 1 in 14,833 with figures ranging from 1 in 9060 in the Tswana, 1 in 9444 in the Swazi, 1 in 15,705 in the Zulu, 1 in 16,858 in the Xhosa, and 1 in 46,991 in the Tonga ethnic group. Although their method of ascertainment was somewhat inadequate because women for varying reasons do not all present for registration and judging skin and hair color from photographs could be quite unreliable, their figures (which represent minimum estimates) are useful for purposes of comparison with those found in a later study (Kromberg and Jenkins, 1982). When comparing the frequencies in the different ethnic groups by means of chi square analysis, Hitzeroth and Hofmeyer found that there was a significant difference between Zulu/Xhosa and the Tswana and between Xhosa and Swazi. They stated that there was no evidence to support the hypothesis that a differential mutation rate between ethnic groups might explain the varying frequencies. But they added that marriage customs might differ and these could influence the frequency of albinism. In several Sotho ethnic subgroups (e.g., Tswana) the custom was for a man to marry the daughter of his maternal uncle, whereas in the Nguni groups (e.g., Zulu, Xhosa and Swazi) exogamous marriages were the rule (Schapera, 1937). They were puzzled, however, that of the four groups with the highest frequencies of albinism (Tswana, South Sotho, Ndebele, and Swazi), one (the Swazi) was of Nguni origin. This matter was clarified later, and higher rates were shown to be associated with higher rates of consanguineous marriages, in a study by Kromberg and Jenkins (1982). Hitzeroth and Hofmeyer, however, suggested that pedigrees should be collected to estimate inbreeding and to determine the selective advantage under which people with albinism reproduce. Further, they recommended that the social status of affected people should be investigated to assess their reproductive chances and popularity (if they are considered to be of supernatural origin) or otherwise (if the condition is considered an abnormality) as a marriage partner.

1973–84

A small study conducted by Dogliotti in 1973 on the treatment of solar keratoses in people with albinism included probably the first assessment of tyrosinase enzyme activity status in African subjects. Hair bulbs from nine affected people were sent to Professor CJ Witkop in Minnesota, United States, for this purpose and the results suggested that three subjects had tyrosinase-negative (this result was subsequently proved to be unlikely, King et al., 1980) and six tyrosinase-positive albinism. This was the first attempt at classifying affected African people by biochemical means at a time when it was assumed that the absence (or malfunction) of tyrosinase could be the cause of the condition.

Probably the first systematic psychological evaluation of people with albinism in Africa was undertaken by Manganyi et al. (1974). They investigated intellectual maturity (since no validated reliable IQ testing instruments were available for intelligence testing in local Black people) and body image boundary differentiation in 28 affected South Africans and a carefully matched normally pigmented control group. Individuals with albinism demonstrated slightly more intellectual maturity than controls and had a more clearly defined, although still rather diffuse, body image boundary. Compared to the controls they appeared to have a negative self-evaluation, which tended to corroborate Burrell's description that many were recluses.

A series of Nigerian people with albinism was studied with a view to registering them early in life, counseling their families, and advising on methods of treatment and prevention of associated problems (Okoro, 1975). Okoro found that skin cancer was a major problem and consequently life expectancy was reduced. There appeared to be a higher prevalence in the South of Nigeria than in the North. However, in the South consanguineous marriages were favored to preserve the presumed superior qualities of the people living there. He also observed that the proportion of married affected women was very low, that children with albinism were often taunted, they dropped out of school because of both uncorrected refractive errors and discouragement, and later had difficulty in obtaining employment. Socially, life was very difficult for them.

Okoro's work was followed by three subsequent Nigerian studies. The first described skin cancer (particularly squamous carcinoma) in Ibo people with albinism (Witkop et al., 1977). The second investigated chromosome breaks and sister chromatid exchanges in 14 Nigerian people with albinism and found that there was no difference in the frequency of these breaks and exchanges compared with a small sample of normal controls (Cervenka et al., 1979). The third study involved the investigation and classification of 79 Nigerian people with albinism (King et al., 1980). Tyrosinase-positive albinism was found in 56 subjects (none was found to have tyrosinase-negative albinism) and the remaining 23 had brown albinism. The latter type of albinism had been described by Barnicot (1952), Pearson et al. (1913) (who had presented one early description of a person with brown albinism, provided by the explorer Sir Richard Burton and of other similarly affected people in a group of subjects with what was then called "Xanthism"), and Stannus (1913) who had classified them as a separate group of "incomplete albinos." The people with brown albinism, however, had not been characterized in detail until the study of King et al. (1980). It is noteworthy that no tyrosinase-negative individuals were found in this sample of Nigerians. This may have been by chance because only 79 subjects were examined. Nevertheless, the finding indicates that tyrosinase-negative albinism is either rare in Nigeria or that the sun's effects (in an equatorial region) and/or their eye problems are so severe in this type of albinism that affected individuals neither leave the shelter of their huts nor present themselves for investigation and/or have a shorter life expectancy.

Also, in the 1970s, the adjustment of 35 young South African people with albinism and 35 normally pigmented controls was explored by the present author, and attitudes toward affected people were examined (Kromberg, 1977). The two groups showed similar general levels of adjustment. However, the specific problems of affected individuals were associated with the physical side effects of their condition, with their difficulties in obtaining employment, and with the widely held community belief that persons with albinism do not die natural deaths. There was much ignorance regarding the cause of albinism, but attitudes were generally found to be positive, although the word "nkau" (or monkey, as mentioned above) was frequently used when referring to them.

Albinism in Cameroon was investigated by Aquaron et al. (1978) and later again by Aquaron (1980). The former group of workers reported on three pairs of twins with albinism and two pairs with unialbinism (one twin being affected and the other unaffected) in the Bamileka tribe, in which there appeared to be a high rate of consanguineous marriage. Blood groups were studied and the zygosity of the twins determined. Aquaron (1980) examined 216 people with albinism and calculated the prevalence rate in one area inhabited by the Bamileka tribe to be 1 in 3800. There were more males than females affected and this difference appeared to be significant when compared with the sex ratio in the general population. Affected people had a shortened life span and the tyrosinase-positive type appeared to predominate. In a subsequent report on this group, a family containing twins with albinism who had a mother and sibling with OCA3/rufous albinism was described (Aquaron et al., 1981). This report is worth noting because it supports the previous reported findings of these two conditions occurring together in one family, from other parts of Africa (Pearson et al., 1913; Loewenthal, 1944; Rose, 1973). Biochemical studies were carried out on the Cameroonian subjects, and studies on blood groups showed no significant difference between them and the controls (Mallett and Aquaron, 1983). In investigating the history of the Bamileka ethnic group, Aquaron (1980) found that around the beginning of the 20th century there were two polygamous chiefs who both had albinism and had many children, suggesting that the high prevalence of the condition in this group might be associated with founder effect.

A small study on skin cancer in people with albinism, treated between 1973 and 1979 in Tanzania, was reported by Alexander and Henschke (1981). They reviewed 10 cases of advanced squamous cell carcinoma and found that radiotherapy could produce acceptable therapeutic results, but that education of patients, concerning avoidance of sun exposure particularly during peak intensity (10.00–15.00 h), the wearing of protective clothing, and the use of sun-screening agents would perhaps be the most practical method of preventing skin cancer in affected people.

Two further papers, during this period, dealt with albinism in South Africa (Kromberg and Jenkins, 1982, 1984). The first paper, which resulted from some of the findings of an extensive doctoral research project (Kromberg, 1985), reported on the prevalence of albinism in a large sample of urban Black people, which showed that it occurred in 1 in 3900 people in Johannesburg. In the second article the issues to be covered in genetic counseling sessions were outlined and discussed. The findings of the latter study indicated that young affected individuals needed counseling concerning the cause of albinism, the associated physical problems, the prevailing myths about the condition, and the discrimination against them especially as marriage partners and employees. The implementation of a broadly defined genetic counseling program as advocated by Fraser (1974), but also designed to include discussion on these issues, was recommended (see Chapter 10).

This review of the older historical studies on albinism in Africa indicates that several different aspects had been investigated up to 1984, some in greater detail than others. The prevalence was estimated in Nigeria, Cameroon, Transkei, and in an urban area of South Africa, and rates appeared to be similar in these countries (i.e., around 1 in 4000). Several forms of albinism have been described over the last century, but studies suggested that the tyrosinase-negative types (now called OCA1) were rare and the tyrosinase-positive and brown types (now called OCA2 and OCA2B, respectively) were more common in Africa. Skin cancer was documented as the major problem of people with albinism, and it was probably associated with a shortened life span (found in Nigeria and Cameroon but not in the Transkei district of South Africa). Skin color in the families of persons with albinism was commented on, in passing, in various reports and these suggested that it was lighter than in the general population, but no studies on skin color, using measuring instruments, had yet been conducted. It was also suggested that a light skin color was looked on as beautiful in some areas of Africa, as far apart as Somalia and the Transkei, for many decades. Fertility was only mentioned briefly, although the few reports gave conflicting results, and the person with albinism has been reported as impotent, infertile, or normally fertile. Psychosocial reports were rare and only presented findings concerning the average intellectual ability, negative self-evaluation, and the generally normal level of adjustment of young people with albinism in a Southern African cohort.

1985–Date

A psychological study on maternal-infant bonding, where the mother was black and the infant had albinism, was carried out in South Africa by Kromberg et al. (1987). Such bonding was found to be delayed in comparison with a matched control group of unaffected dyads (where the black mother had an unaffected child). Later, a clinical study on albinism and skin cancer, documenting the high risk of such cancers, was published (Kromberg et al., 1989). These studies involved a great deal of field work (see Fig. 1.3). Subsequently, a sample of people with OCA3/rufous albinism, living in southern Africa, was examined and described in detail (Kromberg et al., 1990). At the same time, Aquaron (1990) finalized his work on albinism in the Cameroon and published his 15 year follow-up study.

In the late 1980s, research on albinism in South Africa began to focus on molecular genetic studies (e.g., Heim, 1988), and in the 1990s Kedda et al. (1994) found that the OCA2 (previously called tyrosinase-positive albinism) gene localized to chromosome 15q11-q13, with evidence of multiple mutations accounting for the conditions. The following year Stevens et al. (1995) reported that an intragenic deletion of the P gene (first identified in a triracial isolate in the United States (Durham-Pierre et al., 1994)) was the common OCA2 causing mutation in South Africa. In the Human Genetics Department of the University of the Witwatersrand,

FIGURE 1.3 Undertaking field work on albinism in rural South Africa. *Photograph courtesy of Dr. J. Kromberg, University of the Witwatersrand, South Africa.*

Johannesburg, the major long-term research project on albinism in Africa continued, several PhD projects on the condition were conducted, theses were produced (Kromberg, 1985; Heim, 1988; Kedda, 1993; Stevens, 1995; Manga, 1997; Kerr, 2007), and articles were published. Further molecular work on the OCA2B/brown and OCA3/rufous types of albinism in which the causative genes were identified followed (Manga et al., 1997, 2001). Several more articles on the topic have been written and published by South African research workers, the latest two being on the types of albinism found in Southern Africa (Kromberg et al., 2012) and the biology and genetics of albinism and vitiligo (Manga et al., 2013).

Also in the 1990s two studies were carried out by cell biologists investigating the morphology of melanocytes and melanosomes from individuals with albinism. The first was focused on cells extracted from the skin and hair bulbs of subjects with OCA3/rufous albinism (Kidson et al., 1993), whereas the second involved a study of cultured melanocytes from people with albinism (Kidson et al., 1994). A further study was completed by a dermatologist (Bothwell, 1997) on the unique pigmented skin lesions found in South African patients with OCA2.

At this time Lund initiated studies on the condition in Zimbabwe, determining its distribution (1996) and, together with colleagues, describing a cluster among a Tonga isolate in the country (Lund et al., 1997). More recently, she examined ways of managing the condition in a low-resource setting (Lund, 2005) and, with a team of colleagues, carried out research on the condition in other neighboring Southern African countries. Several of these projects will be discussed, in detail, in later chapters of this book (see Chapters 3 and 4).

A Norwegian team investigated albinism in Malawi and collected information on the local knowledge and beliefs about the condition during a qualitative study (Braathen and Ingstadt, 2006). Their findings showed that individuals with albinism and their families knew very little about the condition and that even though myths abounded, there were also stories of approval of affected people. However, they were considered by the community and themselves to be disabled.

One of the significant articles from this period was likely that of a World Health Organization team who investigated albinism in Africa as a public health issue (Hong et al., 2006). They reviewed the prevalence studies conducted in Africa and concluded: "The prevalence of albinism suggests the existence of tens of thousands of people living with albinism in Africa. This finding reiterates the need for increased awareness of and public health interventions for albinism in order to better address the medical, psychological and social needs of this vulnerable population" (Hong et al., 2006, p. 212).

In another recent article, Cruz-Inigo et al. (2011) stated that although there is presently no cure for albinism, the morbidities associated with the condition could be reduced with improved sun protection and eye care. Also, because of the myths regarding the mystic power of people with albinism, which abound in some sub-Saharan countries such as Tanzania, Malawi, Uganda, and Burundi, body parts of persons with albinism are valued for use in witchcraft and the making of powerful medicines. Affected people living in these areas have thus been put in great danger recently and often need protection. Many attacks on them and their families have been reported. Therefore, in these areas of Central and Southern Africa, people with albinism require improved security, in their own homes, or relocation to a safe place. The authors emphasize the urgent need for relevant education in the community and for eliminating the myths surrounding the condition, particularly those that put the lives of affected people at risk. Dr. A. Sethi, a dermatologist and one of the coauthors on this article, has set up clinics in Malawi and spends much time in running awareness campaigns and educating Malawians on these issues.

McBride (2014) reviewed recent literature on albinism in Africa, focusing on the effects of the disorder on the health and education of affected individuals. She commented that in the United Kingdom children with albinism present early in childhood to hospital eye clinics and that, although the risk of developing skin cancer is increased, life expectancy is not altered. Children and adults are provided with glasses or contact lenses, where necessary, as well as low-vision aids such as large print books, magnifiers, telescopes, and electronic vision enhancement systems. They have access to genetic counseling services and dermatology clinics and their vision is monitored regularly. Also UK charities such as The Albino Fellowship (www.albinism.org.uk) assist with information and peer support. In comparison, for example, in Tanzania the prevalence of albinism is estimated to be 8.5 times higher than that in the United Kingdom. The condition has serious health consequences in Africa, which include the high risk of skin cancer, limited use of sunscreen lotions, delayed presentation (often caused by poverty and undereducation) for treatment of skin lesions, and poor follow-up. Thus skin cancer is a major cause of morbidity and even mortality. Furthermore, visual impairment often leads to poor learning at school and individuals dropping out because the necessary visual aids are seldom available and, if they are, the support provided, the training and follow-up, may be inadequate. The author adds three useful appendices covering recommendations for managing albinism in Africa, advice for affected individuals, and advice for teachers with students with albinism.

In July 2015 the First International Workshop on Albinism in Africa was held in Douala, Cameroon. About 120 interested people from all over Africa, as well as from European countries such as France, Spain, Italy and the United Kingdom, attended. These included scientists, basic researchers, geneticists, clinicians, dermatologists, oncologists, ophthalmologists, pediatricians, social scientists, community workers, anthropologists, psychologists, biologists, experts in human rights issues, students, and members of albinism associations. The mission of the workshop was to discuss the scientific and social aspects of albinism, focusing on relevance to Africa. On the first day there were scientific sessions on the pathophysiology and genetics of albinism, on the dermatological, ophthalmologic, and anthropological issues, and on management and treatment. The sessions on the second day covered children with albinism, effective advocacy, human rights issues, and meetings of various albinism associations. The nature and outcomes of these discussions will be covered more fully in a later chapter (see Chapter 13).

Studies on Albinism Elsewhere in the World

The monograph by Pearson et al. (1913) covers the few small research studies, which had been carried out, and reports on observations made (either on Pearson's initiative or previously) in various European and North and South American countries, as well as in African countries, up to that time. Thereafter research on albinism continued sporadically in various countries. This work was comprehensively described by Witkop et al. (1983) in a chapter on albinism and other disorders of pigment metabolism in the widely used resource book entitled "The Metabolic Basis of Inherited Disease." Some years later, a further detailed review was published by King et al. (1995) and later updated by King et al. (2007).

The fact that albinism occurred with unusual frequency in the American Indians, especially those living in the South West of the United States and Central America, was recognized more than a century ago. Hrdlicka (1908) studied the Native Americans in South Western United States and Northern Mexico and reported that there were 11 people with albinism in 10 sibships in a population of 2000 Hopi Indians. These figures gave a very high prevalence rate of 1 in 182 in this isolate. Later Harris (1926) found 138 affected people among the 20,000 San Blas Indians living in Panama, giving a frequency of 1 in 146. Woolf and Grant (1962) studied the Hopi Indian people with albinism again in the 1960s to assess whether the Hopi attached special religious significance to albinism or whether there was cultural selection associated with the condition. They found a prevalence rate of 1 in 227 people and suggested that cultural selection in past generations was a possible explanation, but that there did not appear to be any special religious significance attached to the condition. In the 1950s and 1960s, Keeler (1963) undertook long-term studies on the Cuna Indian population of San Blas province. He reported on their high incidence of albinism, the possibility of selective inbreeding, the practice of infanticide, and the shortened life span of affected people (see Chapter 3 for further discussion of these studies).

A large study was undertaken by Froggatt (1960a) on albinism in Northern Ireland. He identified 122 people with albinism in 89 families and showed that the frequency was 8.75 per 100,000. Consanguineous marriages were found in 9% of the families. The oldest age groups had proportionately fewer affected people than the younger groups, possibly because of ascertainment bias due to more thorough investigation of the latter groups. This comprehensive study produced some very informative data, and it will be discussed in detail in a later chapter (see Chapter 3).

Dr. Carl Witkop et al. from Minneapolis, Minnesota, United States, carried out many original and extensive research studies on albinism in the Americas. One of their reports that should be mentioned here covered the biochemical and clinical characteristics of people with albinism among the Zuni, a Native American ethnic group, and the triracial Brandywine isolate (Witkop et al., 1972). The authors reported that pigmentation increased in affected people, but their nystagmus and photophobia decreased as they aged. Their hair bulbs were examined and found to develop pigment when cultured in L-tyrosine. Data on fertility in affected males showed that they had as many offspring as pigmented males. In 1994, a group led by Dr. M. Brilliant (Durham–Pierre et al.) identified the first OCA2 mutation, a P gene deletion, in a family from this Brandywine isolate.

In 1989 the First International Colloquium on the Neurological Genetic Aspects of Albinism was held from 27 to 30 September at Keble College, Oxford University, United Kingdom. The event was planned to commemorate the 10th anniversary of The Albino Fellowship, based in Oxford. Researchers from all over the world, including Professor C. Witkop from the United States, Dr. R. Aquaron from France (see Fig. 1.4), and Professor T. Jenkins and Dr. J. Kromberg (the author) from South Africa, were invited to participate. Selected papers presented at the colloquium were published in a special issue of the Journal of Ophthalmic Paediatrics and Genetics (1990 11(3)). The articles covered visual field maps in persons with albinism, childhood albinism, retinal image quality, visual anomalies and albinism, some rare associated syndromes, educational implications, and OCA3/rufous albinism, among other topics.

The classification, clinical characteristics, and recent findings in persons with albinism in the United States have been reviewed by Dr. C.G. Summers (2009). She stated that the

FIGURE 1.4   Professor C. Witkop (on left) and Dr. R. Aquaron at First International Colloquium on the neurological genetic aspects of albinism at Oxford University, 1989. *Photograph courtesy of Dr. J. Kromberg, University of the Witwatersrand, South Africa.*

clinical diagnosis can often be made on the cutaneous phenotype, although genetic testing may be necessary to precisely classify the type of albinism. Reporting on the ocular characteristics, she commented that affected individuals can have delayed visual development and spectacles are frequently required to improve both visual acuity and binocular vision. Neurodevelopment had been measured and was considered to be normal; however, there appeared to be a slightly elevated prevalence of attention deficit hyperactivity disorder in this prospective study (Kutzbach et al., 2008).

A recent useful review on albinism in Europe, covering, mainly, the types of albinism found and their prevalence, was published by Martinez-Garcia and Montoliu in 2013. These authors state that OCA1 is the commonest form of albinism among White patients. Molecular testing confirmed that it is the most frequent type in German, French, Danish, and Italian cohorts. In general, OCA2 cases occur, but they are rarer in Europe than in Africa. Steady immigration from Africa to Europe has, however, resulted in OCA2 being identified more frequently. OCA3, the rufous type of albinism, is very rarely found in European studies on the condition. These authors suggest that genotype–phenotype studies are essential for the full understanding of the mechanisms underlying the clinical traits observed in patients with albinism.

The identification of the genes mutated in the various forms of albinism has facilitated studies on the condition in several other regions of the world, including India, Pakistan, and Japan (Mondal et al., 2012; Jaworek et al., 2012; Okamura et al., 2016).

Many of the epidemiological and prevalence studies undertaken in various countries, in addition to those conducted in African communities, will be discussed in detail in a later chapter (see Chapter 3).

III. ELUCIDATING THE GENETIC AND MOLECULAR BASIS OF ALBINISM

Molecular genetics work on albinism was carried out in the United States, South Africa (e.g., Kedda et al., 1994; Stevens et al., 1995; Manga et al., 2001; Kerr, 2007), and elsewhere in the 1980s and 1990s and further expanded in the last two decades. The gene associated with

OCA1 (tyrosinase-negative albinism) was identified first and then that for OCA2 (tyrosinase-positive albinism). Presently, mutations at more than seven loci have been linked to various types of albinism (e.g., *TYR*, *OCA2*, *TYRP1*, and *MATP*). These genes, some of which have been reviewed by Oetting et al. (1996) and Gronskov et al. (2007), as well as the ongoing work in the field, will be discussed further later in this book. Various types of syndromic albinism have also been investigated and the findings have been reviewed by Kamaraj and Purohit (2014), together with findings on new OCA-associated genes. The molecular genetics of these conditions will also be covered in a later chapter (see Chapter 5).

IV. CONCLUSION

In this chapter a detailed background, as far as can be gleaned from the published literature as well as some unpublished documents and the doctoral thesis of the author (Kromberg, 1985), is given regarding previous studies on albinism. This review shows that albinism was recognized and described at least two millennia ago; the main manifestations of unusual skin and hair color have been accurately reported and the visual problems documented. The monograph by Pearson et al. (1913) gives a fascinating and instructive account of the knowledge on the condition available early in the 20th century, and the insight of Garrod (1908), at that time, into the cause of albinism is remarkable. The recessive inheritance of the condition was also understood and documented early in that century (Davenport and Davenport, 1910).

Many studies and observations were carried out in Africa, mostly by explorers, travelers, missionaries, anthropologists, or medical practitioners from Europe, who were working in various countries there. Both the social status, including to some extent the stigmatization, and the clinical problems were described. Initially, forms of classification, based on detailed reports and clinical examinations, were suggested by Pearson et al. (1913) and Stannus (1913). These were found to be too subjective and impractical and the types were later designated according to their tyrosinase status. However, once modern molecular techniques became available the types of albinism had to be reclassified in accordance with their causative genes.

The nature of the skin cancer experienced by people with albinism was also investigated, the high risks were recognized, and methods of preventing its development were proposed. In addition, a few psychosocial studies were undertaken. Latterly, many of the studies on albinism have focused on mapping the genes responsible for the different types of the condition and attempting to explain the molecular mechanisms involved in the etiology of the disorder.

References

Alexander, G.A., Henschke, U.K., 1981. Advanced skin cancer in Tanzanian albinos: preliminary observations. J. Natl. Med. Assoc. 73, 1047–1054.

Aquaron, R., Giraud, F., Battaglini, P., 1978. Albinism and unialbinism in twin Cameroonian negroes. Prog. Clin. Biol. Res. 24 (c), 71–76.

Aquaron, R., 1980. L'albinisme oculo-cutane au Cameroon. A propos de 216 observations. Rev. Epidemiol. Sante Publique 28, 81–88.

Aquaron, R., Ronge, F., Aubert, C., 1981. Pheomelanin in albino negroes: urinary excretion in 5-S-cysteinylodopa in Cameroonian subjects. In: Seijing, M. (Ed.), Pigment Cell Tokyo. University of Tokyo Press, Tokyo, pp. 97–103.

Aquaron, R., 1990. Oculocutaneous albinism in Cameroon. A 15 year follow-up study. Ophthalmic. Paediatr. Genet. 11, 255–263.

Barnicot, N.A., 1952. Albinism in South West Nigeria. Ann. Eugen. 17, 38–73.

Barnicot, N.A., 1953. Red hair in African negroes: a preliminary study. Ann. Eugen. 17, 211–232.

Bothwell, J.E., 1997. Pigmented skin lesions in tyrosinase-positive oculocutaneous albinos: a study in black South Africans. Int. J. Dermatol. 36, 831–836.

Braathen, S.H., Ingstadt, B., 2006. Albinism in Malawi: knowledge and beliefs from an African setting. Disabil. Soc. 21 (6), 599–611.

Cervenka, J., Witkop, C.J., Okoro, A.N., King, R.A., 1979. Chromosome breaks and sister chromatid exchange in albinos in Nigeria. Clin. Genet. 15, 17–21.

Cohen, L., Shapiro, M.P., Keen, P., Henning, A.J.H., 1952. Malignant disease in the Transvaal: 1. Cancer of the skin. S. Afr. Med. J. 26, 932–939.

Cruz-Inigo, A.E., Ladzinski, B., Sethi, A., 2011. Albinism in Africa: stigma, slaughter and awareness campaigns. Dermatol. Clin. 29, 79–87.

Davenport, G.C., Davenport, C.B., 1910. Heredity of skin pigmentation in man. Am. Nat. XLIV (527), 641–674.

Dogliotti, M., 1973. Actinic keratoses in Bantu albinos. Clinical experiences with the topical use of 5-fluro-uracyl. S. Afr. Med. J. 47, 2169–2172.

Durham-Pierre, D., Gardner, J.M., Nakatsu, Y., King, R.A., Francke, U., Ching, A., Aquaron, R., del Marmol, V., Brilliant, M.H., 1994. African origin of an intragenic deletion of the human P gene in tyrosinase-positive oculocutaneous albinism. Nat. Genet. 7 (2), 176–179.

Emery, A.E.H., 1979. Elements of Medical Genetics, fifth ed. Churchill Livingstone, Edinburgh.

Fraser, F.C., 1974. Genetic counseling. Am. J. Hum. Genet. 26, 636–661.

Froggatt, P., 1960a. Albinism in Northern Ireland. Ann. Hum. Genet. 24, 213–233.

Froggatt, P., 1960b. The legend of the White Native race. Med. Hist. 4, 228–235.

Garrod, A.E., 1908. Inborn errors of metabolism. III. Albinism. Lancet 2, 73–79.

Gronskov, K., Ek, J., Brondum-Nielsen, K., 2007. Oculocutaneous albinism. Orphanet J. Rare Dis. 2, 43.

Harris, R.G., 1926. The San Blas Indians. Am. J. Phys. Anthr. 9, 17–63.

Heim, R., 1988. Restriction Fragment Length Polymorphisms in Southern African Populations and a Search for Linkage to Tyrosinase-Positive Oculocutaneous Albinism (Ph.D. thesis). University of the Witwatersrand, Johannesburg, South Africa.

Hitzeroth, H., Hofmeyer, J.D.J., 1964. A survey of Albinism among the Bantu-speaking tribes of South Africa. Mank. Q. 5, 81–86.

Hong, E.S., Zeeb, H., Repacholi, M.H., 2006. Albinism in Africa as a public health issue. B.M.C. Public Health 6, 212–219.

Hrdlicka, A., 1908. Physiological and Medical Observations Among the Indians of Southwestern United States and Northern Mexico. Bulletin No 34. Bureau of American Ethnology, Washington DC.

Jaworek, T.J., Kauser, T., Bell, S.M., Tariq, N., Maqsood, M.I., Sohail, A., Ali, M., Iqbal, F., Rasool, S., Riazuddin, S., Shaikh, R.S., Ahmed, Z.M., 2012. Molecular genetic studies and delineation of the oculocutaneous phenotype in the Pakistani population. Orphanet J. Rare Dis. 7, 44.

Kamaraj, B., Purohit, R., 2014. Mutational analysis of oculocutaneous albinism: a compact review. Biomed. Res. Int.:905472, 1–10.

Kedda, M.A., 1993. A Search for the Tyrosinase-Positive Oculocutaneous Albinism Gene Using Linkage Analysis (Ph.D. thesis). University of the Witwatersrand, Johannesburg, South Africa.

Kedda, M.A., Stevens, G., Manga, P., Viljoen, C., Jenkins, T., Ramsay, M., 1994. The tyrosinase-positive oculocutaneous albinism gene shows locus homogeneity on chromosome 15q11-q13 and evidence of multiple mutations in southern African negroids. Am. J. Hum. Genet. 54 (6), 1078–1084.

Keeler, C.E., 1953. The Caribe Cuna moon-child and its heredity. J. Hered. 44, 163–171.

Keeler, C., 1963. The incidence of Cuna moon-child albinos. J. Hered. 54, 115–120.

Kerr, R., 2007. Genes in the Aetiology of Oculocutaneous Albinism in Sub-Saharan Africa and a Possible Role in Tuberculosis Susceptibility (Ph.D. thesis). University of the Witwatersrand, Johannesburg, South Africa.

Kidson, S.H., Richards, P.D.G., Rawoot, F., Kromberg, J.G.R., 1993. An ultrastructural study of melanocytes and melanosomes in the skin and hair bulbs of rufous albinos. Pigment Cell Res. 6, 209–214.

Kidson, S.H., Wiggins, T., Kromberg, J.G.R., 1994. Culture and morphology of human albino melanocytes. S. Afr. J. Sci. 90, 354–356.

King, R.A., Creel, D., Cervenka, J., Okoro, A.N., Witkop, C.J., 1980. Albinism in Nigeria with delineation of a new recessive oculocutaneous type. Clin. Genet. 17, 259–270.

King, R.A., Hearing, V.J., Creel, D.J., Oetting, W.S., 1995. Albinism. In: Scriver, C.R., Beaudet, A.L., Sly, W.S., Valle, D. (Eds.), The Metabolic and Molecular Bases of Inherited Disease, seventh ed. McGraw-Hill, New York, pp. 4353–4392.

King, R.A., Oetting, W.S., Summers, C.G., Creel, D.J., Hearing, V.J., 2007. Abnormalities of pigmentation. In: Rimoin, D.I., Connor, J.M., Pyeritz, R.E., Korf, B.R. (Eds.), Emery and Rimoin's Principles and Practice of Medical Genetics, fifth ed. Churchill Livingstone Elsevier, Philadelphia, pp. 3380–3427.

Kromberg, J.G.R., 1977. Albino Youth in Soweto: Some Features of Their Adjustment (MA dissertation). University of the Witwatersrand, Johannesburg, South Africa.

Kromberg, J.G.R., Jenkins, T., 1982. Prevalence of albinism in the South African Negro. S. Afr. Med. J. 61, 383–386.

Kromberg, J.G.R., Jenkins, T., 1984. Albinism in the South African Negro. III. Genetic counselling issues. J. Biosoc. Sci. 16, 99–108.

Kromberg, J.G.R., 1985. A Genetic and Psychosocial Study of Albinism in Southern Africa (Ph.D. thesis). University of the Witwatersrand, Johannesburg, South Africa.

Kromberg, J.G.R., Zwane, M.E., Jenkins, T., 1987. The response of black mothers to the birth of an albino infant. Am. J. Dis. Child. 141 (8), 911–916.

Kromberg, J.G.R., Castle, D.J., Zwane, E., Jenkins, T., 1989. Albinism and skin cancer in Southern Africa. Clin. Genet. 36, 43–52.

Kromberg, J.G.R., Castle, D.J., Zwane, E.M., et al., 1990. Red or rufous albinism in Southern Africa. J. Ophthalmic Paediatr. Genet. 11 (3), 229–235.

Kromberg, J.G.R., Bothwell, J., Kidson, S.H., Manga, P., Kerr, R., Jenkins, T., 2012. Types of albinism in the black southern African population. East Afr. Med. J. 89 (1), 20–27.

Kutzbach, B.R., Summers, C.G., Holleschau, A.M., MacDonald, J.T., 2008. Neurodevelopment in children with albinism. Ophthalmology 115, 1805–1808.

Livingstone, D., 1857. Missionary Travels. John Murray, London.

Loewenthal, L.J.A., 1944. Partial albinism and nystagmus in Negroes. Arch. Derm. Syph. 50, 300–301.

Lund, P.M., 1996. Distribution of oculocutaneous albinism in Zimbabwe. J. Med. Genet. 33, 641–644.

Lund, P.M., Puri, N., Durham-Pierre, D., King, R.A., Brilliant, M.H., 1997. Oculocutaneous albinism in an isolated Tonga community in Zimbabwe. J. Med. Genet. 34, 733–735.

Lund, P.M., 2005. Oculocutaneous albinism in southern Africa: population structure, health and genetic care. Ann. Hum. Biol. 32 (2), 168–173.

Mallett, B., Aquaron, R., 1983. Isolation and purification of ceruloplasmin in oculocutaneous albinism, Menke's disease, Wilson's disease and pregnant women. Clin. Chim. Acta 132, 245–256.

Manga, P., 1997. Identification and Molecular Characterisation of the Genes for Brown and Rufous Oculocutaneous Albinism in Southern Africa (Ph.D. thesis). University of the Witwatersrand, Johannesburg, South Africa.

Manga, P., Kromberg, J.G., Box, N.F., Sturm, R., Jenkins, T., Ramsay, M., 1997. Rufous oculocutaneous albinism in Southern African blacks is caused by mutations in the TYRP1 gene. Am. J. Hum. Genet. 61 (5), 1095–1101.

Manga, P., Kromberg, J.G.R., Turner, A., Jenkins, T., Ramsay, M., 2001. In Southern Africa, brown oculocutaneous albinism (BOCA) maps to the OCA2 locus on chromosome 15q: P-gene mutations identified. Am. J. Hum. Genet. 68, 782–787.

Manga, P., Kerr, R., Ramsay, M., Kromberg, J.G.R., 2013. Biology and genetics of oculocutaneous albinism and vitiligo – common pigmentation disorders in southern Africa. S. Afr. Med. J. 103 (12 (Suppl. 1)), 984–988.

Manganyi, N.C., Kromberg, J.G.R., Jenkins, T., 1974. Studies in albinism in the South African Negro. I. Intellectual maturity and body image differentiation. J. Biosoc. Sci. 6, 107–112.

Martinez-Garcia, M., Montoliu, L., 2013. Albinism in Europe. J. Dermatol. 40, 319–324.

McBride, G.R., 2014. Oculocutaneous albinism: an African perspective. Br. Ir. Orthopt. J. 11, 3–8.

McCrackin, R.H., 1937. Albinism and unialbinism in twin African negroes. Am. J. Dis. Child. 54, 786–794.

Menasche, G., Hsuan, C., Sanal, O., Feldmann, J., Tezcan, I., Ersoy, F., Houdiusse, A., Fischer, A., de Saint Basile, G., 2003. Griscelli syndrome restricted to hypopigmentation results from a melanophilin defect (GS3) or a MYO5A F-exon deletion (GS1). J. Clin. Investig. 112, 450–456.

Mondal, M., Sengupta, M., Samanta, S., Sil, A., Ray, K., 2012. Molecular basis of albinism in India: evaluation of seven potential candidate genes and some new findings. Gene 511 (2), 470–474.

NOAH. The National Organisation for Albinism and Hypopigmentation. www.albinism.org/publications/What_do_you_call_me.html.

Oetting, W.S., Brilliant, M.H., King, R.A., August 1996. The clinical spectrum of albinism in humans. Mol. Med. Today 330–335.

Oetting, W.S., King, R.A., 1999. Molecular basis of albinism: mutations and polymorphisms of pigmentation genes associated with albinism. Hum. Mutat. 13, 99–115.

Oettle, A.G., 1963. Skin cancer in Africa. Natl. Cancer Inst. Monogr. 10,197–214.

Okamura, K., Araki, Y., Abe, Y., Shigyou, A., Fujiyama, T., Baba, A., Kanekura, T., Chenen, Y., Kono, M., Niizeko, H., Konno, T., Hozumi, Y., Suzuki, T., 2016. Genetic analysis of oculocutaneous albinism types 2 and 4 with eight novel mutations. J. Dermatol. Sci. 81 (2), 140–142.

Okoro, A.N., 1975. Albinism in Nigeria. A clinical and social study. Br. J. Dermatol. 92, 485–492.

Pearson, K., Nettleship, E., Usher, C.H., 1913. A monograph on albinism in man. In: Drapers Co. Research Memoirs Biometric Series, VIII. Dulau, London.

Plinius Secundus the Elder, 1942. The Natural History of Pliny. Heinemann, London. Book 7. Tr. by H Rackman.

Rose, E.F., 1973. Pigment variation in relation to protection and susceptibility to cancer. Pigment Cell 1, 236–245.

Rose, E.F., 1974. Pigment anomalies encountered in the Transkei. S. Afr. Med. J. 48, 2345–2347.

Schapera, I., 1937. The Bantu Speaking Tribes of South Africa. An Ethnographic Survey. Routledge, London.

Schrire, T., 1958. Aetiology of facial cancer. A speculative and deductive survey. S. Afr. Med. J. 32, 997–1002.

Sorsby, A., 1958. Noah – an albino. Br. Med. J. 2, 1587–1589.

Stannus, H.S., 1913. Anomalies of pigmentation among natives of Nyasaland. Biometrika 333–365.

Stevens, G., 1995. Linkage to the Tyrosinase positive Oculocutaneous Albinism Gene and Characterisation of Mutations in Bantu-speaking African Negroids (Ph.D. thesis). University of the Witwatersrand, Johannesburg, South Africa.

Stevens, G., van Beukering, J., Jenkins, T., Ramsay, M., 1995. An intragenic deletion of the P gene is the common mutation causing tyrosinase-positive oculocutaneous albinism in Southern African negroids. Am. J. Hum. Genet. 56, 586–591.

Summers, C.G., 2009. Albinism: classification, clinical characteristics, and recent findings. Optom. Vis. Sci. 86 (6), 659–662.

Vallois, H.V., 1950. Sur quelques points de l'anthropologie des Noirs. L'anthropologie 54, 272–286.

Watkins-Pitchford, W., 1910. Pigmentation and cancer. Tvl. Med. J. 6, 23–25.

Witkop, C.J., Nance, W.E., Rawls, R.F., White, J.G., 1970. Autosomal recessive albinism in man: evidence for genetic heterogeneity. Am. J. Hum. Genet. 22, 55–74.

Witkop, C.J., 1971. Albinism. In: Harris, H., Hirschhorn, K. (Eds.), Advances in Human Genetics, vol. 2. Plenum Press, New York, pp. 61–142.

Witkop, C.J., Niswander, J.D., Bergsma, D.R., Workman, P.I., White, J.G., 1972. Tyrosinase positive oculocutaneous albinism among the Zuni and the Brandywine triracial isolate: biochemical and clinical characteristics and fertility. Am. J. Phys. Anthr. 36, 397–406.

Witkop, C.J., King, R.A., Cervenka, J., Okoro, A.N., 1977. Oculocutaneous albinism associated with skin cancer in Enugu, Nigeria (Abst.). In: Amer. Acad. Oral Path. Congress, May 9–13. Proceedings, Portland Oregon.

Witkop, C.J., Quevedo, W.C., Fitzpatrick, T.P., 1983. Albinism and other disorders of pigment metabolism. In: Stanbury, J.B., Wyngaarden, J.B., Fredrickson, D.S., Goldstein, J.L., Browne, M.S. (Eds.), The Metabolic Basis of Inherited Disease. McGraw Hill, New York, pp. 301–346.

Woolf, C.M., Grant, R.B., 1962. Albinism among the Hopi Indians of Arizona. Am. J. Hum. Genet. 14, 391–399.

Clinical Features, Types of Albinism, and Natural History

Jennifer G.R. Kromberg
University of the Witwatersrand and National Health Laboratory Service,
Johannesburg, South Africa

I. INTRODUCTION

In this chapter the clinical features of albinism, as well as some of those occurring in carriers of an albinism gene (heterozygotes), will be outlined and discussed with reference to reports in the literature, as well as to some unpublished studies. The dermatological and ophthalmological abnormalities will be presented only briefly in this discussion as they will be described in more detail in later chapters. Research studies and issues associated with the auditory system, intelligence, fertility, and longevity will be included, but those associated with psychosocial problems will be discussed in detail in a later chapter on that topic (see Chapter 9). Because albinism comprises a heterogeneous group of disorders, the types, as presently delineated, will be characterized with a focus on the clinical characteristics of persons with albinism commonly found in Africa; the molecular findings will be discussed in a later chapter. (see Chapter 5). A clinical guide to differential diagnosis will also be presented. The characteristics and types that occurred in an actual cohort of people with albinism, living in South Africa, will then be specified. Finally, the natural history of the condition will be outlined.

II. CLINICAL CHARACTERISTICS

The clinical features and characteristics associated with a diagnosis of oculocutaneous albinism (OCA) that will be covered in the following sections include: skin pigmentation in homozygotes (affected people with a double dose of a gene for albinism) and their risks of skin cancer, pigmentation in heterozygotes (unaffected people with one gene for albinism and one gene for normal pigmentation), hair and eye pigmentation, visual problems, intelligence, fitness, and fertility in homo- and heterozygotes, and longevity. These features will be discussed with reference to the published reports as well as contentious issues (such as the possible shortened life span of people with albinism). Where there are relevant original data that have never been published, this material will be included.

Skin Pigmentation

The color of the skin and hair is the most obvious of the characteristics of the person with albinism, particularly where the parents and family are unaffected and of Black African origin. The unusual skin and hair (including eye brows and eye lashes) color is apparent soon after birth, in affected children from both Black and White families. The skin and hair color usually darkens slightly with age and sun exposure (except in the case of OCA1) and varies in intensity in an albinism type-dependent fashion in individual cases. However, the pale skin and hair color alone are not sufficient to make a diagnosis of OCA2 in the newborn child. Confirmation is usually possible only when the eyes of the infant open and are exposed to light, and nystagmus becomes apparent, a few days after birth. The other types, being less obvious originally, might not be diagnosed until later.

The skin problems experienced by most people with albinism, particularly those living in tropical and subtropical regions, include the development of erythema (redness), lentigines (pigmented flat spots with irregular borders), freckles (or ephelides, small round pigmented spots), pachyderma (thickening and wrinkling of the skin), and keratoses (premalignant or malignant lesions) (Bothwell, 1997). Affected people in the Black population frequently dislike the appearance of freckles for cosmetic reasons and because, in Africa particularly, they can hardly avoid sun exposure, the solar keratoses often become malignant. Preventive treatment consists of the application of antiactinic-barrier creams and avoidance of sun exposure (see Chapter 6 for further information on this topic).

Skin Cancer Risks

The high risk of developing skin cancer, which is associated with albinism, has been recognized for a long time. In South Africa, Watkins-Pitchford described this susceptibility to skin cancer in 1910, and it was well documented in the Transvaal province of South Africa in 1952 (Cohen et al.). Oettle (1963) studied skin cancer in 17 South Africans with albinism and found there were 14 cases with squamous cell carcinoma and one case each with basal cell and sweat gland carcinoma, and melanoma. The 14 cases with squamous cell carcinoma represented 9.6% of all such skin cancer found in Black South Africans. The commonest site for the cancer was the head (11 cases), followed by the upper limbs (3), neck (1), and back (1).

King et al. (1980), in a study of Nigerian people with albinism, noted that solar keratoses were most frequently found on the forearm and hand, closely followed by the head, and less frequently the neck and back. In a recent review of 134 biopsies from 86 persons with albinism who attended the Regional Dermatology Training Centre in Moshi, Tanzania, from 2002 to 2011, researchers found that squamous cell carcinoma was more prevalent than basal cell carcinoma, and that the sites affected were usually the head and neck (Kiprono et al., 2014).

However, Rose (1974) reported that skin cancer appeared to be infrequent in the Transkei (now in the Eastern Cape province of South Africa) and that Dr Burrell, who worked for many years in the area on cancer epidemiology with Dr Rose, had only noted one case among the 309 people with albinism he examined, and that was a melanoma on the back of the knee of a 50-year-old woman. Rose investigated 27 families with affected members and questioned them regarding cause of death. None reported skin cancer as a cause.

Nevertheless, Okoro (1975) when describing his Nigerian sample, stated that there were no people with albinism over 20 years of age who were free from premalignant or malignant lesions and keratoses and superficial ulcers were present in 50% of his subjects. He suggested that these destructive lesions were caused by years of exposure to actinic irradiation, which resulted in slight tanning at first, followed by sunburn, blisters, solar elastosis, ephelides, facial lentigines, solar keratoses, chronic superficial ulcers, and ultimately squamous and basal cell epitheliomata.

Rippey and Schmaman (1972) catalogued cases of squamous and basal cell carcinoma in Black patients at Baragwanath (now Chris Hani Baragwanath) hospital, Johannesburg, and found 11% of squamous cell patients and 41% of basal cell patients had albinism. In a later study of cancer in urban Blacks at the same hospital, Isaacson et al. (1978) examined cancer records for a 10-year period from 1966 to 1975. This survey yielded 101 cases of squamous cell carcinoma of which seven were in individuals with albinism. Basal cell carcinoma was found in only nine cases and, of these, three had albinism and one was a child with xeroderma pigmentosa. Malignant melanoma was found to be very rare, and the prevalence rate was 1.65 per 100,000 of the population. No cases were reported in people with albinism.

Comorbidity of skin cancer with albinism was later studied by Kromberg et al. (1989). They enrolled 111 people with albinism in a project on the Black population of Johannesburg, South Africa. Altogether 23.4% had skin cancers, and the risk of developing such cancers increased with age. Squamous cell carcinoma was more common than basal cell malignancy, and the site most commonly affected was the head. No melanomas were detected. The identifiable risk factors included: exposure to environmental ultraviolet radiation, absence of freckles (ephelides), or lentigines, and possibly ethnicity (as some ethnic groups had higher rates than others). Because the type of cancer experienced by affected individuals appeared to be particularly aggressive, prevention was strongly advocated.

A review by a Danish group stated that skin cancer is increased in people with albinism, and that due to the high rate of the condition in Africa, it may pose a serious health problem there (Gronskov et al., 2007). This comment suggests that skin cancer is not such a problem in Northern European developed countries, such as Denmark, where exposure to the sun and its radiation is minimal, early diagnosis and treatment is possible, and preventative measures are available and utilized.

Skin Color and Pigmentary Changes in Heterozygotes

Harris (1926) and, later, Keeler (1953) were convinced that families, with a member with albinism in San Blas, contained persons of fairer complexion than that of the average local Indians. Further, Waardenburg et al. (1961) stated that evidence seemed to show that in the dark-skinned ethnic groups heterozygotes for generalized albinism tend to have lighter skin pigmentation than their unaffected fellows. No attempt was made by these researchers to measure skin color, although reflectance spectrophotometry was used by Barnicot (1952) to measure hair color in affected people and their unaffected relatives, and again by Barnicot (1958) to investigate skin color in unaffected Nigerians.

Skin color has been measured by various methods over the years, and data are available, which indicate how skin color is geographically distributed throughout the world (Roberts, 1977). The original Evans Electroselenium Limited (EEL, London) reflectance spectrophotometer was designed for use in industry, but Weiner (1951) decided that, because of its portability and simplicity, it would be useful for anthropological field studies. The results from his trial study showed that reflectance values obtained with the EEL instrument were comparable with those obtained with the cumbersome and elaborate Hardy photoelectric recording spectrophotometer (Hardy, 1936) used successfully by Edwards and Duntley (1939) for the measurement and analysis of the properties of human skin.

Few quantitative studies of skin color have been performed in Southern Africa. The one of most relevance to the present discussion is that of Wassermann and Heyl (1968). They obtained data, by means of spectrophotometry, on skin pigmentation in South African Whites, people of mixed ancestry (called Cape Coloreds at that time), and Black individuals. They found that females tended to be lighter than males in all three groups, and that the individuals of mixed ancestry showed a wide overlap in skin color with both White and Black subjects. The data they obtained on the Black population were from 104 males and 100 females, mostly from the Xhosa ethnic group. There were 19 females who were pregnant and probably should have been excluded because skin color has been shown to darken during pregnancy (Rook, 1969); the results on the females are therefore slightly biased. Nevertheless, the findings are useful for the purposes of comparison with those from later studies.

Tobias (1961) also used an EEL instrument and published quantitative data on skin reflectance in the San ("Bushmen") and San-European hybrids. Weiner et al. (1964) took measurements on the skin color of four different groups: the San; Khoikhoi ("Hottentots"); San-Khoikhoi hybrids; and Okavango Bantu-speaking Negroes, most of whom were living in South West Africa/Namibia. Although skin reflectance can be measured at nine different wave lengths, the reflectance at 685 nm (609 filter) was the only measurement systematically recorded in these samples, probably because reflectance at this wavelength gives a good index of the melanin content of the skin (Harrison and Owen, 1956).

A study on the effect of homozygosity and heterozygosity for the phenylketonuria (PKU) gene on skin color in one Yemeni family was carried out by Roberts (1977). PKU results from mutations in phenylalanine hydroxylase, which catalyzes the reaction that converts phenylalanine to tyrosine, the amino acid used in pigment synthesis. Serum levels of tyrosine in PKU are low and may result in decreased skin pigmentation. Roberts found that the affected child had appreciably lightened skin pigmentation, in comparison with his father and mother, when measured at nine different wavelengths. This work appears to be the only one in which a recessive genetic disorder and skin color were studied in homo- and heterozygotes.

In an attempt to assess whether skin pigmentation in obligate heterozygotes (unaffected people with a child with albinism) for an albinism gene and random subjects is significantly different and whether heterozygotes could be identified in this way, Roberts et al. (1986) set up a research project in Swaziland. Pigmentation was measured by reflectance spectrophotometry in 210 Swazi subjects of whom 43 had albinism, 44 were their close relatives, and 123 were random controls. The findings showed that the mean reflectance of the subject groups diminished in the sequence: affected subjects, obligate heterozygotes, other unaffected family members, and random subjects. The differences among the unaffected groups were highly significant. However, although a discriminant function fitted to the obligate heterozygotes and random controls was also highly statistically significant, it was not efficient enough to be applied to the diagnosis of the carrier/noncarrier state.

From these studies it seems that although research workers have investigated various means of assessing skin color to identify carriers of a gene for albinism, this was not possible until molecular genetic testing became available, and the genes for the common types of albinism were identified.

Hair and Eye Pigmentation

Hair and eye pigmentation can vary widely in people with albinism, in Africa and elsewhere (King et al., 2007). Furthermore, both can change and darken with age. Depending on the ethnic group of origin and the type of albinism found in an individual the hair can be milky white, pale yellow, or corn colored, light brown, or reddish. Similarly, eye color may be pale blue, blue–gray, dark blue, hazel, light brown, or darker brown. Some of these differences have been correlated with variants at pigment loci such as melanocortin 1 receptor (Preising et al., 2011).

Various studies have attempted to examine the eye color of heterozygotes for the albinism gene to assess whether or not pigmentary changes occurred in them. Waardenburg (1947) found that in obligatory heterozygotes, the iris was translucent in some but not all cases with a history of oculocutaneous or ocular albinism, and he suggested that this information was of practical use for genetic purposes. Froggatt (1960) also found that although abnormally translucent irides were not a constant finding in all carriers, they were significantly more frequent in them than in controls. Witkop et al. (1973), however, believed that iris translucency was not a reliable indicator of heterozygosity for OCA2. In a small series of American Black

OCA1 (termed tyrosinase-negative at that time) obligatory heterozygotes they found only about half the subjects had translucent irides, showing that it is not even a consistent finding in that group. Froggatt (1960) noted the eye and hair color of the parents and children of affected people and suggested that blue eyes and fair hair were more common in this group than in the general population in Northern Ireland. Additional studies are needed to assess these features in obligatory heterozygotes with OCA1 and OCA2, following confirmation of mutations by molecular genetic testing.

Vision

Clinically, the person with albinism has several problems associated with the hypopigmentation of the eyes. During embryonic development decreased melanin synthesis (or reduced tyrosinase activity) in the retinal pigment epithelium results in foveal hypoplasia and dysregulation of adjacent retinal ganglion cells and consequently abnormal decussation of the nerve fibers connecting the retina to the brain at the optic chiasm (Jeffery, 1997). Individuals with OCA thus experience reduced visual acuity, nystagmus (the eyes move in a more or less rhythmical manner from side to side, or in a rotary manner from the original point of fixation, MacNalty, 1965), photophobia (abnormal intolerance of, or sensitivity to, light, MacNalty, 1965), and strabismus (squinting). Dark glasses and occasionally tinted contact lenses are used to reduce photophobia. Because corrected vision is reasonably good in most people with albinism (except perhaps those with OCA1) they should be able to cope in mainstream schools, using appropriate visual aids and with the sympathetic help of the teacher.

Auditory System

The lack of pigmentation in people with albinism also causes minor subtle changes in the auditory system (King et al., 2007). Animals with an absence of melanin pigment have been shown to be susceptible to hearing loss caused by noise, although in affected humans such exposure to noise may only cause a prolonged temporary threshold shift. Such shifts occur in all humans but are greatest in people with little pigment.

Other effects, due to this lack of pigment in humans with albinism, appear to have little clinical significance and to be so minor as to not yet have been identified by the medical professionals or scientists studying people with the condition.

Intelligence

The intelligence of people with albinism in the United States has been shown to be within the normal range (Beckham, 1946) as has the intellectual maturity in affected people in South Africa (Manganyi et al., 1974). However, psychological problems have been shown to occur (Stewart and Keeler, 1965), and assumptions are sometimes made regarding their abilities, which may or may not be justified (Cameron, 1979). Counseling and educational advice should be provided, where necessary, and community education on the nature of albinism and the fact that intelligence is within the normal range should be offered more widely (see Chapter 9).

Fitness and Fertility

The biological fitness of individuals is generally assessed by the number of their offspring who reach reproductive age (infant deaths and stillbirths are therefore excluded). Fitness is unity (or 100%), if an individual and spouse have two such offspring (Emery, 1979). Fertility, on the other hand, is indicated by the total number of offspring produced, whether or not they survive.

Neither the fitness nor fertility of people with albinism has been thoroughly assessed. Their fertility is affected by the fact that many do not marry, an observation made in both White (Froggatt, 1960) and Black (Vallois, 1950; Barnicot, 1952; Okoro, 1975; Kromberg and Jenkins, 1984) communities. Among African societies marriage is virtually universal, and almost every woman marries at least once in her lifetime (Rinehart, 1979), so when people with albinism do not marry this fact is very striking. Froggatt (1960) estimated that the chance of an affected person marrying was reduced in comparison with unaffected people in Northern Ireland. He found that in his sample only 12 of 30 (40%) affected females and 20 of 41 (49%) affected males, in the over 20-year-old age group, were married. However, fertility, in those who married, did not seem to be compromised, and they had a similar number of offspring, on average, to comparable groups in the general population. Okoro (1975) stated that in Nigeria it was very difficult for a person with albinism to marry, and he found that a very small proportion of affected women in his sample were married. Further, in Barnicot's (1952) Nigerian study only five of eight females and eight of 13 males, aged 20 years or older, were married. Barnicot added that he could not consider the question of fertility because of the limited available data and diverse ages of the subjects in his group.

Freire-Maia et al. (1978) discussed fertility in a small group of people with albinism who came from an isolate on Lencois Island (a Northern Brazilian island). The 10 affected people in the sample had on average 2.90 children per person, in comparison with 37 of their normally pigmented relatives who had 2.46 children per person. There was therefore no reason to conclude, from their findings, that fertility was reduced in the people who had albinism. They did not, however, assess fitness of their subjects in terms of survival of their children to reproductive age. Witkop et al. (1972) studied a Zuni and Brandywine American Indian isolate (in the United States), with a high rate of the condition, and found that data on fertility showed that males with albinism had a lower average fertility, but it was not significantly different from that of pigmented males.

In a study on 128 South African Black people with albinism Kromberg (1985) collected pedigrees on each family. The data on children, siblings, and parents' siblings were analyzed and compared with the relevant figures available for the general population. Altogether 35.9% of the sample had offspring; however, there were fewer males (22.6%) than females (48%) with children. The mean age for all subjects was 29.1 years, but for females with children it was 33.7 years, whereas for males with children it was 40.9 years. The data are shown in Table 2.1.

The lifetime infertility rate has been defined as "not ever having children and being 30 years or older" (Irwig, personal communication, 1984). This rate was estimated for males, females, and the total group. The findings are presented in Table 2.2.

The infertility rate of 210 per 1000 women with albinism was compared with figures for women in the Xhosa ethnic group living in a rural area of South Africa (the only group for which comparable figures were available at the time). The rate in this group for females was 57 per 1000

TABLE 2.1　Number of Females and Males With Albinism Who Had Children

Ages (Years)	Females		Males		Total	
	Total No	With Children No (%)	Total No	With Children No (%)	Total No	With Children No (%)
15–19	17	6 (35.3)	16	0 (0)	33	6 (18.2)
20–29	30	11 (36.7)	22	3 (13.6)	52	14 (26.9)
30–39	8	6 (75.0)	10	2 (20.0)	18	8 (44.4)
40–49	4	3 (75.0)	12	7 (58.3)	16	10 (62.5)
50–59	3	3 (100)	1	1 (100)	4	4 (100)
60–69	2	1 (50)	1	1 (100)	3	2 (66.7)
70+	2	2 (100)	0	0	2	2 (100)
Total	66	32 (48)	62	14 (22.6)	128	46 (35.9)
Mean age (years)	28.9	33.7	29.2	40.9	29.1	35.9

After Kromberg, J.G.R., 1985. A Genetic and Psychosocial Study of Albinism in Southern Africa (Ph.D. thesis). University of the Witwatersrand, Johannesburg, South Africa.

TABLE 2.2　Lifetime Infertility Rate for Females and Males With Albinism (Aged 30–59 Years)

	Subjects		Lifetime Infertility
	Total No	Without Offspring No (%)	Per 1000 People
Females	19	4 (21.0%)	210
Males	24	13 (54.2%)	542
Total	43	17 (39.5%)	395

After Kromberg, J.G.R., 1985. A Genetic and Psychosocial Study of Albinism in Southern Africa (Ph.D. thesis). University of the Witwatersrand, Johannesburg, South Africa.

women (Irwig, personal communication, 1984). The infertility rate for women with albinism was therefore significantly higher ($P = .02$, Binomial test, Siegel, 1956). The data showed that women with albinism were less likely than the unaffected Xhosa women to have children in every age group, except the youngest group (15–19 years) in which the numbers were very small.

The number of children per subject with albinism was 0.89 per female and 0.63 per male (Kromberg, 1985). Although strictly comparable data were not available, and some subjects might still have had further children, these figures suggest that fertility is reduced in both affected males and females.

Fertility was also assessed in a group (N = 36) of obligatory heterozygotes (i.e., parents of an affected child) and an age matched control group (N = 36, couples with unaffected children and no history of albinism). The subjects in the obligatory group had the same number of children (2.08 per couple) as the controls (2.08), although the number in both groups was small. The data collected on 65 heterozygote x normal unions (fertility rate 4.22) compared with 59 control unions (fertility rate 3.63) also showed that there was no significant difference between the two groups, but again the sample was small (Kromberg, 1985).

In general, the data from these studies suggest that fertility may be compromised in people with albinism, but that this may be because of sociocultural reasons (which vary in different communities), rather than physiological reasons, for which there is no evidence. However, the fertility rates do not appear to be compromised in heterozygote unions, nor in those between heterozygotes and normal controls.

Life Expectancy

As long ago as 1729 Wafer stated that people with albinism were but "short lived" (Pearson et al., 1913), and other reports have suggested that their life expectancy is reduced (e.g., Okoro, 1975; Schrire, 1958; Aquaron, 1980). Much depends, however, on where they live and how much care they receive. King et al. (1980) stated that in Nigeria no OCA2 subjects over 40 years of age were found in a sample of 79 affected people, and skin malignancies appeared to be the major factor in causing a limited life span. It is possible that people with OCA2, OCA2B, or OCA3 live longer than those with the OCA1 type because the former groups have some little protection from the sun in the form of the pigment, which develops, but no relevant data seem to be available. Okoro's (1975) data on 1000 unclassified people with albinism also suggested that life expectancy is decreased. In the East Central State of Nigeria Okoro found that only 10% of affected people were aged 31–60 years, whereas 20% of unaffected dermatology clinic patients were in that age group. Okoro, however, admitted that his methods of ascertainment might have caused the older patients to be missed.

Barnicot (1952) on the other hand found that the ages of his 40 Nigerian subjects were not markedly divergent from those of the general population. There were six, or 22.5%, of his sample and 20% of the general population in the age range 35 years and older. These results, however, are subject to the limitations of his small sample. Burrell (unpublished papers) and Oettle (1963) also found that the ages of the individuals in their groups of affected people in the Transkei (Eastern Cape, South Africa) showed a normal distribution.

Froggatt (1960) studied 122 people with albinism in Northern Ireland and compared their age distribution with that of the general population. From the figures presented it is apparent that: 43% of his subjects and 23% of the general population were in the age range 0–14 years; 47% of both groups were aged 15–49 years; and 10% of his subjects and 24% of the general population were in the age group 50 years or older. It was evident, therefore, that for affected people, the younger group was overrepresented and the older group underrepresented. Froggatt suggested four potential reasons to account for the finding. First, the life expectancy of the person with albinism might be reduced compared with the general population, but clinical examination did not suggest systemic weaknesses or the susceptibility to any life-threatening conditions; nevertheless albinism might still impact negatively on life expectancy. Second, the frequency of albinism might have been increasing, either due to a change in mating patterns, or an increase in the mutation rate, both of which seemed unlikely. Third, the criteria of acceptance into the study might have led to the inclusion of more young than old people. It was possible that older people were missed because if they had white hair they would not attract attention, and old people with albinism might no longer have nystagmus (since it is said to improve with age), the presence of which was one of the criteria for inclusion in the study. Fourth, the methods of ascertainment meant that the study was more comprehensive for the younger age group.

A large study was carried out by Kromberg (1985) over a 12-year period on 254 people with albinism in a Black community living in Johannesburg. During this time only three deaths were reported. The three who died included two females aged ~1 and 61 years, respectively, and a male aged 44 years. Because the number of deaths was so few, it was impossible to construct a life table for the subjects in the sample. Therefore, to obtain a general idea of how long the affected people were living their ages were categorized according to the age groups used in the South African census (1980), and the figures for these groups were compared with those from the census data for the local Black population. The distribution of the ages of the subjects was found to be significantly different from that of the general population ($P < .001$, Kolmogorov–Smirnov test, Siegel, 1956). The group of people with albinism tended to be younger than the general population, the mean age of the two groups being 24.2 and 28.2 years, respectively. Those affected were overrepresented in three groups: those aged 4 years or less; 15–24 years; and 65–74 years (see Table 2.3). It is possible that the ascertainment methods led to more complete ascertainment in the very young age group (<4 years) because this study ran concurrently with another study during which infants (<12 months) with albinism were sought for a maternal-infant bonding research project on mother-infant dyads (Kromberg et al., 1987). Furthermore, the initial prevalence project was conducted mainly

TABLE 2.3   Ages of a Cohort of 254 People With Albinism Compared With the Ages of the General Population in Soweto, Johannesburg

Age (Years)	People With Albinism Expected Number[a]	People With Albinism Observed Number (%)	General Population Number (%)
<4	26	34 (13.5)	103,480 (10.0)
5–14	40	37 (14.6)	162,260 (14.6)
15–24	54	82 (32.3)	218,340 (21.2)
25–34	53	49 (19.3)	211,600 (20.5)
35–44	37	28 (11.0)	147,180 (14.3)
45–54	26	8 (3.1)	106,120 (10.3)
55–64	14	7 (2.7)	57,140 (5.6)
65–74	5	8 (3.1)	19,000 (1.8)
75+	2	1 (0.4)	6,580 (0.6)
Total	257	254 (100)	1,031,700

[a] *Calculated using the prevalence rate of 1:4000 (Kromberg and Jenkins, 1982).*
After Kromberg, J.G.R., 1985. A Genetic and Psychosocial Study of Albinism in Southern Africa (Ph.D. thesis). University of the Witwatersrand, Johannesburg, South Africa.

through visits to schools in the decade 1971–81, so that these subjects were in the 15–24 age group at the time the data on ages were analyzed. It is, however, difficult to explain the excess of subjects in the 65–74-year-old group. The numbers of observed subjects compared with those expected were found to be fewer in the 35–54-year age group. Only 14.1% of the subjects with albinism belonged to this group, whereas 24.5% of subjects were expected. The subjects in this age group either did not come forward to participate in the study or they had a higher mortality rate than other groups. In the 45 years and older age group 47 (18.3%) affected people were expected, but only 24 (9.3%) were observed. These findings suggest that people with albinism may actually have a shorter life span than their peers.

In the Kromberg (1985) study the differences in the age structure of the affected male and female populations, when compared with the general population, were investigated. The data on 132 males with albinism showed that the distribution of their ages in comparison with that of males in the general population was highly significantly different ($P < .001$). Specifically, males were underrepresented in the over 45-year group, in which about 25 were expected, but only 10 were found. These results suggest that these males might not live as long as their unaffected counterparts (however, it is possible that they moved to join their relatives in the rural areas, as they grew older). Similarly, the information regarding 117 affected females showed that there were fewer females in the 45+ group than were expected (14 compared with 21, respectively). However, the results showed only marginal evidence, which was not significant, of a difference in the distribution of ages in females with albinism and the general population ($P > .10$).

Although some ascertainment bias is possible in this study (Kromberg, 1985), these findings suggest that people with albinism were overrepresented in younger age groups and underrepresented in older age groups, and that their life expectancy may be reduced. The age distribution was particularly skewed at two specific points; there were more affected individuals than were expected in the 15–24 year old group and less than expected in the 45+ group. This pattern suggests that affected people might be sent from the rural to the urban areas for educational and employment reasons, whereas other unaffected people, finding limited opportunities in the urban areas, might move as adolescents to the rural areas; then occasionally after age 35 years, but generally after 45, many affected people either die of skin cancer or leave the city for their rural village. Those who do not die at this stage, it seems, can live to an old age and, in the 65+ group although the sample was very small (8 subjects), people with albinism were again overrepresented. This latter group presumably represents those who were well cared for, economically comfortable, nourished, sheltered, throughout their lives, avoided sun exposure, used antiactinic creams and/or were not susceptible to malignancies (or had them detected and treated early) at a young age.

Okoro (1975) stated that the age distribution of his sample of Nigerians with albinism showed a rapid decline in numbers after age 30, and although he admitted that this might, also, have been because of his methods of ascertainment, he nevertheless concluded that affected people probably had a shorter life span. Oettle (1963), however, presented age statistics for the Transkei area of South Africa, which indicated that there was no discrepancy between those for people with albinism and for the general population, and he concluded that the life expectancy of affected people did not seem to be reduced in that area. This finding fits well with that observed during a study on skin cancer in Soweto, Johannesburg (Kromberg et al.,

1989) because Xhosa people with albinism (who generally live in the Transkei and Ciskei, now the Eastern Cape province of South Africa) were noted to have markedly reduced rates of skin cancer when compared with affected people from some of the other ethnic groups. It appears, therefore, from Oettle's (1963) study that those affected in the Transkei, can expect a normal life span because they do not appear to succumb to skin cancer.

Aquaron (1980) claimed that in the Cameroon people with albinism had a shortened life span. Furthermore, in Nigeria no people with albinism older than 40 years were seen by King et al. (1980). The different types of OCA, however, have not been investigated separately for longevity, and it might be assumed that the milder the type the less effect the condition will have on life span. Barnicot's (1952) claim that longevity was not compromised in affected people in Nigeria is at variance with the findings of King et al. (1980), but it is possible that regional differences could explain these two findings since Barnicot worked in a coastal region and King et al. in the East Central region of Nigeria. In addition the lifestyle that existed in the 1950s, when Barnicot was investigating the condition may have undergone radical changes in the intervening 30 years. From the findings of the Kromberg (1985) study and the Oettle (1963) study, which also vary, regional differences may also occur in South Africa.

The data from the South African study (Kromberg, 1985) show that both affected males and females were overrepresented in the younger age groups and underrepresented in the older groups compared with the general population, which supports the finding that affected subjects are likely to have a reduced life span. For males, the difference was highly significant, whereas for females it was borderline not significant. Expectation of life for local Black males and females was estimated at 55 years and 60 years, respectively, in 1980 (Mostert and van Tonder, 1982). (It is lower now in the early 21st century, due to the HIV/AIDS epidemic in South Africa). The results showed that some of the affected people could reach the expectation of life of the general population. However, the difference in life expectancy for men and women might be explained by their differing lifestyles. It is probable in the local situation that men with albinism are subjected to more sun exposure, especially where they have limited education, restricted employment opportunities and, consequently, need to work as manual laborers (often out of doors). Women, however, although in the rural areas they may tend the fields, may be able to avoid sun exposure more successfully. Furthermore women, if they do not marry, may remain sheltered by their biological family for life, whereas the men may leave their parental home in search of independence and employment.

There is therefore still some debate regarding the life expectancy of both White and Black people with albinism. There is also recognition that many factors affect the life span, these include cultural, lifestyle, health and medical care, gender, and social class, as well as psychosocial and economic factors.

III. TYPES OF ALBINISM

Several decades ago 10 different types of albinism, differentiated by their clinical, biochemical, and genetic characteristics, were described by Witkop et al. (1983). These included five common types (then named tyrosinase-negative, tyrosinase-positive, yellow mutant, brown, and rufous) and five rarer types, which included those associated with syndromes (Hermansky–Pudlak, Chediak–Higashi, and Cross), as well as the autosomal dominant OCA, and the Black Locks Albinism syndrome. However, in 1985 King and Olds published a paper in which they used different classes and types determined by the tyrosinase activity in each type. This classification was superseded again, in the late 1980s and 1990s, due to the dramatic findings identifying some of the causative genes by the molecular geneticists. Types of albinism due to *tyrosinase* gene mutations were classified as OCA1, *OCA2* gene mutations as OCA2 with subtypes OCA2A (OCA2 without pigmented freckles/ephelides), OCA2AE (OCA2 with pigmented freckles/ephelides), OCA2B (brown albinism), and OCA3 (rufous albinism) due to *tyrosinase-related protein 1* gene mutations. In 1996 Oetting et al. wrote a review on the clinical spectrum of albinism in humans. They provided evidence that multiple loci were associated with human albinism and stated that mutations associated with slow accumulation of pigment in the hair and eyes over time, while the ocular defects remain consistent, were of particular interest. Investigation of these mutations, they suggested, might provide insights into the interaction between the pigment system and the development of the optic system.

There is still some difficulty in classifying an individual with albinism unless all the relevant symptoms and signs are manifest and full information on the characteristics and the necessary test results (such as molecular genetic studies) are available. Without this information one type may appear very similar to another, and the brown and rufous categories, for example, show some clinical overlap. Molecular testing is the method of choice to confirm a clinical diagnosis, if required, and to categorize the type of OCA since the genes involved in the common forms of albinism have now been identified. However, where molecular testing is not available (as is the case in many African countries), clinical characteristics may be useful, although it is well recognized that they do not constitute a fully reliable method, in determining the type of albinism in an affected individual. Furthermore, while a few mutations account for most pathogenic alleles not all mutations have been identified. For example, although it has been shown that brown albinism is caused by *OCA2* mutations, and these individuals have been shown to carry the common 2.7 kb deletion, a second mutation is yet to be identified (Manga et al., 2001).

The classification of albinism in Online Mendelian Inheritance in Man (OMIM) includes nine different types of albinism. The genetic characteristics of three types, OCA1–3, which have been described in Africa, are detailed in Table 2.4.

OCA2 is possibly the commonest form of albinism worldwide, due mostly to its high frequency in sub-Saharan Africa. It is also the commonest autosomal recessive genetic condition in Black populations in Southern Africa. Sickle cell anemia is the most common recessively inherited disorder in other areas of Africa, where high carrier frequencies occur because of heterozygote advantage where malaria is endemic. OCA affects at least 1 in 3900 black individuals in South Africa (Kromberg and Jenkins, 1982) and in some Southern African areas, frequencies are as high as 1 in 1300 (Botswana) (Kromberg, 1985).

OCA1 (OMIM 203100) is the result of *tyrosinase* gene (*TYR*, localized to chromosome 11) mutations that significantly reduce the amount of functional tyrosinase enzyme. The most severe form was initially referred to as tyrosinase-negative OCA because there is complete loss of tyrosinase activity (Fig. 2.1 shows a child with OCA1). Type 2 OCA (OMIM 203200), which has a slightly milder phenotype, with normal to high tyrosinase activity (tyrosinase-positive OCA), results from mutations in the *OCA2* gene (formerly the *P* gene, localized to

TABLE 2.4   Genetic Characteristics of Three Types of Albinism Observed in Africa

Type of Albinism	Inheritance	Locus	Chromosome	OMIM[a] No
OCA1 (previously tyrosinase-negative)	Autosomal recessive	TYR	11q14-21	203100
OCA2 (previously tyrosinase-positive)	Autosomal recessive	*P*	15q11-13	203200
OCA2B—brown	Autosomal recessive	*P*	15q11-13	203200
OCA3—rufous	Autosomal recessive	*TYRP1*	9p23	203290

OCA, oculocutaneous albinism.

[a] *Online Mendelian Inheritance in Man (OMIM) Reference number.*

Adapted From Kromberg, J.G.R., Bothwell, J., Kidson, S.H., Manga, P., Kerr, R., Jenkins, T., 2012. Types of albinism in the Southern African population. E. Afr. Med. J. 88 (4), 124–131.

FIGURE 2.1   A child with oculocutaneous albinism type 1. *Photograph by Rick Guidotti, Positive Exposure. Courtesy of the National Organization for albinism and hypopigmentation (http://www.albinism.org/).*

chromosome 15q). Brown OCA (OCA2B) is also linked to the *OCA2* locus, and it is therefore a subtype of OCA2. Type 3 OCA (OMIM 203290), also known as red or rufous OCA, is caused by mutations in the *TYRP1* gene on chromosome 9p. These and the other conditions associated with albinism will be discussed in more detail in the chapter on molecular genetics (see Chapter 5).

Types of Albinism Found in Southern Africa

The focus of this section is on the types of albinism most frequently observed in Southern Africa, viz. OCA2 (including Brown OCA) and OCA3/Rufous OCA. Because the different types found in other parts of Africa have not yet been specifically identified and described, and there may be some differences between the affected populations living in the Northern and Southern parts of the continent, the discussion will focus on those types occurring in Southern Africa.

Studies on types of albinism and their clinical features were conducted by Kromberg (1985) and Kromberg et al. (2012) to determine which types were most commonly found in Southern Africa, what features they had and what the implications were for the affected individuals and their families. The initial study was a long-term one, and subjects were ascertained over a 15 year period, by various means (Kromberg, 1985). Community surveys and fieldwork were carried out in Johannesburg (South Africa's largest city) and the neighboring countries of Lesotho, Botswana, and Swaziland. Visits were made to schools, health centers, clinics, and hospitals to inform local communities about albinism and the ongoing research projects, as well as the services available for affected people. As affected individuals were identified, they received genetic counseling, health education, leaflets on albinism, free antiactinic cream, and referrals to a dermatology clinic and to the Albinism Society of South Africa. Also, they were informed about ongoing research projects and were invited to participate if they wished to do so.

Families of affected individuals were also approached and many volunteered to participate in various studies, some of which have been reported previously (for example see Kromberg and Jenkins, 1982; Kromberg, 1985; Kromberg et al., 1989, 1990, Ramsay et al., 1992; Bothwell, 1997; Manga et al., 2001). The three cohorts, all from the Black African population, who presented with albinism and were included in the study on types of albinism, were the following:

1. Affected individuals (96) identified through community surveys in Soweto and Johannesburg. Individuals were classified by means of clinical examinations, and in some cases classification was confirmed later by molecular studies;
2. Affected individuals (62) from the greater Johannesburg area who presented at a hospital dermatology clinic where they were examined by a dermatologist for skin problems;
3. Affected individuals ascertained from the greater Johannesburg area, as well as during a community survey in Lesotho, specifically for studies on the OCA2B (11) and OCA3 (20) types of albinism (a few of whom were also included in group 1 and 2 above).

Individuals in the three groups were examined and data collected to describe and classify (initially using the criteria of Witkop et al., 1983) the common types of albinism found locally. Clinical examinations were performed and documented by several different medical practitioners, who completed a standard form specifically compiled for the study, and skin examinations were carried out by a dermatologist. Blood samples were collected for molecular studies, from subjects who were willing and gave informed consent, and the results of these studies have been reported elsewhere (for example see Ramsay et al., 1992; Manga et al., 1997, 2001; Kerr, 2007). Statistical analysis was used to identify the significance of differences between groups of data (a P value of $<.05$ was considered significant).

Classification: Initial clinical examination of the first group of 96 subjects showed that the majority (82%) were classified as having OCA2. No subjects were classified as having OCA1 because all had some form of pigmentation. Details of the classification are shown in Table 2.5. This sample of subjects cannot be said to be representative of the total group of people with albinism as they were volunteers who happened to present to our research team, but the results indicate a trend suggesting that OCA2 occurs more frequently than either OCA2B or OCA3.

TABLE 2.5   Clinical Types of Albinism Found in 96 Subjects

Type	No (%)
OCA 1	0 (0)
OCA 2	79 (82)
OCA2B/brown OCA	11 (12)
OCA3/rufous OCA	6 (6)
Total	96 (100)

OCA, oculocutaneous albinism.
Adapted From Kromberg, J.G.R., Bothwell, J., Kidson, S.H., Manga, P., Kerr, R., Jenkins, T., 2012. Types of albinism in the Southern African population. E. Afr. Med. J. 88 (4), 124–131.

OCA2: clinical features and subtypes. Two distinct subtypes became apparent in the OCA2 group (excluding OCA2B) during clinical examinations: subjects who had no pigmented freckles (OCA2A) and subjects who had such freckles or ephelides (OCA2AE) (Figs. 2.2 and 2.3 show two individuals with OCA2, one without and one with freckles). Of the 79 OCA2 subjects, 73 were examined for freckles and 41 (56%) had none, while 32 (44%) had freckles on skin of sun-exposed areas. To assess whether freckles developed in all affected members of a family, 22 sibships were examined, and the results are presented in Table 2.6. Because freckles were noted to be generally absent in children under the age of 10 years, this group was excluded from the analysis. Siblings were usually concordant for freckling with concordance occurring more often than discordance ($\chi^2 = 6.6$, $P < .01$).

On examination of clinical information available for the OCA2 subjects (N = 33 without freckles and N = 23 with freckles) those without freckles had more nevi and keratoses than did the subjects with freckles (see Table 2.7).

Molecular studies show that a 2.7 kb deletion accounts for 78% of OCA2 chromosomes (Stevens et al., 1995). The remaining mutations remain somewhat elusive (Kerr et al., 2001). Those without freckles and those with them cannot be differentiated on a molecular level at present.

OCA2B (Brown albinism, subtype of OCA2). Eleven individuals with OCA2B were identified and examined (some have been included in an article reporting comparison of OCA3 and OCA2B, Kromberg et al., 1990). Brown OCA also maps to the *OCA2* locus. Affected individuals are heterozygous for the common *OCA2* 2.7 kb deletion, but the second pathogenic mutation is yet to be identified. Affected individuals were clinically distinct from those with OCA2 because of their light cream colored skin, ginger to brown hair, which darkened with age, eyes, which were often hazel to brown, and ability to tan without too much concomitant skin damage (see Fig. 2.4). In comparison to individuals with OCA2, those with OCA2B have

FIGURE 2.2   A young adult with oculocutaneous albinism type 2A. *Photograph courtesy of Dr. J. Kromberg, University of the Witwatersrand, South Africa.*

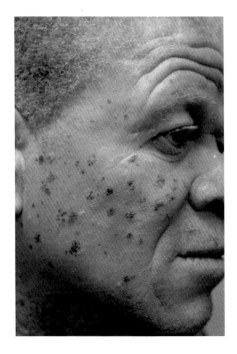

FIGURE 2.3   An individual with oculocutaneous albinism type 2AE. *Photograph courtesy of Dr. J. Kromberg, University of the Witwatersrand, South Africa.*

TABLE 2.6   Freckles in Sibships With Oculocutaneous Albinism Type 2

	Sibships
Freckles	No (%)
Present in all siblings	8 (36)
Absent in all siblings	9 (41)
Present in some, absent in some	5[a] (23)
Total	22 (100)

[a] includes one 3 child sibship in which 2 siblings had freckles and one did not.
Adapted From Kromberg, J.G.R., Bothwell, J., Kidson, S.H., Manga, P., Kerr, R., Jenkins, T., 2012. Types of albinism in the Southern African population. E. Afr. Med. J. 88 (4), 124–131.

TABLE 2.7   Skin and Eye Defects in Oculocutaneous Albinism Type 2 Subjects (N = 56) Without (N = 33) and With Freckles (N = 23)

Defect	Subjects Without Freckles	Subjects With Freckles	No Information	Total
	No (%)	No (%)	No	No (%)
Naevi present	31 (94)	18 (78)		49 (87)
Elastosis	32 (97)	23 (100)	1	55 (98)
Keratosis	28 (85)	14 (61)		42 (75)
Carcinomas	3 (9)	2 (9)	1	5 (9)
Keloids	1 (3)	0 (0)	1	1 (2)
Nystagmus	33 (100)	23 (100)		56 (100)
Translucency	11 (33)	3 (13)	1	14 (25)

Adapted From Kromberg, J.G.R., Bothwell, J., Kidson, S.H., Manga, P., Kerr, R., Jenkins, T., 2012. Types of albinism in the Southern African population. E. Afr. Med. J. 88 (4), 124–131.

a darker skin color, fewer visual problems, such as nystagmus, and more of them have brown eyes. The characteristics of people with OCA2B and OCA3 have been compared here, and the findings (presented in Table 2.8) show that these two types have different pigmentation.

OCA3, rufous/red albinism. This type of albinism (see Fig. 2.5) is visually striking. Despite being more heavily pigmented than people with OCA2 and OCA2B, when requests are circulated during field studies asking people with albinism to report to a specific place, individuals with the OCA3 also present themselves (see Kromberg et al., 1990). Evidently, members of the community see such individuals as different and rejection at birth has been reported in at least one case where the mother was unhappy at the sight of her unusually pigmented infant.

Affected individuals usually have a light brick red skin and gold hair, which often darken with age to red-brown and ginger, respectively, reduced visual acuity, and/or photophobia, and some have nystagmus and/or strabismus, but none has blue eyes.

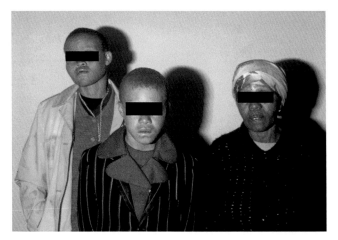

FIGURE 2.4   Two sisters with oculocutaneous albinism type 2B and their mother. *Photograph courtesy of Dr. J. Kromberg, University of the Witwatersrand, South Africa.*

TABLE 2.8   Skin Pigment, Hair and Eye Color, and Visual Defects in 11 Individuals With Brown Albinism (Oculocutaneous Albinism Type 2B) and 20 Individuals With Rufous Albinism (Oculocutaneous Albinism Type 3)

Characteristic	OCA2B		OCA3	
	No (%)	Total Examined	No (%)	Total Examined
Skin				
Light tan	11 (100)	11		
Red			19 (100)	19
Naevi	7 (60)	11	3 (33)	10
Freckles present			0 (18)	18
Hair Color				
Ginger	8 (73)	11	17 (94)	18
Light brown	3 (27)	11		
Reddish			2 (11)	18
Iris Color				
Blue	2 (18)	11	0 (0)	20
Hazel	3 (27)	11	6 (30)	20
Brown	6 (55)	11	14 (70)	20
Nystagmus	5 (55)	9	8 (42)	19
Strabismus	2 (22)	9	1 (10)	10

OCA, oculocutaneous albinism.
Adapted From Kromberg, J.G.R., Bothwell, J., Kidson, S.H., Manga, P., Kerr, R., Jenkins, T., 2012. Types of albinism in the Southern African population. E. Afr. Med. J. 88 (4), 124–131.

FIGURE 2.5   Two siblings with oculocutaneous albinism type 3 with their unaffected sister. *Photograph courtesy of Dr. J. Kromberg, University of the Witwatersrand, South Africa.*

Differentiating Between Albinism Types

It was generally possible to distinguish between the four types of albinism commonly found in Southern Africa (including the three OCA2 subtypes, namely OCA2A, OCA2AE and OCA2B, as well as OCA3 or rufous OCA) on clinical grounds, based on the characteristics presented above. This classification was not confirmed by molecular testing in every subject. However, when those in the rufous category, for example, were subsequently tested for *TYRP 1* mutation, their classification was verified 100% (Manga et al., 1997), providing some credibility for our assessment in the field. For the purpose of clinical classification, therefore, Table 2.9 provides all the characteristics documented for each type, including the molecular findings, in a form similar to that presented by Witkop et al. in 1983, King et al. in 1995, and Kromberg et al. in 2012.

Thus the four common types of albinism among Black Southern Africans can be clinically distinguished from each other (with little overlap), using a few simple guidelines. These findings are similar to those of Stannus (1913), who described various grades of albinism, more than 100 years ago, in Nyasaland (now Malawi).

Results from these studies show that the vast majority of Black Southern African individuals with albinism have OCA2, and the remainder has OCA3 or rufous OCA. Although ascertainment bias is a confounding factor, observations of affected members of over 600 families (enrolled in many different South African research projects) support this finding. At present, the proportion of individuals with OCA2B and OCA3 cannot be estimated with more certainty than 12% and 6%, respectively, based on the group presented here. These may be underestimates because individuals with these forms of albinism are probably less likely to present for research purposes since they experience fewer problems than those with OCA2 and are unlikely to be similarly motivated. Furthermore, because of the heterogeneity of the local population and the broad range of skin color, associated with both San and White admixture, people with OCA2B or OCA3 are not as easily detected in the general population

TABLE 2.9 Characteristics of the Types and Subtypes of Albinism Found Among the Black People of Southern Africa

Characteristic	OCA2A (Without Freckles)	OCA2AE (With Freckles)	OCA2B (Brown OCA)	OCA3 (Rufous OCA)
Hair color	Light yellow–gold, darkens with age	Yellow–gold darkens with age	Light brown to dark brown, darkens with age	Gold, ginger, or reddish—darkens with age
Skin color	White to cream	White to cream	Cream to light tan	Reddish brown
Freckles	Absent	Present ++	May be present	Absent
Susceptibility to skin cancer	+++	++	+	May be present
Reflectance spectrophotometry	Lighter skin color than caucasoids	Lighter skin than caucasoids		? No characteristic spectrum
Eye color	Blue to light brown	Blue to brown	Blue (few) to hazel or brown	Hazel to brown
Iris translucency	++	+	+	Unusual
Red reflex	May be present	Rarely present	Rarely present	Absent
Fundal pigment	Little	Little	Little?	Normal to reddish
Nystagmus	+++	++	+	May be present
Photophobia	+++	++	+	May be present
Visual acuity	Infants severe adults better, but still poor	Infants severe, adults better, but still poor	May be defective	Defective to normal
Strabismus	May be present	May be present	Occasional	Rare
Tyrosinase assays	Normal levels	Normal levels	Normal levels	Normal to high levels
EM studies	Present	Present	Present	Present
Melanosomes	Type I–III	Types I–III (type IV present in freckles)	Types I–III	Types I–IV
Visual evoked potential	Decussation defect present	Decussation defect present	Decussation defect may be present	Decussation defect may be present in some
Genetics	AR[a], P gene mutations on chromosome 15q11	AR[a], P gene mutations on chromosome 15q11	AR[a], P gene mutations on chromosome 15q11	AR[a], TYRP1 gene mutations on chromosome 9p23
Other	In all population groups	Possibly recognized only in black Africans?	Possibly recognized only in Africans?	Common in Africa, Papua New Guinea and the Pacific Islands

OCA, oculocutaneous albinism.
[a] AR, autosomal recessive.
Adapted From Kromberg, J.G.R., Bothwell, J., Kidson, S.H., Manga, P., Kerr, R., Jenkins, T., 2012. Types of albinism in the Southern African population. E. Afr. Med. J. 88 (4), 124–131.

as are individuals with the classic OCA2. A study of OCA3 in the Johannesburg area, of South Africa, found a frequency of about 1 in 8500 (Kromberg et al., 1990), supporting the idea that the 6% estimate presented above is very likely to be an underestimate. Barnicot (1952) in his Nigerian study suggested that OCA3 occurred quite frequently there. The condition has also been described in Cameroon (Aquaron et al., 1981).

The situation with individuals with the OCA2 subtype brown OCA is equally confounding, and differential diagnosis between this type of albinism and children of mixed ancestry as well as those with malnutrition induced depigmentation can be difficult in the early years. Although no further data are available, the 12% proportion of the group presented here for OCA2B is also likely an underestimate. Based on the characteristics of OCA2B and OCA3 albinism it is obvious, in most cases, that the individuals with OCA3 have more pigment than those with OCA2B. A few of the OCA2B subjects (2/11) had blue eyes, but none (0/20) of the OCA3 subjects had them. Also, half (55%) those with OCA2B had nystagmus, while slightly fewer (42%) of those with OCA3 had it.

In cases where individuals present with an uncharacteristic or intermediate phenotype, the possibility of interplay between the different genes involved in pigment production should be considered. Such a situation has been described in Southern African siblings, presenting with the unusual phenotype of OCA2 but reddish skin and hair (Manga et al., 2001). At the molecular level these siblings were found to be homozygous for *TYRP1* mutations, and therefore affected with OCA3, but were found, in addition, to be heterozygous for the 2.7 kb *P* gene deletion mutation, which caused a further reduction in pigmentation. The genetic basis of OCA types 1–4 has been reviewed by Rooryk et al. (2009) and will be further discussed in a later Chapter (see Chapter 5).

It is hoped that, for practical purposes, the findings from these Southern African studies will assist those working in the field (many of whom do not have access to molecular diagnostics), in Africa, to understand the characteristics of people with the different common types of albinism.

IV. PHYSICAL CHARACTERISTICS IN A SPECIFIC AFRICAN COMMUNITY: COMPARATIVE DATA

To obtain a picture of the range of characteristics found in a group of affected African people in South Africa, a study of the sample of 96 people with albinism (mentioned above), which included nine sibships, living in Johannesburg, was undertaken. All the subjects were examined clinically by two medical practitioners, and their physical characteristics were noted (Kromberg, 1985). The data were reanalyzed recently (in 2014, by the present author) and the participants classified, retrospectively, into three groups (OCA2, the OCA2B/brown subgroup and OCA3 or rufous group) according to these characteristics because no molecular tests were available at the time of the initial investigation. The findings regarding skin and hair color are presented in Table 2.10. In one sibling pair, one brother (the older of the two, aged 38 years) had very pale yellowish white hair, whereas his younger brother (aged 22 years) had hair with a distinct yellow tinge. The accumulation of pigment also varied within sibships. Hair color appeared to darken with age in two of nine sibships, in a further five the siblings of different ages had more or less the same hair color, and in the remaining two sibships the younger siblings had darker hair than the older ones.

TABLE 2.10   Hair Color and Skin Pigment in 96 Subjects Classified According to Type of Albinism

Characteristic	OCA2 No (%)	OCA2B Brown OCA No (%)	OCA3 Rufous OCA No (%)
Hair Color			
Light yellow	28 (35.4)		
Yellow	34 (43.1)		
Dark yellow/gold	14 (17.7)		
Light yellow brown	2 (2.5)	8 (72.7)	4 (66.8)
Red brown	0		1 (16.6)
Brown	0	3 (27.3)	
No information	1 (1.3)		1 (16.6)
Total	79 (100)	11 (100)	6 (100)
Skin Pigment			
Freckles[a]			
Present	32 (40.5)	7 (64)	
Absent	41 (52.0)	2 (18)	3 (50)
No information	6 (7.5)	2 (18)	3 (50)
Total	79 (100)	11 (100)	6 (100)

OCA, oculocutaneous albinism.
[a] *Includes ephelides and lentigines.*
After Kromberg, J.G.R., 1985. A Genetic and Psychosocial Study of Albinism in Southern Africa (Ph.D. thesis). University of the Witwatersrand, Johannesburg, South Africa.

In the group of people with OCA2B the majority of subjects (8/11, 72.7%) had fair or light brown hair. A similar hair coloring was found in some affected individuals in the OCA3 group, where in five subjects examined four had fair hair and only one had reddish-brown hair. A few affected people had dyed their hair black, and it was difficult to assess their original hair color.

Skin pigment, in the form of unusually large freckles, was found only in OCA2 and OCA2B. The freckles were generally about half to one centimeter in diameter, dark brown in color with a ragged edge. There was one subject with OCA2 in which the freckles were unusually small (+/−3mm in diameter), and they were almost confluent over the face and other exposed parts of the body. The three OCA3 subjects (aged 3, 8, and 15 years) examined for freckles had none at all.

The color of the irides of the subjects and the presence of nystagmus and strabismus were also noted wherever possible during the medical examination, and the findings are reported in Table 2.11. There were few subjects who could be said to have pure blue eyes. The majority of those with blue eyes had some form of yellow or light brown color radiating out from the

TABLE 2.11   Ocular Findings in 96 People With Albinism

Type Characteristic	OCA2	OCA2B Brown albinism	OCA3 Rufous albinism
	No (%)	No (%)	No (%)
Iris Color			
Blue	12 (15.2)	2 (18.2)	
Blue + yellow rays	49 (62.0)		
Hazel	12 (15.2)	3 (27.3)	1 (16.7)
Brown		4 (36.3)	3 (50.0)
No information	6 (7.6)	2 (18.2)	2 (33.3)
Total	79 (100)	11 (100)	6 (100)
	N/NE[a] (%)	N/NE[a] (%)	N/NE[a] (%)
Nystagmus	73/73 (100)	5/9 (55)	0/3 (0)
No information	6	2	3
Strabismus	30/59 (50.8)	2/9 (22)	0/3 (0)
No information	20	2	3
Total	79	11	6

OCA, oculocutaneous albinism.
[a] *Number with characteristic/total number examined (% with characteristic).*
After Kromberg, J.G.R., 1985. A Genetic and Psychosocial Study of Albinism in Southern Africa (Ph.D. thesis). University of the Witwatersrand, Johannesburg, South Africa.

pupil. Blue eyes, however, were found in a few subjects in all categories, except in the small sample of individuals with OCA3. Hazel eyes were observed in only 15% of the subjects in the OCA2 group, but they were more frequent in the OCA2B and OCA3 groups. In the latter two groups brown eyes, somewhat lighter in color than those of the general population were found quite commonly.

Nystagmus was observed in all subjects in the OCA2 group. In the OCA2B subgroup, however, only 55% had the condition and none of the small sample of OCA3 individuals was affected. The severity varied from a mild type, which only occurred in response to light, to a severe continuous roving nystagmus.

Strabismus was a common finding among subjects, and it was observed in half the OCA2 subjects and 22% of OCA2B subjects. This finding was not consistent among siblings and in several cases one sibling had severe strabismus, whereas the other one or two siblings were not affected at all. In the nine sibships examined there was only one where both siblings had strabismus.

Although this material was collected in a somewhat subjective manner, it presents a comprehensive and detailed picture of the wide range of physical characteristics, which can occur in a group of Black African individuals with albinism, and suggests how they can differ, even in affected siblings.

V. NATURAL HISTORY

The infant with albinism is often only diagnosed in the first few days of life unless the mother knows of a history in the family and recognizes the condition at birth. The medical personnel in the maternity ward will then generally give the parents some explanation concerning the cause and genetics of the disorder, the problems, which can be expected regarding the eyes and skin, how to manage these problems, and, ideally, referral to an ophthalmology, dermatology, and genetic counseling clinic. The parents generally take the infant home and, initially, bonding may only develop slowly because of the differences in skin and hair color between mother, father and child, the difficulty the parents have in identifying with the child, the nystagmus and the reduced visual acuity present in the child (Kromberg et al., 1987). Also, the nature and timing of the bonding, as well as acceptance of their situation, might differ between mother and father and might take longer in one or the other.

Later, integrating an affected child into the community may not be easy, and explanations regarding the condition, which may be difficult for the parents to provide, are often necessary (Morris et al., 2015). The hair and skin color of the children may darken with age, even in those of European extraction (King et al., 2007). However, every child with albinism needs to be protected from the sun, initially, taught to avoid sun exposure as much as possible, preferably to keep the hair longish (shaving the head, which is locally popular, should be avoided), wear a hat, long sleeved cotton shirts, long pants or dresses, and to use antiactinic creams on exposed areas of skin.

In the first 2 years the child may be slower to crawl and walk than other unaffected children because of the poor visual acuity (Kromberg et al., 1987). As the child grows older he/she may experience some rejection and teasing by peers and other members of the community. Parents should be aware of this situation, support the child, and bolster self-esteem. Parents need to take their small child to the dermatology and ophthalmology clinics for advice and checks on the skin and vision and to the low vision clinic to discuss spectacles and appropriate visual aids, prior to the child going to school. They also should attend the genetic counseling clinic (Morris et al., 2015) for an explanation of the nature, genetics, prognosis, and management of the condition, as well as the genetic risks associated with it (see Chapter 10).

When the child first enters school, the parents should discuss the condition with the school teacher and explain the need for the child to sit at the front of the class and away from the glare of windows. The teacher should also be informed that the child is normal in every respect, apart from visual problems and sun sensitivity, and that intelligence is within the normal range. At school, outdoor ball games may be difficult for the child; however, some sporting activities, such as athletics and swimming (bearing in mind that sun exposure should be limited), are possible. Occasionally affected children are sent to schools for the blind and partially sighted, sometimes for their own protection (especially if they are being stigmatized and/or abused in their own community), but this should not be necessary, unless the visual defect is very severe, and remaining with supportive parents, in their community and at the local school, is preferable (Lund, 2001).

After schooling is completed children with albinism can proceed to an institution of higher education, if their academic standard is high enough and their school leaving certificate qualifies them for university entrance. Several universities (such as the University of the

Witwatersrand, Johannesburg, who set up their disability office in 1986) have a disability office with specialized assistants to help students with various disabilities. They are also able to assist partially sighted students with assistive devices, IT equipment (such as Zoom text, which can be used to magnify the text on the computer screen), visual aids, editing lecture notes and other material, managing tests and exams, as well as information and psychosocial support (A. Sam, personal communication, 2017). Other young adults with albinism can seek places in technical colleges, or obtain employment, but preferably they should avoid those jobs which involve daily sun exposure or require excellent vision. Depending on the extent of their visual impairment, many will not be able to drive a car.

Marriage and permanent relationships might be difficult for both affected men and women. However, many have succeeded in finding a suitable partner, starting a family, and producing healthy children. Genetic counseling should be offered to the affected individual and the couple, so that the recurrence risks and choices are understood, and decisions can be made about carrier testing, reproduction, child bearing, and prenatal diagnosis. The couple should also understand how to manage the condition, so that if they have affected children, they will provide them with the best possible quality of life, and the best chance of achieving their potential.

As affected individuals age they should be aware that early treatment of their solar keratoses is essential to prevent the development of skin cancers, and that the use of sun barrier creams is always necessary and life-long. They should know about, and reject, the myths that are common in sub-Saharan African communities regarding their unusual death (Kromberg, 1992). They need to be capable of facing these myths and reassuring themselves, and others that their only difference from other human beings is their skin color and visual limitations, also, that their bodies are made of skin, bone, tissue, and blood, like others are, and so they will die, at the end of their lives, as others do.

VI. CONCLUSION

In conclusion, many of the characteristics of people with the differing types of OCA are similar, basically affecting the eyes and skin to varying degrees. In many ways, the necessity for differentiating between the types seems minor. Some affected people will want to know which type they have, for scientific reasons, or out of pure curiosity. However, the treatment, on the clinical level, remains similar, as does the genetic counseling, although more emphasis should be placed on the essential nature of such treatment when pigment is minimal.

The previously little covered issues, associated with longevity (life span may be compromised in certain situations) and fertility (which may be somewhat reduced because of the presence of sociocultural rather than physiological factors), which are outlined in this chapter, were included here for the sake of completeness. However, they still require further investigation and research in different communities and at different times. Also, the nature of these characteristics, which can be influenced by many different individual psychosocial and cultural factors, probably changes together with the various changing environmental, community, social, and health issues, which surround the affected person.

Many of the topics covered briefly in this chapter will be picked up and discussed in more detail and more comprehensively in the clinical chapters that follow, particularly those on the ophthalmology and dermatology effects of having OCA (see Chapters 6 and 7). Also, detailed chapters on the psychosocial issues associated with the condition, how best to manage it and how to prevent the development of complications will be presented (see Chapters 9 and 13).

References

Aquaron, R., 1980. L'albinisme oculo-cutane au Cameroun. A propos de 216 observations. Rev. Epidemiol. Sante Publique 28, 81–88.

Aquaron, R., Ronge, F., Aubert, C., 1981. Pheomelanin in albino negroes: urinary excretion in 5-S-cysteinylodopa in cameroonian subjects. In: Seijing, M. (Ed.), Pigment Cell. University of Tokyo Press, Tokyo, pp. 97–103.

Barnicot, N.A., 1952. Albinism in South West Nigeria. Ann. Eugen. 17, 38–73.

Barnicot, N.A., 1958. Reflectometry of the skin in Southern Nigerians and some mulattoes. Hum. Biol. 30, 150–160.

Beckham, A.S., 1946. Albinism in negro children. J. Genet. Psychol. 69, 199–215.

Bothwell, J., 1997. Pigmented skin lesions in tyrosinase-positive oculocutaneous albinos: a study of black South Africans. Int. J. Dermatol. 36, 831–836.

Cameron, D., 1979. On being an albino: a personal account. Br. Med. J. 1, 28–29.

Cohen, L., Shapiro, M.P., Keen, P., Henning, A.J.H., 1952. Malignant disease in the Transvaal: 1. Cancer of the skin. S. Afr. Med. J. 26, 932–939.

Edwards, E.A., Duntley, S.Q., 1939. The pigment and color of living human skin. Am. J. Anat. 65, 1–33.

Emery, A.E.H., 1979. Elements of Medical Genetics. Churchill Livingstone, Edinburgh.

Freie-Maia, N., de Andrade, F.L., de Athayde-Neto, A., Cavalli, I.J., Oliviera, J.C., Marcallo, F.A., Coelho, A., 1978. Genetic investigations in a Northern Brazilian island. II. Random drift. Hum. Hered. 28, 401–410.

Froggatt, P., 1960. Albinism in Northern Ireland. Ann. Hum. Genet. 24, 213–233.

Gronskov, K., Ek, J., Brondum-Nielsen, K., 2007. Oculocutaneous albinism. Orphanet J. Rare Dis. 2, 43–50.

Hardy, A.C., 1936. Handbook of Colorimetry. The Technology Press, Cambridge, Mass.

Harris, R.G., 1926. The San Blas Indians. Am. J. Phys. Anthropol. 9, 17–58.

Harrison, G.A., Owen, J.J.T., 1956. The application of spectrophotometry to the study of skin color inheritance. Acta Genet. 6, 481–484.

Isaacson, C., Selzer, G., Kaye, V., Greenberg, M., Woodruff, J.D., Davies, J., Ninin, D., Vetten, D., Andrews, M., 1978. Cancer in the urban blacks of South Africa. S. Afr. Cancer Bull. 22, 49–84.

Jeffrey, G., 1997. The albino retina: an abnormality that provides insight into normal retinal development. Trends Neurosci. 20 (4), 165–169.

Keeler, C.E., 1953. The Caribe moon child and its heredity. J. Hered. 44, 163–177.

Kerr, R., 2007. Genes in the Aetiology of Oculocutaneous Albinism in sub-Saharan Africa and a Possible Role in Tuberculosis susceptibility (Ph.D. thesis). University of the Witwatersrand, Johannesburg, South Africa.

Kerr, R., Stevens, G., Manga, P., Salm, S., John, P., Haw, T., et al., 2001. Identification of P gene mutations in individuals with oculocutaneous albinism in sub-Saharan Africa. Hum. Mutat. 15, 166–172.

King, R.A., Creel, D., Cervenka, J., Okoro, A.N., Witkop, C.J., 1980. Albinism in Nigeria with delineation of a new recessive oculocutaneous type. Clin. Genet. 17, 259–270.

King, R.A., Olds, D.P., 1985. Hairbulb tyrosinase activity in oculocutaneous albinism: suggestions for pathway control and block location. Am. Med. Genet. 20, 49–55.

King, R.A., Hearing, V.J., Creel, D.J., Oetting, W.S., 1995. Albinism. In: Scriver, C.R., Beaudet, A.L., Sly, W.S., Vale, D. (Eds.), The Metabolic and Molecular Basis of Inherited Disease. McGraw-Hill, New York, pp. 4353–4392.

King, R.A., Oetting, W.S., Summers, C.G., Creel, D.J., Hearing, V.J., 2007. Abnormalities in pigmentation. In: Rimoin, D.L., Connor, J.M., Pyeritz, R.E., Korf, B.R. (Eds.), Emery and Rimoin's Principles and Practice of Medical Genetics, fifth ed. Churchill Livingstone, Elsevier, Philadelphia, pp. 3380–3427.

Kiprono, S.K., Chaula, B.M., Beltraminelli, H., 2014. Histological review of skin cancers in African albinos: a 10-year retrospective review. BMC Cancer 14, 157.

Kromberg, J., 1992. Albinism in the South Africa negro. IV. Attitudes and the death myth. Birth Defects Orig. Artic. Ser. 28 (1), 159–166.

Kromberg, J.G.R., Jenkins, T., 1982. Prevalence of albinism in the South African negro. S. Afr. Med. J. 61, 383–386.

Kromberg, J.G.R., Jenkins, T., 1984. Albinism in the South African negro. III. Genetic counselling issues. J. Biosoc. Sci. 16, 99–108.

Kromberg, J.G.R., 1985. A Genetic and Psychosocial Study of Albinism in Southern Africa (Ph.D. thesis). University of the Witwatersrand, Johannesburg, South Africa.

Kromberg, J.G.R., Zwane, M.E., Jenkins, T., 1987. The response of black mothers to the birth of an albino infant. Am. J. Dis. Child 141, 911–916.

Kromberg, J.G.R., Castle, D., Zwane, E., Jenkins, T., 1989. Albinism and skin cancer in Southern Africa. Clin. Genet. 36, 43–52.

Kromberg, J.G.R., Castle, D.J., Zwane, E.M., Bothwell, J., Kidson, S., Bartel, P., Phillips, J.I., Jenkins, T., 1990. Red or rufous albinism in Southern Africa. Ophthalmic. Paediatr. Genet. 11 (3), 229–235.

Kromberg, J.G.R., Bothwell, J., Kidson, S.H., Manga, P., Kerr, R., Jenkins, T., 2012. Types of albinism in the Southern African population. E. Afr. Med. J. 88 (4), 124–131.

Lund, P.M., 2001. Health and education of children with albinism in Zimbabwe. Health Educ. Res. 16 (1), 1–7.

MacNalty, A.S. (Ed.), 1965. Butterworths Medical Dictionary. Butterworths, London.

Manga, P., Kromberg, J.G.R., Box, N.F., Jenkins, T., Ramsay, M., 1997. Rufous albinism in SA blacks is caused by mutations in the TYRP1 gene. Am. J. Hum. Genet. 61, 1095–1101.

Manga, P., Kromberg, J.G.R., Turner, A., Jenkins, T., Ramsay, M., 2001. Southern Africa, brown oculocutaneous albinism (BOCA) maps to the OCA2 locus on chromosome 15q: P-Gene mutations identified. Am. J. Hum. Genet. 68, 782–787.

Manganyi, N.C., Kromberg, J.G.R., Jenkins, T., 1974. Studies on albinism in the South African negro. I. Intellectual maturity and body image differentiation. J. Biosoc. Sci. 6, 107–112.

Morris, M., Glass, M., Wessels, T.-M., Kromberg, J.G.R., 2015. Mothers' experiences of genetic counselling in Johannesburg, South Africa. J. Genet. Couns. 24, 158–168.

Mostert, W.P., Van Tonder, J.L., 1982. Moontlike bevolkingsgroei in Suid-Afrika tot die middle van die 22e eeu. Raad vir Geesteswetenskaplike Navorsing, Pretoria, pp. 5–83. RGN verslag.

Oetting, W.S., Brilliant, M.H., King, R.A., August 1996. The clinical spectrum of albinism in humans. Mol. Med. Today 330–335.

Oettle, A.G., 1963. Skin cancer in Africa. Natl. Cancer Inst. Monogr. 10, 197–214.

Okoro, A.N., 1975. Albinism in Nigeria. A clinical and social study. Br. J. Dermatol. 92, 485–492.

Online Mendelian Inheritance in Man (OMIM), http://www.ncbi.nlm.nih.gov/Omim/.

Pearson, K., Nettleship, E., Usher, C.H., 1913. A Monograph on Albinism in Man. Drapers Co., Research Memoirs Biometric Series, vol. VIII. Dulau, London.

Preising, M.N., Gonzer, M., Lorenz, B., 2011. Screening of TYR, OCA2, and MC1R in patients with congenital nystagmus, macular hypoplasia and fundus hypopigmentation indicating albinism. Mol. Vis. 17, 939–948.

Ramsay, M., Colman, M.-A., Stevens, G., Zwane, E., Kromberg, J., Farrell, M., Jenkins, T., 1992. The tyrosinase-positive oculocutaneous albinism locus maps to chromosome 15q11.2-q12. Am. J. Hum. Genet. 51, 879–884.

Rinehart, W., (Ed.), 1979. Age at Marriage and Fertility: Africa, vol. 7, Population Reports, pp. 130–138.

Rippey, J.J., Schmaman, A., 1972. Skin tumours of Africans. In: Marshall, J. (Ed.). Essays in Tropical Dermatology, vol. 2. Excerpta Medica, Amsterdam, pp. 98–115.

Roberts, D.F., 1977. Human pigmentation: its geographical and racial distribution and biological significance. J. Soc. Cosmet. Chem. 28, 329–342.

Roberts, D.F., Kromberg, J.G.R., Jenkins, T., 1986. Differentiation of heterozygotes in recessive albinism. J. Med. Genet. 23, 323–327.

Rook, A., 1969. The ages of man and their dermatoses. In: Rook, A., Wilkinson, D.S., Ebling, F.J.C. (Eds.), Textbook of Dermatology, vol. 1. second ed. Blackwell, Oxford, pp. 138–153.

Rooryk, C., Morice, F., Lacombe, D., Taeib, A., Arveiler, B., 2009. Genetic basis of oculocutaneous albinism. Expert Rev. Dermatol. 4, 611–622.

Rose, E.F., 1974. Pigment anomalies encountered in the Transkei. S. Afr. Med. J. 48, 2345–2347.

Schrire, T., 1958. Aetiology of facial cancer. A speculative and deductive survey. S. Afr. Med. J. 32, 997–1002.

Siegel, S., 1956. Nonparametric Statistics. McGraw-Hill, New York.

Stannus, H.S., 1913. Anomalies of pigmentation among natives of Nyasaland. Biometrika 9, 333–365.

Stevens, G., van Beukering, J., Jenkins, T., Ramsay, M., 1995. An intragenic deletion of the P gene is the common mutation causing tyrosinase-positive oculocutaneous albinism in southern African negroids. Am. J. Hum. Genet. 56, 586–591.

Stewart, H.F., Keeler, C.E., 1965. A comparison of the intelligence and personality of moon-child albino and control cuna Indians. J. Genet. Psychol. 106, 319–324.

Tobias, P.V., 1961. Studies on skin reflectance in Bushman-European hybrids. In: Proceedings of the 2nd International Congress on Human Genetics, Rome, pp. 461–471.

Vallois, H.V., 1950. Sur quelques points de l'anthropologie des noirs. L'anthropologie 54, 272–286.

Waardenberg, P.J., 1947. Recognising heterozygote heredity in ocular and universal albinism by the permeability of the iris to light (Dutch). Ned. Tijdschr. Geneeskd. 91, 1863–1866.

Waardenberg, P.J., Klein, D., Franschetti, A., 1961. Genetics and Ophthalmology. Royal van Gorcum, Assen.

Wassermann, H.P., Heyl, T., 1968. Quantitative data on skin pigmentation in South African races. S. Afr. Med. J. 42, 98–101.

Watkins-Pitchford, W., 1910. Pigmentation and cancer. Tvl. Med. J. 6, 23–25.

Weiner, J.S., Harrison, G.A., Singer, R., Harris, R., Jopp, W., 1964. Skin color in Southern Africa. Hum. Biol. 36, 294–307.

Weiner, J.S., 1951. A spectrophotometer for measurement of skin color. Man 253, 152–153.

Witkop, C.J., Niswander, J.R., Bergsma, D.R., Workman, P.I., White, J.G., 1972. Tyrosinase positive oculocutaneous albinism among the Zuni and Brandywine triracial isolate: biochemical and clinical characteristics and fertility. Am. J. Phys. Anthr. 36, 397–406.

Witkop, C.J., Hill, C.W., Desnick, S., Thies, J.K., Thorn, H.L., Jenkins, M., White, J.G., 1973. Ophthalmologic, biochemical, platelet, and ultrastructural defects in the various types of occulocutaneous albinism. J. Investig. Dermatol. 60, 443–456.

Witkop, C.J., Quevedo, W.C., Fitzpatrick, T.P., 1983. Albinism and other disorders of pigment metabolism. In: Stanbury, J.B., Wyngaarden, J.B., Frederickson, D.S., Goldstein, J.L., Brown, M.S. (Eds.), The Metabolic Basis of Inherited Disease. McGraw-Hill, New York, pp. 301–346.

Epidemiology of Albinism

Jennifer G.R. Kromberg

University of the Witwatersrand and National Health Laboratory Service,
Johannesburg, South Africa

I. INTRODUCTION

The epidemiology of albinism has been studied in numerous countries over many decades. These studies, however, have been of various quality, stringency, and sample size. Some have been based in urban areas, others in rural areas, some undertaken by explorers or casual observers, others by medical professionals, anthropologists, scientists, or community workers, and a few by epidemiologists. Sometimes the same population has shown different rates over time or in changing environmental circumstances, and various subpopulations and/or age groups of the same population might also show differing rates. Most of these investigations are older studies and very few newer ones based on molecular findings have been undertaken.

Many factors affect the reliability and validity of epidemiological studies. The diagnostic criteria used for inclusion of the subjects into the study group may vary depending on how rigorous the study is and the expertise of the researchers. Local health, environmental, and social and cultural conditions (such as preference for consanguineous unions) can determine the changes in epidemiological findings, which occur over time and in various geographical areas. Also, the ascertainment methods and data collection approaches used in the survey, e.g., hospital records based, region based (urban or rural), postal or community based, can give differing results.

In this chapter, several of the international epidemiological studies on albinism will be reviewed. Some studies with specific value and relevance and many of those based in Africa will be covered in more detail.

II. INTERNATIONAL EPIDEMIOLOGICAL STUDIES

Worldwide Prevalence

Albinism, which occurs in people of all ethnicities, affects about 1 in 17,000 people worldwide (Witkop, 1979), suggesting that 1 in 70 people carries a gene for one or other of the types of albinism (Gronskov et al., 2007).

In general, the prevalence of oculocutaneous albinism varies according to population group (see Table 3.1). In Europe, the overall frequency is about 1:20,000 with estimates ranging from 1:10,000 in Norway (Magnus, 1922) to 1 in 15,000 in the Netherlands (Van Dorp, 1987) and 1:29,000 in Italy (Raseri, 1879; cited in Pearson et al., 1913). In contrast, the frequency among the Cuna Indian isolates of San Blas Province, Lower Panama, is 1:200 (Keeler, 1964) and among the Hopi American Indians of Arizona it is 1:227 (Woolf and Grant, 1962). In Africa, prevalence estimates range from 1:5000 in Nigeria (Barnicot, 1952) to 1:2875 in one ethnic group in Cameroon (Vallois, 1950) and higher in a few isolates, which will be discussed below. The figures in Table 3.1 are best estimates for all types of albinism together, at the time. They are based on diagnosis by phenotype as no new published data based on large-scale mutation testing appear to be available. Furthermore, many of these figures originate from cursory studies and observations in the field by research workers and their colleagues.

Various suggestions have been put forward to explain the exceptionally high rates in groups such as the Hopi American Indians in the United States. One such proposal is based on cultural habits of the local people (Woolf and Dukepoo, 1969). Mansell (1972) proposed that, previously, in some South American villages affected people were regarded as holy beings unsuited for normal village life. He added that there they were revered and lead lives of luxury. Another possible cause is a founder mutation with a high prevalence in a small population, as has been found recently in the Cuna (also called Kuna) Indian Americans of Panama (Currasco et al., 2009). Other reasons, such as a cultural preference for consanguineous unions and selective advantage for carriers of the gene, have been suggested to account

TABLE 3.1 Frequency of Albinism in Different Populations

Population	Frequency	Source
EUROPE		
Denmark	1:14,000	Gronskov et al. (2009)
Italy	1:29,000	Pearson et al. (1913)
The Netherlands	1:15,000	Van Dorp (1987)
Northern Ireland	1:10,000	Froggatt (1960)
Norway	1:9,650	Magnus (1922)
Russia	1:100,000	Pearson et al. (1913)
Scotland	1:12,000	Pearson et al. (1913)
AMERICA		
Panama, Cuna	1:213	Stout (1946)
The United States	1:16,000	Witkop et al. (1983)
Hopi, Arizona	1:227	Woolf and Grant (1962)
Navajo, Arizona	1:1500–2000	Yi et al. (2003)
African American	1:10,000	Witkop et al. (1983)
White	1:19,000	Witkop et al. (1983)
Canada		
British Columbia	1:20,600	McLoed and Lowry (1976)
AFRICA		
Botswana (isolate)	1:1307	Kromberg (1985)
Cameroon (Badjoue)	1:2875	Vallois (1950)
Cameroon (Bamileke)	1:7900	Aquaron (1990)
Malawi	1:3400	Bar (1993)
Nigeria	1:5000	Barnicot (1952)
South Africa		
Transkei[a] (Xhosa)	1:3759	Oettle (1963)
Soweto, Johannesburg	1:3900	Kromberg and Jenkins (1982)
Swaziland	1:1951	Kromberg (1985)
Zimbabwe: Shona	1:4182	Lund (1996)
Tonga (isolate)	1:1000	Lund et al. (1997)
ASIA		
Borneo	1:5–10,000	Abrahams (1972)
China: Han population	1:18,000	Gong et al. (1994)
Japan	1:47,000	Neel et al. (1949)

[a] *Now in the Eastern Cape province of South Africa.*

for the high rates in Africa. A high rate of consanguinity has also been proposed to explain the increased rate of albinism found on the island of Borneo (Abrahams, 1972).

Figures are not available for prevalence in many countries and those for Russia, for example, are estimates seemingly based on educated guess work (Pearson et al., 1913). No epidemiological studies appear to have been carried out in the Middle East or Asian countries, such as India, although Chaki et al. (2006) undertook molecular work on 25 affected Indian families and found that OCA1 was the commonest type, occurring in 50% of their sample. Furthermore, little epidemiological research has been done in the countries of the Far East. However, in China, the Han population has been studied and found to have a prevalence rate of about 1 in 18,000 and in the Shandong province 3.83% of the population were estimated to be carriers (Gong et al., 1994); later, again, the most common form found was OCA1 (in 70% of cases, Wei et al., 2010). Also, an old study, reported by Neel et al. (1949), stated that a prevalence rate of 1:47,000 had been found in the Gifu prefecture of Japan and cursory observations made in Borneo suggested that it might be as common as 1 in 5000 to 1 in 10,000 there (Abrahams, 1972). By 2008, mutations causing OCA4 had been identified in Turkish, German, Dutch, French, Belgian, Moroccan, Indian, Korean, Japanese, and Brazilian patients, and mutations for OCA1, OCA2, and OCA3 had been identified in patients from many different countries (see the study carried out by Rooryck et al., 2008, in France, on multinational patients). Fig. 3.1 shows the prevalence of albinism in different countries worldwide.

Prevalence in Europe

The Northern Ireland (Froggatt, 1960) prevalence study is one of the most comprehensive and detailed of those carried out in Europe, and, therefore, although it was conducted prior to molecular testing becoming available, it is presented in some detail here. Froggatt aimed to present a genetic and statistical appraisal of the condition based on the data derived from a complete ascertainment in Northern Ireland. His objective was to describe the appearance of the typical European person with albinism so that more stringent criteria for inclusion into the affected group could be adopted. His criteria included presence of the following:

1. hypopigmented fundi oculi
2. congenital nystagmus
3. translucent irides
4. white, straw colored or fair hair, with white hair at birth.

Froggatt had multiple methods of ascertainment, which included reports from general practitioners and ophthalmologists, local health and welfare authorities, special schools and institutions, and opticians. In this way 122 individuals were ascertained; Froggatt estimated that ascertainment was 80% complete and that missed cases were likely to be of high social class, older in age, and in families with few affected members. All affected individuals were visited and examined by the author and a family and genetic history (including any consanguineous unions) were collected. Where possible, parents of affected individuals were examined for any pigmentary anomalies (particularly of the fundus).

On analysis, Froggatt's data supported the autosomal recessive gene hypothesis for the condition. The frequency of albinism calculated from 122 individuals in a population of 1,393,800 was found to be 8.75 per 100,000 (about 1 in 12,000). He refined this estimate for various groups

FIGURE 3.1 Shows the prevalence of albinism in different countries worldwide. *Modified map from https://stock.adobe.com/. File #: 13610494.*

in the population: in the counties of Londonderry (urban) and Tyrone (rural) the rates were 15.1 and 12.8, respectively, per 100,000; in the age group of 1–14 years the rate was 13.1; and assuming an ascertainment rate of only 80%, the rate was 10.9 per 100,000 or about 1 in 9174 people, in the total population. He suggested that the excess in the younger age groups was due to ascertainment being better in these age groups, and/or that life expectancy for people with albinism was less than that of the general population and/or the condition was becoming more common, and/or the criteria for acceptance were more appropriate for younger than older people. Froggatt found an excess of affected males (81) over females (57), and it is possible that in his study, males came forward more often than females for social reasons, as suggested previously by Haldane (1938), in his study on the frequency of recessive conditions in man. The consanguinity rate in Froggatt's study was 9% (half these unions being between first cousins and half between second cousins), whereas the rate in the general population was between 0.5% and 3.3%. The rate of 4.5% for first cousins was lower than that of 10.7% reported in Holland (Sanders, 1938) and 12% in the pedigrees reported by Pearson et al. (1913).

Furthermore, Froggatt considered the possibility that the gene had some expression in the heterozygous form and he examined some of the unaffected parents and children of people with albinism. However, although iris translucency was more common in these obligatory heterozygotes (70% were affected) than in the controls, it was not a consistent finding. Also, only a few minor pigment anomalies were reported and three subjects had white patches of hair, while one had white eyebrows and eyelashes (similar findings were reported by Sanders, 1938).

An earlier, extensive but less detailed study on albinism was undertaken in the Netherlands (Sanders, 1938). In 140 families with 702 children, 216 were found to have albinism. Of these, 115 were boys and 101 girls; thus there was no significant gender-based difference. Among all the male children 31% were affected and among all the female children 30% were affected. The prevalence rate was estimated as 1 in 20,000 people. The consanguinity rate in the parents of children with albinism was found to be ~10% compared with <1% in the general population. Extensive family pedigrees, showing the types of consanguinity and in some cases the family history of affected relatives, were compiled. Many affected individuals were also investigated and their features, such as iris color, astigmatism, and visual acuity, were recorded in detail. In some cases, the unaffected parents were also examined and minor pigmentary and other abnormalities noted.

This study was followed by one undertaken by Van Dorp (1985) working in Amsterdam as part of a doctoral project. She studied many families and suggested that the prevalence rate was higher than that found by Sanders and was probably nearer to 1 in 16,000. She also found that 10% of the children in the local schools for the visually handicapped had albinism (Van Dorp et al., 1983). Later Van Dorp (1987), having completed a 4-year study, stated that the prevalence for all forms of albinism in the Netherlands was 1 in 15,000, but that about 10% of these were cases of X-linked ocular albinism. Furthermore, she admitted that diagnosis was often difficult on clinical grounds because some patients with this condition could appear to be normally pigmented, whereas other patients with X-linked ocular albinism could have greatly reduced pigmentation compared with family members. She suggested that the albinism definition should be revised in a situation such as this one, that the word oculocutaneous should be abandoned and the term autosomal recessive albinism should be used instead (however, this still did not include the X-linked form).

Later a Danish study was carried out by Gronskov et al. (2009). They planned to investigate the birth prevalence of patients with autosomal recessive albinism based on the data collected retrospectively in a compulsory national birth register. Altogether 218 patients were identified from this register giving a minimum prevalence rate of 1 in 14,000. However, among these cases 55% appeared to have recessive OCA, whereas 45% apparently had autosomal recessive ocular albinism, suggesting that the actual prevalence rate of the types of OCA, alone, was very low. Gronskov et al. (2009) added that the clinical spectrum of albinism varies greatly, and in a population with a light complexion it may become especially difficult to differentiate between those with OCA and those with a form of autosomal recessive ocular albinism.

From the brief review of these few European studies, it seems that rates of albinism in Europe range from about 1 in 10,000 to 1 in 20,000. Presently, these figures cannot be further refined (until molecular studies are carried out) and no comparisons can be made because the criteria used for the types of albinism included in these surveys often varied. Also, patients with ocular albinism have sometimes been included, while such cases have generally been excluded in the studies emanating from Africa. Nevertheless, in general terms, the rates in Europe appear to be at least two or three times lower than those reported for African communities.

Prevalence in Central America

As Woolf and Grant (1962) state, the variation in the population frequency of albinism is an intriguing problem, especially considering the low rate in Europe (1 in 17,000 in newborns, including all known forms, Martinez-Garcia and Montoliu, 2013) and the high rates reported for Central America (e.g., 1 in 213 in the San Blas American Indians of Panama, Stout, 1946).

In 1946 Stout investigated the incidence and sociological position of people with albinism among the San Blas Cuna of Panama. He stated that the earliest reference to Cuna people with albinism was that of Salcedo, in 1640, and he based his information on the observations of a missionary Adrian de Santo Tomas. Stout (1946) found 98 affected people in a population of 20,831, giving an incidence of about 0.47% (or 1:200). However, the incidence was higher on islands where missionaries were active and consequently infanticide of affected infants, although still carried out in other areas, was reduced.

In 1953 Keeler studied albinism in the San Blas province on the Caribbean coastal islands of Panama and stated that the highest incidence in the world occurred there. Later, Keeler (1964) investigated the incidence of the condition among the Cuna in more detail and tried to give reasons for the high rate. He explained that the name moon-child was used for affected people because the Cuna believed albinism was caused by either the mother or father looking at the moon too much during gestation. Keeler found that the local customary practice of infanticide reduced the rate of albinism for many years. However, when a supreme chief, who was a prestigious leader, taught that people with albinism sinned less than others, were on better terms with the Sun-God, and should be kept alive to tell the people more about their God, the rate of infanticide decreased. Nevertheless, after his death the numbers of people with albinism declined again, especially in the rural areas. At the time Keeler was investigating the condition (1950s and 1960s), he was informed that there was a considerable amount of infanticide. He found that in the most civilized town, where the killing of newborn children was frowned on, there were 144 people with albinism per 10,000 of the population. Keeler was told by older people in the community that life span was reduced in affected people, partly because of the development of aggressive skin cancer and infections. Affected females seldom married and lived a more protected and therefore longer life than males. He stated that the high rate of the condition could also be partly ascribed to the selective inbreeding that occurred. However, Currasco et al. (2009) studied five Cuna individuals with OCA2. All were homozygous for an intron 17 acceptor splice site mutation suggesting a common pathogenic mutation.

Woolf and Grant (1962) carried out studies on albinism in the Hopi in Arizona among whom a high rate (Hrdlicka, 1908, 1 in 182), and also a possible religious significance associated with being affected, had been reported previously. Their research showed that the frequency was 1 in 227 people, but that there was no specific religious significance attached to having the condition and affected people were well integrated into the community. However, Woolf and Dukepoo (1969) investigated the reasons for the high rate of albinism in the Hopi and found that there was some genetic drift (random fluctuations of frequency), as well as a probable founder effect; the heterozygote rate was 12.4% in some groups, and there was an increased rate of inbreeding. They added that there were cultural issues associated with the high-homozygote rate, affected males were not expected to work in the fields, and they had more time and opportunity to engage in sexual activity than other men. Together these factors could explain the high frequency of the condition.

A similar high rate of 1 in 240 people (23 people with albinism among 5500 people) was found among the Zuni, an Amerindian isolate living in New Mexico and the South-Western parts of the United States (Witkop et al., 1972; Woolf, 1965). Testing of these affected people, at that time, showed that they had tyrosinase-positive (ty-pos) albinism (OCA2) and they developed increased pigmentation in the hair and eyes and decreased nystagmus and photophobia as they aged.

A review of the literature of OCA2 in American Indians was presented by Woolf in 2005. He raised the question of why the condition should be so widely distributed when being affected is so detrimental to the well-being of the individual. Also, fitness would have been reduced in people with OCA in early nomadic hunter-gatherer populations because of their poor eyesight and sun sensitivity. He proposed a hypothesis for the variable frequencies of the gene in terms of chance processes (such as founder effect, bottleneck effect, and genetic drift) in small populations and/or natural or cultural selection, and/or the interaction of these processes in some generations.

Prevalence by Type of Albinism

Prior to molecular genetic testing becoming available, the prevalence rates of the so-called tyrosinase-negative (ty-neg, OCA1) and tyrosinase-positive (ty-pos, OCA2) types of albinism were investigated in a few populations and generally the former was found to be rarer than the latter type, particularly in Africa. According to Witkop et al. (1970) from all subjects screened (100 subjects were tested by the assessment of hair bulb tyrosinase levels, which is considered a somewhat unreliable method today, particularly because some tyrosinase mutations are now known to cause reduced but not complete loss of function) in the United States, at that stage, 27% were ty-neg and 73% ty-pos. There appeared, however, to be ethnic group variations and in the same study Witkop and colleagues found that in 21 North American White subjects with the condition 67% were ty-neg and 33% ty-pos, whereas in 37 North

American Black subjects 27% were ty-neg and 73% ty-pos. Because many of Witkop's subjects were selected from schools for the partially sighted, his figures for the White group may have been biased in favor of the ty-neg type who experience more severe eye defects. McLoed and Lowry (1976) suggest that the ty-neg type is probably less frequent than the ty-pos type in Whites and Blacks. In their sample of 46 affected Whites in British Columbia, they found that 17 (37%) were ty-neg and 29 (63%) ty-pos. Similarly, Van Dorp et al. (1983) found, in the Netherlands, that among 78 patients, 17 (22%) were ty-neg and 61 (78%) were ty-pos.

Among Nigerians, ty-neg albinism appeared to be much rarer and none was found in a sample of 79 people with albinism studied there (King et al., 1980). In South Africa, this type of albinism is also very unlikely to occur. However, in one old study (Dogliotti, 1973) hair bulb samples from nine patients were sent to Professor C.J. Witkop in Minnesota, United States, for tyrosinase testing. Because such testing has been shown to be unreliable, the results showing that 3/9 cases were ty-neg are somewhat suspect and no further ty-neg (OCA1) cases have been identified in South Africa. Ty-neg albinism has been described in the Cameroon where Aquaron et al. (1981) identified one case, with no apparent Caucasian admixture. This middle-aged man was seen by the author (JGRK) at the recent International Workshop on Albinism in Africa, held in Douala, Cameroon, in July 2015. He was certainly paler in skin and hair color than the many other people with albinism at the meeting. Molecular testing has confirmed his OCA1 status (Badens et al., 2006).

The findings for many of these old studies depended on the methods of ascertainment of the subjects, but nevertheless, ty-neg (OCA1) albinism appeared to be less frequent than ty-pos (OCA2) in Whites and to occur very rarely in Blacks, while, at that time (1970s–80s), ty-pos (OCA2) albinism was thought to be more common in the United States, Canada, and Africa.

Regarding the rarer types, initially, the OCA2B (brown OCA) type of albinism was only observed in Africa and New Guinea, and OCA3 (red or rufous OCA) also appeared to be much more common in these two areas of the world than elsewhere (Witkop et al., 1983; Barnicot, 1953). The prevalence of some of the other rarer types of albinism is only now being investigated as their molecular causes become better understood.

Recent information from molecular studies has provided more clarification on the prevalence of the different types and the distinction between them, as well as delineating further types. One of the latest review articles suggests that OCA1 is the most common subtype in Caucasians, and it probably accounts for about 50% worldwide, whereas OCA2, together with BOCA, only accounts for 30% worldwide, but it is most common in Africa (Kamaraj and Purohit, 2014). Furthermore, research in Japan showed that OCA1 was the commonest type found in a series of people with albinism, followed by OCA4 (which is rare in the rest of the world) and then OCA2, while OCA3 was not found at all (Suzuki and Tomita, 2008). These studies, and the issues arising, will be discussed in more detail, later, in the chapter covering molecular genetics (see Chapter 5).

III. PREVALENCE RATES IN AFRICA

In 2006 a World Health Organization (WHO) team investigated the prevalence of albinism in Africa and the issues concerning vulnerable populations (Hong et al., 2006). They undertook a comprehensive literature search and distributed a pilot survey on albinism to African countries, through African WHO regional offices. They found that good epidemiological data were only available for South Africa (Kromberg and Jenkins, 1982; Venter et al., 1995), Zimbabwe (Kagore and Lund, 1995; Lund, 1996; Lund et al., 1997), Tanzania (Luande et al., 1985), and Nigeria (Okoro, 1975). Prevalence rates were estimated as ranging from 1:5000 to 1:15,000.

Prevalence in Southern Africa

Southern African is here defined as South Africa together with its neighboring countries (excluding those to the north of South Africa, such as Zimbabwe, Namibia, and Mozambique but including Botswana, on the west, Lesotho, which lies within the boundaries of South Africa, and Swaziland, which, except for a small Eastern boundary with Mozambique, is surrounded by South Africa. The term "Southern" is often used, in other contexts, to include several countries further north).

Urban Studies in South Africa

The South African study undertaken by Kromberg and Jenkins (1982) is worth reviewing here because it was one of the few comprehensive African cross-sectional studies, identified by Hong et al. (2006), with multiple methods of ascertainment and a significant sample size. Also, most of the affected individuals were personally contacted by the researchers and were seen and interviewed. The objectives of this study were to determine the following:

1. prevalence of albinism in an urban population and in various ethnic groups
2. consanguinity rate among parents of children with albinism
3. sex ratio in affected individuals

The fieldwork was carried out, over a 10-year period, in the urban area of Soweto, a satellite city adjacent to Johannesburg, with its large population of ~803,000 Black people (Census, 1970). This population was made up of representatives of most of the major South African ethnic groups, at the time, as well as some groups from neighboring countries (Kromberg and Jenkins, 1982). Seven methods of ascertainment were used: visits to schools; visits to health clinics and two local hospitals (one a general hospital, the other an eye hospital); welfare organizations and social workers were approached for information; and lastly families with a member with albinism offered information on other affected individuals. Each affected person was then visited, a checklist documenting demographic details was completed and a pedigree was drawn up. The subjects were also invited to attend a special clinic. A medical examination was performed, photographs were taken, and individuals were provided with antiactinic sun barrier cream. Where the affected person was willing and volunteered, informed consent was obtained and a blood sample was collected and stored for future research. However, testing for tyrosinase status was not undertaken, and the prevalence figures included all the known forms of the condition together.

In this way 213 people with albinism were identified, in 126 families, and 206 (97%) of them were interviewed. The remaining seven (3%) individuals were away from home and unavailable at the time of the interview. The checklist was usually fully completed, except for the item on consanguinity, which, in the absence of some of the parents, could not always be reported and the information was only obtained on 90/126 (71%) families.

The prevalence rate in this study was found to be 1:3900 using the available total population figure from the 1970 Census (Kromberg and Jenkins, 1982). From this information, a carrier rate of 1:32 and a gene frequency of 0.0160 could be calculated. This prevalence was significantly higher than any reported for Caucasian populations but similar to that found in Nigeria (1:5000, Barnicot, 1952) and the Transkei region (now in the Eastern Cape province) of South Africa, which is inhabited mostly by Xhosa ethnic groups (1:3759, Oettle, 1963).

The data for the major South African ethnic groups (Nguni and Sotho) and their subgroups showed that prevalence varied (see Table 3.2). The highest prevalence was found in the South Sotho group, while the Sotho groups together (mostly Southern Sotho and Northern Sotho/Tswana) showed significantly higher prevalence rates than the Nguni (mostly Zulu, Xhosa, and Swazi) ($P < .01$). Similarly, Hitzeroth and Hofmeyer (1964), in their study on albinism in black women in South Africa, found the highest rates were in the Tswana (a Northern Sotho subgroup), other Sotho groups, and Swazi, and that there was a highly significant difference between the Nguni and Sotho population rates.

Although this study was based on a large sample, the prevalence estimated here could still be an underestimate, either because of people with albinism remaining in the rural areas, where they can maintain a relatively sheltered existence, or because they have been sent to those areas to be cared for by relatives. Alternatively, they may be overrepresented in urban areas, where they might be sent for schooling or might migrate to for employment opportunities, or, possibly, the family might find that stigmatization is less in urban areas. Furthermore, they and their families might consider them unsuited for work in the rural areas where manual labor and working in the fields, with its concomitant and detrimental sun exposure, are required. It has been reported that people with albinism among the Cuna fitted much better into a community where a variety of indoor employment opportunities was available (Woolf, 2005).

The data on consanguinity rates among the parents of children with albinism provided some interesting insights into the reasons for the differing prevalence rates. Altogether the consanguinity rate calculated for the 90 families was 24% (see Table 3.3). This figure appears high, but the consanguinity rate for parents of an affected child in Japan was nearer 44% (Neel et al., 1949). When the rates in the various ethnic groups were analyzed,

TABLE 3.2 Prevalence of Albinism in Soweto, South Africa, by Ethnic Group

Ethnic Group	No. of Subjects	Total Population	Prevalence
Nguni			
Zulu	55	245,248	1:4459
Xhosa	17	81,511	1:4794
Swazi	11	29,872	1:2716
Sotho			
South Sotho	53	108,190	1:2041
Tswana	42	146,184	1:3481

Adapted from Kromberg, J.G.R., 1985. A genetic and psychosocial study of albinism in Southern Africa (Ph.D. thesis). University of the Witwatersrand, Johannesburg, South Africa.

TABLE 3.3   Consanguinity Among Parents of People With Albinism in Soweto, Johannesburg

Ethnic Group	Consanguineous Cases	No. of Families	Consanguinity Rate (%)
Nguni			
Zulu	1	23	4.3
Xhosa	2	10	20.0
Swazi	2	6	33.3
Sotho			
South Sotho	6	22	27.3
Tswana	10	24	41.7
Other[a]	1	5	20.0
Total	22	90	24.4

[a] Includes Pedi, Venda, and Shangaan groups.
Adapted from Kromberg, J.G.R., 1985. A genetic and psychosocial study of albinism in Southern Africa (Ph.D. thesis). University of the Witwatersrand, Johannesburg, South Africa.

the Tswana and South Sotho groups showed rates of 42% and 27%, respectively, whereas the Zulu rate was 4%. The only significant difference ($P < .02$) between the groups was that between the Zulu and Tswana groups. The data showed a high prevalence of albinism in the South Sotho and Tswana groups, which both had high consanguinity rates, compared with the Zulu group, which had a low consanguinity rate. It is possible that the gene frequency is similar in different groups but the prevalence differs because of the consanguinity rate.

To determine whether albinism was more common in males or females, the sex ratio was calculated. Although this ratio was found to be 1.21 and more males (113) than females (93) were observed in the sample, the difference was not significant ($P > .10$). Also, it is possible that more females than males were sent to boarding schools for the visually impaired, for their protection, and therefore some of the females might not have been identified in this study. Alternatively, females might have been less likely than males to present themselves to the research team for inclusion in the project.

Rural Prevalence Studies

To compare and contrast the urban and rural situation, four rural prevalence studies were conducted by Kromberg (1985), in selected areas of two neighboring countries, Botswana and Swaziland, and two areas of South Africa, Kwazulu/Natal and the Transkei region (now within the Eastern Cape province) (see map in Fig. 3.2). The findings from these four studies are summarized below.

BOTSWANA

The large village of Mochudi, in Botswana, had a Tswana population of 18,300 people and there were 14 people (in nine families) with albinism living there (in 1980) (Kromberg, 1985). Of this group 10 were seen personally, by the author (JK) and interviewed in their homes,

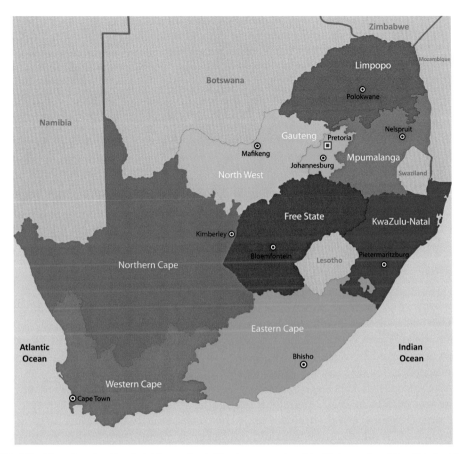

FIGURE 3.2 Map showing South Africa and neighboring countries. *Modified map from https://stock.adobe.com/ by Iryna Volina. File #: 56914311.*

while the remaining four were not seen but were included because they were described by reliable witnesses (an anthropologist and medical practitioner), who had experience of albinism. The results showed that the prevalence in this village was 1:1307 and the carrier rate 1 in 19 (see Table 3.4).

This prevalence rate was approximately three times higher than that (1:3481) found for the Tswana ethnic group living in the urban area of Johannesburg. The consanguinity rate, in the rural village, was 11%, i.e., one of the nine couples with an affected child had a consanguineous union. However, due to the small sample, this may not reflect the true rate because in the much larger urban sample the consanguinity rate was 41%.

A family history was reported in 5/6 families, for whom information was available, and one family had multiple affected members in the extended family. Another family was found to have 3/9 children with an unusual reddish yellow skin, brown eyes, and fair hair and they were considered to have OCA3.

TABLE 3.4 Prevalence of Albinism in Botswana and Swaziland

	Botswana	Swaziland
No. of subjects	14	82
No. of families	9	36
Total population	18,300	160,000
Prevalence	1 in 1307	1 in 1951
Gene frequency	0.0537	0.0442
Carrier rate	1 in 19	1 in 23

Adapted from Kromberg, J.G.R., 1985. A genetic and psychosocial study of albinism in Southern Africa (Ph.D. thesis). University of the Witwatersrand, Johannesburg, South Africa.

SWAZILAND

During the study of albinism in the Hhohho district (population size 160,000) of Swaziland, 49 people with albinism from 36 families were identified and interviewed (Kromberg, 1985). In these families, a further 28 affected relatives who were living in the district were reported. Apart from these 77 individuals, 2 individuals were observed in the street and 3 further individuals were reported by reliable witnesses and included. The final total of affected people was therefore 82 in 160,000, giving a prevalence figure of 1:1951 (see Table 3.4).

The prevalence rate in this large rural Swazi sample was higher than expected and verified the high rate (1 in 2716 in 29,872 people) found in the small sample of Swazi people living in the urban area (Kromberg and Jenkins, 1982). The consanguinity rate for the 36 rural families was found to be 17%, although there could have been some underreporting as consanguineous marriages are not strictly culturally permitted in this ethnic group. However, the condition is so common in Swaziland that 58% of the families reported a strong family history.

Altogether five children in two Swazi families were examined and found to have the characteristics of OCA3. These children had similar physical characteristics, including light copper-colored skin and fair gold hair, to the children with OCA3 albinism seen in Botswana. Only one of these affected children had nystagmus, none had obvious photophobia, two had brown eyes, one light brown eyes, and two a gray–green eye color.

KWAZULU/NATAL PROVINCE

This study was based in the Nongoma district of the Province of Kwazulu/Natal, on the North Eastern coastline of South Africa (Kromberg, 1985). Altogether 15 individuals with albinism in 11 families were identified in the population of ~90,000 people (unpublished data from the Nongoma Magistrate's Court files, assembled by Klopper, 1979, personal communication). Of this group, 13 were seen and interviewed at the local hospital or at a local school, either alone or with a relative. Two were living in an inaccessible area, too far away to be contacted. Because of the mountainous area and poor roads, ascertainment was probably incomplete. A better estimate of prevalence was provided by the study of the children in 48 schools in the district (see Table 3.5) than by the general population study.

TABLE 3.5 Prevalence of Albinism in Nongoma District

	Nongoma District	School Children
No. of subjects	15	7
No. of families	11	7
Total population	90,000	18,036[a]
Prevalence	1 in 6000	1 in 2576
Gene frequency	0.0255	0.0386
Carrier rate	1 in 39	1 in 26

[a] In 48 schools.

Adapted from Kromberg, J.G.R., 1985. A genetic and psychosocial study of albinism in Southern Africa (Ph.D. thesis). University of the Witwatersrand, Johannesburg, South Africa.

There were no consanguineous unions in the parents of these affected individuals, as far as the informants knew. However, there is a local taboo on marriages within the fathers', mothers', or grandmothers' clans (Clegg, 1979, personal communication), so that marriages are generally exogamous, and consanguineous marriages if they occurred might neither be reported nor admitted.

TRANSKEI

The fieldwork on albinism in selected districts of the Transkei (now in the Eastern Cape Province) resulted in the identification of 43 affected people (Kromberg, 1985). These individuals were all seen and interviewed and they described a further 15 people with albinism. On examination 39/43 were found to have the typical form of albinism (13 were seen at the local school for the blind), and three had light reddish-brown skin and pale-yellow hair.

Although clinics, drawing patients from 11 of the 28 different magisterial districts of the region, were visited, the prior arrangements were inadequate and the ascertainment was poor. The prevalence rates therefore were found to be much lower than expected (the highest was 1 in 10,000 in the district with the highest ascertainment) and not worth documenting here. The consanguinity rate for the sample was zero (0/43), but 6.3% said they came from the same family and 15.6% from the same village. However, this ethnic group holds the same taboos as the Zulu (both being of Nguni extraction), and therefore relevant information was difficult to collect reliably.

The three affected individuals who had the reddish skin and fair hair had a history of affected sibs and normally pigmented parents, which supported the theory that this form of albinism was also recessively inherited, as was suspected in the Botswana family. Also, a marriage between two people with albinism was reported. The affected father was interviewed and two of his five affected children were seen (the father stated that all their offspring were affected and the mother had died of tuberculosis (TB)) (see Fig. 3.3, undertaking fieldwork in rural southern African villages).

FIGURE 3.3 Undertaking field-work on a hazy day in rural southern African villages. *Photograph courtesy of Dr. P. Manga, New York University School of Medicine.*

Discussion of Southern African Prevalence Rates

The prevalence rate for albinism found in the urban area of Soweto, Johannesburg, was 1:3900 (Kromberg and Jenkins, 1982). This rate is remarkably high being approximately four times higher than that reported for many White populations (Pearson et al., 1913; Sanders, 1938; Witkop et al., 1983). It is likely to be a minimum rate and only an estimate because ascertainment in Soweto was probably incomplete and the fluctuations in the Soweto population made accurate enumeration difficult.

It is not clear why the rate should be raised in a subtropical country with high daily sunshine rates and an environment generally detrimental to the well-being of affected individuals. It is possible a founder effect has contributed to the high carrier rate (1:32) in this population. Alternatively, there could be some form of heterozygote advantage (as suggested by Kromberg, 1987). Such advantage could be related to immunity to one of the major killers in the tropics, as has been found for sickle cell anemia carriers. Resistance to malaria would have given carriers an advantage in the past in tropical Africa. Further investigations of endemic diseases in carriers are required to address this query regarding albinism. Another form of advantage was mentioned briefly by Oettle (1963) who suggested that if the light skin color he observed in carriers was considered a sign of beauty, they would have a selective advantage. Kerr (2008) investigated the possibility that carrier status provides some immunity to developing TB in African people. However, her findings showed that presence or absence of the common OCA mutation (a 2.7 kb deletion mutation on chromosome 15q) causing OCA2 in Africa did not appear to influence susceptibility to TB. Similarly, Tuli et al. (2012) suggested that OCA2 carriers were less susceptible to developing leprosy; however, the number of OCA2 mutation carriers in a cohort of individuals with leprosy in Tanzania was not significantly reduced.

Selection in favor of homozygotes might have occurred if people with albinism had been considered special or honored. Such an attitude does not seem to have been prevalent locally, however, because the killing of affected children in Southern Africa was reported by travelers, such as Livingstone (1857) and Stannus (1913). Also, according to Turner (in Pearson et al., 1913), individuals with the condition were considered undesirable as wives or husbands. Considering that fertility may be reduced (Kromberg, 1985), it seems unlikely the homozygote advantage could account for the raised prevalence rates found, in the past and present, locally. It seems more likely that there is a heterozygote advantage, as yet unidentified. The local high consanguinity and high carrier rate, together with genetic drift, might also partly explain the prevalence rate.

Although there are many reports on albinism in Africa, not many systematic epidemiology studies have been performed. The available rates for the condition vary from 1:2858 to 1:5000 (see Vallois, 1950; Barnicot, 1953; Oettle, 1963; Aquaron, 1980; Lund, 1996) and approximate those found in the Soweto study, suggesting that albinism might occur at a similar rate throughout Africa. The African rates, however, are not as high as those that occur in certain isolates (e.g., 1 in 227 in the Hopi in Arizona, Woolf and Grant, 1962). The exceptions are the Botswana village (Mochudi), with a population of 18,300, described by Kromberg (1985), (prevalence 1:1300) and the cluster described by Lund et al. (1997) in a Zimbabwe valley, with a population of about 11,000, (prevalence 1 in 1000). Currently, further studies are being conducted in Namibia and Tanzania, and the Zimbabwe data are being reanalyzed, and the findings will be presented in the following chapter (see Chapter 4).

The urban prevalence rate reported above might have been affected by various local social factors. The urban Soweto population figure used to determine the prevalence consisted of people who had been in the city for, at most, two generations. It is possible therefore, that the actual prevalence is higher because affected people and their families might not migrate to the cities where competition for employment and resources is strong. They might prefer to remain in the rural areas where they can live within the supportive extended family. Alternatively, individuals with the condition may be taken or sent out of the urban area (to be cared for by rural relatives) and this pattern of behavior would reduce the urban prevalence. On the other hand, the rate may be too high if affected people tend to seek good education and indoor industrial employment, only found in urban areas, rather than having to submit to rural occupations frequently associated with excessive sun exposure. These factors, however, could counterbalance each other and the estimates would then seem reasonable.

When examining prevalence by ethnic group, in South Africa, the rates for albinism vary, the Zulu and Xhosa (two of the Nguni groups) having significantly lower rates than the Southern Sotho group. The latter group, however, had a consanguineous marriage rate more than twice as high as the combined Zulu/Xhosa rate. An investigation of local customs confirmed this finding and Sotho culture condones and even promotes such marriage between relatives, particularly between first cousins, such as cross-cousins (first cousins, e.g., a man with his mother's brother's daughter) or parallel cousins (e.g., a man with his father's brother's daughter) because this arrangement keeps the bridewealth in the family and, in those of noble birth, preserves their elite status (Hammond-Tooke, 1993). However, in Zulu/ Xhosa culture there are strict rules about clan exogamy and consanguineous marriages are generally taboo (Hammond-Tooke, 1993). Specific details on the types of consanguineous marriages in affected families in the Southern African study (Kromberg, 1985) and, for comparison, in the general population were not available, so that calculations to take into account the effect of the consanguinity rate on the gene frequency could not be completed. The data, however, suggest that the heterozygote rate might be similar in all ethnic groups (all are originally Bantu-speaking peoples), but the homozygote rate might be increased in certain groups because of their cultural practices.

The prevalence in males and females was found to be similar in the South African study, although some authors have found higher rates in males than females (Barnicot, 1952; Froggatt, 1960; Aquaron, 1980). Barnicot (1953) suggested that the excess of males in his Nigerian sample was due to his ascertainment methods and to local social pressure, which might have encouraged males to participate in research projects, whereas females might have been less likely to volunteer.

The four studies on rural prevalence indicated a wide range of local rates, which were higher than those found in urban Soweto, Johannesburg. The highest was 1 in 1307, found in Mochudi (in rural Botswana) where the acknowledged consanguinity rate was estimated at only 11%. It is possible that the urban Tswana send their affected family members to live with rural relatives, and one definite case of this behavior occurring was described during the study (Kromberg, 1985). It is also possible that this large village might be considered an isolate where inbreeding, though neither admitted nor perhaps even recognized, may frequently occur. The rural rate far exceeded the urban rate of 1 in 3481 (with a consanguinity rate of 41%).

In Kwazulu/Natal the prevalence rate was found to be 1 in 6000 for the general population and 1 in 2576 for schoolchildren (compared with 1 in 4459 for all urban Zulus). Barnicot (1952) also reported a higher rate in schoolchildren in Nigeria. It may be, in a community where not everyone can attend school, that children with albinism are preferentially selected (by their parents) for schooling, over their unaffected siblings, or the affected children are more likely to persevere at their lessons because they generally prefer indoor to outdoor activities. Alternatively, the ascertainment in schools may be more complete. Also, if affected people have shorter life spans, they will be overrepresented in the younger age groups. Kagore and Lund (1995) reported a high rate of 1 in 2833 in urban schoolchildren in Harare, Zimbabwe, while Lund (1996) reported a lower rate of 1 in 4728 in a large sample of rural Zimbabwe school children. These findings again suggest that children with albinism may be sent to the urban areas for schooling, unless affected children were more likely to be missed in the rural study than in the urban study, or the findings from the urban study were biased in some way.

The prevalence figures for the Transkei area in the Kromberg study (1985) were somewhat unreliable because ascertainment was incomplete as shown by the finding of an unusually low rate (<1 in 10,000). Oettle (1963), who undertook research on skin cancer in the area, suggested that, from his observations, he could estimate a rate of 1 in 3759. Later, Rose (1974), who also worked with cancer patients in the area, stated that her figures showed a prevalence rate of about 1 in 3000. These figures are more in keeping with the estimated prevalence of 1 in 4794 found for this ethnic group in the Soweto study (Kromberg and Jenkins, 1982).

In Swaziland, the lines of communication were effective and well established and ascertainment was consequently quite easy. Both the prevalence (1 in 1951) and the consanguinity rate (17%) were found to be higher than in other Nguni groups (and this finding is supported by the results from the study of Hitzeroth and Hofmeyer (1964)). The Swazis appear to be an exception and although the Nguni generally have taboos against consanguineous marriages, the Swazi have tended to promote such marriages, especially among the ruling families. The urban Swazi rate of 1 in 2716 was also high compared with those in the other Nguni ethnic groups. From discussions in Swaziland, it appears that people with albinism are generally well accepted there, even as marriage partners, and the superstitions, which surround the condition elsewhere (Kromberg and Jenkins, 1984), are not so widely believed. Such social factors together with the increased consanguineous marriage rate could have contributed to the high rate in this ethnic group.

The findings of ethnic differences in prevalence rates in this South African study substantiate the results from a previous study (Hitzeroth and Hofmeyer, 1964). In that study, too, the highest rates were found in the Tswana (a Sotho group) and significantly lower rates in the Zulu and Xhosa (Nguni groups).

In summary, the urban prevalence of albinism in South African Blacks was ~1 in 3900, which is similar to that found in other African countries. Males and females were equally affected. The Nguni groups (e.g., Zulu and Xhosa) generally showed both a low prevalence and a low consanguinity rate. The exception to this rule was the Swazi, who are of Nguni extraction, but had both a high prevalence and a high consanguinity rate. The Sotho groups (e.g., Tswana and Southern Sotho) who favored consanguineous marriage also had higher rates. It is therefore possible that the gene frequency is similar in all the local ethnic groups, but the homozygous frequency is affected by the localized consanguinity rates, perhaps by the possible selective advantage of heterozygotes, which, if based on lighter skin color, could differ from group to group and/or by genetic drift.

Prevalence in Other African Countries

Cameroon

Vallois (1950), a French anthropologist, visited Cameroon and heard that albinism was quite common there. He then identified eight people with albinism in some Badjoue villages, in which 23,000 people lived. He therefore estimated the prevalence rate to be about 1 in 2875 in the Badjoue ethnic group.

Later, Aquaron (1990), working in Cameroon during several visits over a 15-year period, examined 190 people with albinism, in an area inhabited by the Bamileke ethnic group. Using an estimate of 1,500,000 for the Bamileke population, he calculated a prevalence rate of 1 in 7900. He suggested that the rate of inbreeding was quite high in this ethnic group because marriages tended to occur within the same clan. He added that around the beginning of the 20th century the group had had two polygamous chiefs with albinism (Aquaron, 1980); they had many wives and many children, so that founder effect was clearly illustrated.

Malawi

Stannus (1913) described many people with albinism in Nyasaland (now Malawi) in his doctoral thesis on the subject. He stated that albinism was common in the country, but he did not attempt to calculate the prevalence.

However, in 1993 Bar reported that, in his study on albinism in Malawi, he had found a prevalence rate of 1 in 3400. He also found a high rate of skin cancer, including both squamous and basal cell carcinoma, but no cases of malignant melanoma.

Nigeria

Barnicot (1952) appears to have completed the first scientific study to estimate the frequency of albinism in Africa. In Nigeria, he found a frequency of one child with albinism in every 2858 schoolchildren. Furthermore, in a Nigerian township the rate was 1 in 5000. In a later study Barnicot (1953) investigated and described 23 people, with a type of albinism he called xanthous, who had brownish-yellow hair and copper-colored skin. He maintained that this unusual type of albinism was quite common in Nigeria, although he did not attempt to estimate a prevalence rate.

In 1975, Okoro studied a series of people with albinism. He then estimated a prevalence of only 1 in 15,000 for East-Central Nigeria. He added that there appeared to be a higher prevalence in Southern Nigeria than in Northern Nigeria and suggested that this was because of the preference in the South for consanguineous unions. These figures illustrate again the variations in rates that can occur in one large country with a large population.

Tanzania

In 1985 Luande et al. suggested that the prevalence rate in Tanzania might be as high as 1 in 2500. However, their study was focused on the skin of persons with albinism and the nature of the skin cancer found in the 350 affected people attending their Tanzania Tumor Center in Dar es Salaam. Because the data were extracted from a register of these patients, it was probably somewhat biased, and the denominator might have been difficult to calculate correctly. Lookingbill et al. (1995) also examined actinic damage and skin cancer in 164 patients, in the Moshe region of Northern Tanzania, but they could not obtain the necessary population statistics to estimate the prevalence in the area. Another study is currently underway in Tanzania and this will be discussed in the following chapter (see Chapter 4).

Zimbabwe

Albinism has been studied in some detail in Zimbabwe. Initially, Kagore and Lund (1995) surveyed schoolchildren in a postal survey of all the secondary schools in the capital, Harare, and 1 in 2833 children were reported to have albinism. Then Lund (1996) carried out a national survey of schools. By means of a postal questionnaire to the principals of all the secondary schools, outside Harare, and heads of primary schools in 4 of the 8 provinces, a total of 1,314,358 school children, aged 6–23 years, were screened. Altogether, 278 affected children were identified, giving an albinism prevalence rate of 1 in 4728. The frequency in certain areas, particularly, the rural eastern province of Manicaland (Lund, 1996) and the urban area of Harare (Lund, 2005), was significantly higher than that found in other areas. In general, however, the prevalence of OCA among schoolchildren was significantly increased in urban areas when compared with that in rural areas.

The following year Lund et al. (1997) investigated the condition in a Tonga isolate in a valley in Northern Zimbabwe. They found 11 people with albinism, in a population of about 12,000 people, and interviewed 5 affected adults and 1 school girl. The incidence there was

1 in 1000 people and the carrier rate was 1 in 16. The five cases on which molecular studies were carried out all had OCA2. These studies will be discussed in more detail in the following chapter (see Chapter 4).

IV. CONCLUSION

The epidemiological studies presented in this chapter show that high rates of the condition occur in Africa, although, in general, these are not as high as the rates that are reported from some of the population isolates existing elsewhere in the world, e.g., in Arizona, United States. The rates reported from different African countries vary only slightly but locally rural and urban rates can differ (as shown in the South African and Zimbabwean studies). Not many studies have commented on the sex ratio of affected people, but in general a significant difference in the numbers of affected males and females is not expected, although more people of one sex or the other may present themselves (or be identified) during epidemiological studies due to local social and cultural factors.

The types of OCA (presently mostly determined by clinical examination and not by molecular investigation) found in different countries across the world vary, with OCA2 being much more common in Africa than elsewhere. Consanguinity rates influence the prevalence of OCA and, for example, the rate shown in one South African community-based study (in Soweto, Johannesburg) was 24%, whereas in an Irish study it was 9%. However, because rates tend to change with various current cultural and social forces, more recent studies are required to determine if these rates still apply to the present situation regarding preferred mate selection.

Further studies on the epidemiology of albinism are required. Adequate prevalence studies have only been performed in a few countries, and data for countries such as those in the Far East, including China and Japan, as well as India, are limited. In general, the prevalence for the peoples of Africa is estimated to be around 1 in 4–5000, whereas Northern European countries have lower rates, but isolated populations, particularly those in the Americas, have much higher rates. High rates are associated with consanguinity in some cases, with founder effect or selective breeding or cultural advantage for people with albinism or the carriers of the gene, in other cases, and selective advantage for heterozygotes may still be a possibility. Also, rates may vary over time, with varying environmental factors, in various geographical, social, and cultural situations and from generation to generation.

References

Abrahams, P.H., January 8, 1972. Albinos in Borneo. Lancet 101–102.

Aquaron, R., 1980. L'albinisme oculo-cutane au Cameroun. A propos de 216 observations. Rev. Epidemiol. Sante Publique 28, 81–88.

Aquaron, R., Ronge, F., Aubert, C., 1981. Pheomelanin in albino Negroes. Urinary Excretion of 5-S-cysteinyldopa in Cameroonian Subjects. In: Seijing, M. (Ed.). Pigment Cell. Tokyo University Press, Tokyo, pp. 97–103.

Aquaron, R., 1990. Oculocutaneous albinism in Cameroon. A 15 year follow-up study. Ophthalmic Paediatr. 11 (4), 255–263.

Badens, C., Courrier, S., Aquaron, R., 2006. A novel mutation (del AACT) in the tyrosine gene in a Cameroonian black with Type 1A oculocutaneous albinism. J. Dermatol. Sci. 42 (2), 121–124.

Bar, G., 1993. Albinism in Malawi. Malawi Med. J. 9, 10–12.

Barnicot, N.A., 1952. Albinism in South Western Nigeria. Ann. Eugen. 17, 38–73.

Barnicot, N.A., 1953. Red hair in African Negroes: a preliminary study. Ann. Eugen. 17, 211–232.

Chaki, M., Sengupta, M., Mukhopadhyay, A., Subba Rao, I., Majumder, P.P., Das, M., Samanta, S., Ray, K., 2006. OCA1 in different ethnic groups of India is primarily due to founder mutations in the tyrosinase gene. Ann. Hum. Genet. 70, 623–630.

Currasco, A., Forbes, E.M., Jeambrun, P., Brilliant, M.H., 2009. A splice site mutation is the cause of the high prevalence of oculocutaneous albinism type 2 in the Kuna population. Pigment Cell Melanoma Res. 22, 645–647.

Dogliotti, M., 1973. Actinic keratoses in Bantu albinos. clinical experiences with the topical use of 5-fluro-uracil. S. Afr. Med. J. 47, 2169–2172.

Froggatt, P., 1960. Albinism in Northern Ireland. Ann. Hum. Genet. 24, 213–230.

Gronskov, K., Ek, J., Brondum-Nielsen, K., 2007. Oculocutaneous albinism. Orphanet J. Rare Dis. 2, 43.

Gronskov, K., Ek, J., Sand, A., Scheller, R., Bygum, A., Brixen, K., Brondum-Nielsen, K., Rosenberg, T., 2009. Birth prevalence and mutation spectrum in Danish patients with autosomal recessive albinism. Investig. Ophthalmol. Vis. Sci. 50, 1058–1064.

Gong, Y., Shao, C., Zheng, H., Chen, B., Guo, Y., 1994. Study on genetic epidemiology of albinism. J. Genet. Genom. 21 (3), 169–172.

Haldane, J.B.S., 1938. Estimation of frequencies of recessive conditions in man. Ann. Eugen. Lond. 8, 255.

Hammond Tooke, D., 1993. The Roots of Black South Africa. An Introduction to the Traditional Culture of the Black People of South Africa. Jonathan Ball, Johannesburg.

Hitzeroth, H.W., Hofmeyer, J.D.J., 1964. A survey of albinism among the Bantu of South Africa. Mank. Q. 5, 81–86.

Hrdlicka, A., 1908. Physiological and Medical Observations Among the Indians of South-western United States and Northern Mexico. Bureau of American Ethnology Bulletin No. 34, Washington, DC.

Hong, E.S., Zeeb, H., Repacholi, M.H., 2006. Albinism in Africa as a public health issue. BMC Public Health 6, 212–219.

Kagore, F., Lund, P.M., 1995. Oculocutaneous albinism among school children in Harare, Zimbabwe. J. Med. Genet. 32, 859–861.

Kamaraj, B., Purohit, R., 2014. Mutational analysis of oculocutaneous albinism: a compact review. Biomed. Res. Int. 1–8.

Keeler, C.E., 1953. The Caribe Cuna moonchild and its heredity. J. Hered. 44, 163–171.

Keeler, C.E., 1964. The incidence of Cuna moon-child albinos. J. Hered. 55, 115–118.

Kerr, R., 2008. Genes in the Aetiology of Oculocutaneous Albinism in Sub-Saharan Africa and a Possible Role in Tuberculosis Susceptibility (Ph.D. thesis). University of the Witwatersrand, Johannesburg, South Africa.

King, R.A., Creel, D., Cervanka, J., Okoro, A.N., Witkop, C.J., 1980. Albinism in Nigeria with delineation of a new recessive oculocutaneous type. Clin. Genet. 17, 259–270.

Kromberg, J.G.R., Jenkins, T., 1982. Prevalence of albinism in the South African negro. S. Afr. Med. J. 61, 383–386.

Kromberg, J.G.R., Jenkins, T., 1984. Albinism in the South African negro. III. Genetic counselling issues. J. Biosoc. Sci. 16, 99–108.

Kromberg, J.G.R., 1985. A genetic and psychosocial study of albinism in Southern Africa (Ph.D. thesis). University of the Witwatersrand, Johannesburg, South Africa.

Kromberg, J.G.R., 1987. Albinism in Southern Africa: why so common in blacks. S. Afr. J. Sci. 83, 68.

Livingstone, D., 1857. Missionary Travels. John Murray, London.

Lookingbill, D.P., Lookingbill, G.L., Leppard, B., 1995. Actinic damage and skin cancer in albinos in northern Tanzania: findings in 164 patients enrolled in an outreach skin care program. J. Am. Acad. Dermatol. 32, 653–658.

Luande, J., Henschke, C.I., Mohammed, N., 1985. The Tanzanian human albino skin. Cancer 55, 1823–1828.

Lund, P.M., 1996. Distribution of oculocutaneous albinism in Zimbabwe. J. Med. Genet. 33, 641–644.

Lund, P.M., Puri, N., Durham-Pierre, D., King, R.A., Brilliant, M.H., 1997. Oculocutaneous albinism in an isolated Tonga community in Zimbabwe. J. Med. Genet. 34, 733–735.

Lund, P.M., 2005. Oculocutaneous albinism in southern Africa: population structure, health and genetic care. Ann. Hum. Biol. 32 (2), 168–173.

Magnus, V., 1922. Albinism in man (abstract) J. Am. Med. Ass. 79, 780.

Mansell, M.A., 1972. Albinos in South America. Lancet 1, 265.

Martinez-Garcia, M., Montoliu, L., 2013. Albinism in Europe. J. Dermatol. 40, 319–324.

McLoed, R., Lowry, R.B., 1976. Incidence of albinism in British Columbia (BC). Separation by hairbulb tests. Clin. Genet. 9, 77–88.

Neel, J.V., Kodani, M., Brewer, R., Anderson, R.C., 1949. The incidence of consanguineous matings in Japan. With remarks on the estimation of comparative gene frequencies and the expected rate of appearance of induced recessive mutations. Am. J. Hum. Genet. 6, 156–178.

Oettle, A.G., 1963. Skin cancer in Africa. Natl. Cancer Inst. Monogr. 10, 197–214.

Okoro, A.N., 1975. Albinism in Nigeria. A clinical and social study. Br. J. Dermatol. 92, 485–492.

Pearson, K., Nettleship, E., Usher, C.H., 1913. A monograph on albinism in man. In: Research Memoirs Biometric Series, vol. VIII. Drapers Co, Dulau, London.

Rooryck, C., Morice-Picard, F., Elcioglu, N.H., Lacombe, D., Taeib, A., Arveiler, B., 2008. Molecular diagnosis of oculocutaneous albinism: new mutations in the OCA1-4 genes and practical aspects. Pigment Cell Melanoma Res. 21, 583–587.

Rose, E.F., 1974. Pigment anomalies encountered in the Transkei. S. Afr. Med. J. 48, 2345–2347.

Sanders, J., 1938. Die hereditat des albinismus. Genetica 20, 97–120.

Stannus, H.S., 1913. Anomalies of pigmentation among Natives of Nyasaland: a contribution to the study of albinism. Biometrika 9, 333–365.

Stout, D.B., 1946. Further notes on albinos among the san Blas Cuna, Panama. Am. J. Phys. Anthr. 4, 483–490.

Suzuki, T., Tomita, Y., 2008. Recent advances in genetic analyses of oculocutaneous albinism types 2 and 4. J. Dermatol. Sci. 51, 1–9.

Tuli, A.M., Valenzuela, R.K., Kamugisha, E., Brilliant, M.H., 2012. Albinism and disease causing pathogens in Tanzania: are alleles that are associated with OCA2 being maintained by balancing selection? Med. Hypotheses 979 (6), 875–878.

Vallois, H.V., 1950. Sur quelques points de L'anthropologie des Noirs. L'anthropologie 54, 272–286.

Van Dorp, D.B., Van Haeringen, N.J., Delleman, J.W., Apkarian, P., Westerhof, W., 1983. Albinism: phenotype or genotype. Doc. Ophthalmol. 56, 183–194.

Van Dorp, D.B., 1985. Shades of Grey in Human Albinism (Ph.D. thesis). Vrije Universiteit te Amsterdam, The Netherlands.

Van Dorp, D.B., 1987. Albinism, or the NOACH syndrome. Clin. Genet. 31, 228–2421.

Venter, P.A., Christianson, A.L., Hutamo, C.M., Makhura, M.P., Gericke, G.S., 1995. Congenital anomalies in rural black South African neonates – a silent epidemic? S. Afr. Med. J. 85, 15–20.

Wei, A., Wang, Y., Long, Y., Wang, Y., Guo, X., Zhou, Z., Zhu, W., Liu, J., Bian, X., Lian, S., Li, W., 2010. A comprehensive analysis reveals mutational spectra and common alleles in Chinese patients with oculocutaneous albinism. J. Invest. Dermatol. 130, 716–724.

Witkop, C.J., Nance, W.E., Rawls, R.F., White, J.G., 1970. Autosomal recessive albinism in man: evidence for genetic heterogeneity. Am. J. Hum. Genet. 22, 55–74.

Witkop, C.J., Niswander, J.D., Bergsma, D.R., Worman, P.I., White, J.G., 1972. Tyrosinase positive oculocutaneous albinism among the Zuni and the Brandywine triracial isolate: biochemical and clinical characteristics and fertility. Am. J. Phys. Anthropol. 36, 397–406.

Witkop, C.J., 1979. Albinism: hematologic-storage disease, susceptibility to skin cancer, and optic neuronal defects shared in all types of oculocutaneous and ocular albinism. Ala. J. Med. Sci. 16, 327–330.

Witkop, C.J., Quevedo, W.C., Fitzpatrick, T.P., 1983. Albinism and other disorders of pigment metabolism. In: Stanbury, J.B., Wyngarden, J.B., Fredrickson, D.S., Goldstein, J.L., Browne, M.S. (Eds.), The Metabolic Basis of Inherited Disease. McGraw Hill, New York, pp. 301–346.

Woolf, C.M., Grant, R.B., 1962. Albinism among the Hopi Indians in Arizona. Am. J. Hum. Genet. 14, 391–400.

Woolf, C.M., 1965. Albinism among Indians in Arizona and New Mexico. Am. J. Hum. Genet. 17, 23–35.

Woolf, C.M., Dukepoo, F.C., 1969. Hopi Indians, inbreeding and albinism. Science 164, 30–37.

Woolf, C.M., 2005. Albinism (OCA2) in Amerindians. Yearb. Phys. Anthropol. 48, 118–140.

Yi, Z., Garrison, N., Cohen-Barak, O., Karafet, T.M., King, R.A., Erickson, R.P., Hammer, M.F., Brilliant, M.H., 2003. A 122.5-kilobase deletion of the P gene underlies the high prevalence of oculocutaneous albinism type 2 in the Navajo population. Am. J. Hum. Genet. 72 (1), 62–72.

Prevalence and Population Genetics of Albinism: Surveys in Zimbabwe, Namibia, and Tanzania

Patricia M. Lund[1], Mark Roberts[2,3]

[1]Coventry University, Coventry, United Kingdom; [2]African Institute for Mathematical Sciences (AIMS), Bagamoyo, Tanzania; [3]University of Surrey, Guildford, United Kingdom

O U T L I N E

Albinism in Africa
http://dx.doi.org/10.1016/B978-0-12-813316-3.00004-0

I. INTRODUCTION

Very few large-scale surveys on oculocutaneous albinism (OCA) have been conducted in African populations. Given the very distinctive hypopigmentary phenotype, which is readily recognizable (Fig. 4.1), studies of this condition provide rare opportunities to gather accurate frequency data on a genetic condition that shows an autosomal recessive pattern of inheritance and is known to have a relatively high frequency throughout sub-Saharan Africa. Similar opportunities are provided by censuses. Concerns that have grown in recent years about the health and security challenges faced by people with albinism in Africa mean that albinism is now beginning to be included in national census questionnaires by African governments. In addition to providing more accurate frequency estimates, large-scale surveys and censuses can also reveal surprising patterns of variation and may cast some light on the underpinning population genetics. However, results need to be interpreted with considerable care, given the sensitive nature of issues concerned with albinism.

In this chapter, the results of a large-scale survey of school children in Zimbabwe in 1994–95 and data from national censuses in Namibia and Tanzania in 2011 and 2012, respectively, are reviewed and discussed from the perspective of population genetics.

FIGURE 4.1   Shop owner with albinism in Tanzania. *Photograph courtesy of Dr. P. Lund, Coventry.*

II. ZIMBABWE: A POSTAL SURVEY OF SCHOOLS

The largest ever (noncensus) survey of albinism in schools in an African country was conducted in 1994–95 in Zimbabwe, a country located in the central part of southern Africa (see Fig. 1.2, for a map of Africa). Zimbabwe has a relatively homogenous population, with the majority belonging to the Shona group. At the time when this study was undertaken it had a population of about 11 million people, a reliable postal service, and a well-regarded schooling system with well-trained teachers. Children with albinism were educated in mainstream schools.

Survey Methodology

A pilot survey was conducted in the capital city of Harare in 1994 (Kagore and Lund, 1995). A questionnaire in English was sent to head teachers at all secondary schools in the city, with the prior permission of the Ministry of Education and Culture and the regional education office. The latter provided the researchers with the names and addresses of the 69 schools falling within their region. The questionnaire requested information on the total number of pupils at each school and their sex and age range. Head teachers were asked for details of any pupils with albinism, including their sex, age, and ethnic origin. An accompanying letter gave a description of the phenotype of albinism and the vernacular names for people with albinism. It stressed that replies were required from schools where there were no pupils with albinism and from those educating pupils with this condition. A stamped and preaddressed envelope was included for return of the completed data sheet. The head was asked to tick a box to indicate the location of the school: urban (cities, towns, and peri-urban) or rural (including communal lands and commercial farms). The questionnaire was accompanied by a copy of the letter of permission from the Ministry of Education. A second mailing was sent after 3 months to schools which had not responded (21.7%); those that failed to respond to either mailing were visited in person to collect the data. Most of the cases of albinism were confirmed by visits from one of the researchers and eight were interviewed in their homes. In a parallel survey data were collected from primary schools in Harare.

Following the pilot study in Harare, a more extensive survey of primary and secondary schools across the country was undertaken, using the same methodology as described above, to investigate the geographical distribution of OCA in the country as a whole, covering both urban and rural areas (Lund, 1996). The questionnaire was mailed to the head of every secondary school outside the capital Harare (1447 schools) and to primary schools in four of the eight provinces outside Harare (1747 schools). The response rate for this survey was 90.8%, with replies from 2899 of the 3194 schools sampled. The authors attribute this extraordinarily high response rate to the inclusion of the letter from the Ministry giving permission to collect the data. Head teachers would have considered it a requirement to provide this information if requested to do so by the Ministry. The final return rate was slightly higher than this, with some envelopes returned even after the results were analyzed and the journal article published, perhaps due to a head teacher finding the questionnaire, at the bottom of a pile of papers at the end of a busy school year! It was not feasible to confirm all the classifications of OCA made by the head teachers, but the author visited a total of 50 pupils in seven different provinces to confirm they had OCA.

Survey Results

The main results of the surveys are summarized in Table 4.1. The national prevalence of OCA in the school-going population was 1 in 4182, but the distribution across the country was not even. The highest frequencies were in the capital Harare: 1 in 2792 in primary schools and 1 in 2661 in secondary schools. In primary schools in the four provinces outside Harare that were sampled, rates ranged from 1 in 3843 in Mashonaland East to 1 in 7539 in Matabeleland South. In secondary schools, they ranged from 1 in 2956 in Manicaland to 1 in 7311 in Mashonaland Central province.

About one-third of the schools in Harare (24/69 schools; 34.8%) had at least one pupil with albinism, whereas outside the capital the proportion was much lower: 157 pupils with OCA were being educated in 126 different primary schools and 121 in 101 secondary schools, with a total of 7.8% of schools educating pupils with OCA.

The authors of this chapter have revisited these data and conducted an additional chi-squared analysis, which revealed no significant difference between the recorded frequencies of albinism in primary schools and those in secondary schools in any of the five provinces for which both sets of data were obtained. This remains true when the data from the five provinces are combined (Table 4.1). These results suggest that pupils with albinism are as likely to progress to secondary level schooling as their unaffected peers.

Sex Parity

The autosomal recessive inheritance of OCA predicts that males and females are equally likely to be affected. Lund (1996) reported a male to female sex ratio of 1.18 for all pupils with albinism identified in the postal survey, not significantly different from the expected ratio of 1. Further analysis by the current authors showed no significant differences between the recorded frequencies of albinism in the male and female populations at either primary or secondary level or when these levels were combined (see Table 4.2).

In Zimbabwe the proportion of females proceeding to secondary schools is generally less than that of males, regardless of whether or not they have albinism. However, the finding that the frequency of albinism among females in secondary schools is not significantly different from the frequency in males suggests that females with albinism are not disadvantaged any more than females without albinism.

Urban–Rural Prevalence

Lund (1996) noted that, for the country as a whole, the prevalence of albinism in schools in urban areas was significantly higher than in rural areas ($P < .01$), with 65.8% pupils with albinism being educated in urban locations. Reanalysis of the data shows that, if the Harare data are excluded, the frequency for urban areas is *lower* than for rural areas, though not significantly so. This suggests that the national urban–rural difference is a consequence of the significantly higher frequency in the almost entirely urban Harare.

TABLE 4.1 Frequency of Albinism in Zimbabwe[a]

Province	Primary Schools			Secondary Schools			Primary and Secondary Combined		
	Total Pupils	Pupils With OCA	Frequency 1 in x	Total Pupils	Pupils With OCA	Frequency 1 in x	Total Pupils	Pupils With OCA	Frequency 1 in x
Harare	203,795	73	2792	87,817	33	2661	291,612	106	2751
Manicaland		No data		91,619	31	2955	91,619	31	2955
Masvingo		No data		91,324	25	3653	91,324	25	3653
Mashonaland East	242,094	63	3843	80,898	13	6223	322,992	76	4250
Matabeleland North		No data		62,690	14	4478	62,690	14	4478
Mashonaland West	215,401	47	4583	56,592	9	6288	271,993	56	4857
Midlands		No data		88,151	17	5185	88,151	17	5185
Mashonaland Central	172,016	28	6143	36,553	5	7311	208,569	33	6320
Matabeleland South	143,247	19	7539	33,773	7	4825	177,020	26	6808
Total	976,553	230	4246	629,417	154	4087	1,605,970	384	4182

[a] Data from Kagore, F., Lund, P.M., 1995. Oculocutaneous albinism among schoolchildren in Harare, Zimbabwe. J. Med. Genet. 32 (8), 859–861 and Lund, P.M., 1996. Distribution of oculocutaneous albinism in Zimbabwe. J. Med. Genet. 33, 641–644.

TABLE 4.2   Frequency of Male and Female Pupils With Albinism in Zimbabwe Excluding the Capital, Harare[a]

	Total Population	Population With Albinism	Frequency of Albinism (1 in x)
PRIMARY SCHOOLS			
Male	393,105	76	5172
Female	378,602	81	4674
Total	771,707	157	4915
SECONDARY SCHOOLS			
Male	300,031	74	4054
Female	240,294	47	5113
Total	540,325	121	4465
TOTAL: PRIMARY AND SECONDARY SCHOOLS			
Male	693,136	150	4621
Female	618,896	128	4835
Total	1,312,032	278	4720

[a] Data from Kagore, F., Lund, P.M., 1995. Oculocutaneous albinism among schoolchildren in Harare, Zimbabwe. J. Med. Genet. 32(8), 859–861 and Lund, P.M., 1996. Distribution of oculocutaneous albinism in Zimbabwe. J. Med. Genet. 33, 641–644. The reported data do not give numbers for males and females in Harare for primary schools so the capital was excluded from this table.

Geographic Distribution

The frequency data in Table 4.1 and their pairwise chi-squared comparisons shown in Table 4.3 reveal the following:

1. The highest frequencies at both primary and secondary levels are in Harare. When primary and secondary levels are combined, the Harare frequency is significantly higher than those in each of the three nearest provinces, Mashonaland East, Mashonaland West, and Mashonaland Central ($P < .01$).
2. The provinces with the four highest frequencies when primary and secondary levels are combined, namely Harare, Manicaland, Masvingo, and Mashonaland East, form a single block in the east of the country. The only significant difference between frequencies within this block is that between Harare and Mashonaland East noted in the previous point.
3. Similarly, the provinces with the five lowest frequencies form a block in the north and west of Zimbabwe. There are no significant differences between the frequencies for provinces in this block. There are, however, a large number of significant frequency differences between provinces in this block and those in the "eastern block" described above.

These observations suggest two factors that could be influencing the distribution of albinism in Zimbabwe. The first is the singular nature of Harare with its exceptionally high frequencies compared with the rest of the country. Lund (1996) speculated that this could be

TABLE 4.3 Comparison of Frequencies of Albinism in Combined Populations of Primary and Secondary Pupils in Different Provinces of Zimbabwe[a]

	Manica	Masvingo	Mash East	Mat North	Mash West	Midlands	Mash Central	Mat South
Harare	ns	ns						
Manica		ns	ns	ns				
Masvingo			ns	ns	ns	ns	ns	
Mash East				ns	ns	ns		
Mat North					ns	ns	ns	ns
Mash West						ns	ns	ns
Midlands							ns	ns
Mash Central								ns

Manica, Manicaland; *Mash*, Mashonaland; *Mat*, Matabeleland.

[a] *Significant differences are in black for P < .001, dark gray for P < .01, and pale gray for P < .05, while "ns" indicates no significant difference. See the text for a discussion of the splitting of the provinces into two blocks.*

caused by migration of schoolchildren with albinism to Harare from nearby provinces. If, as anecdotal evidence suggests, this was particularly true for secondary school students, then it might account for the fact that the frequency of albinism among secondary school students is *higher* than among primary school students in Harare but *lower* in each of the nearby provinces Mashonaland East, Mashonaland West, and Mashonaland Central (see Table 4.1).

The second factor is the approximate east–west divide. More data and further work are needed to explore possible reasons for this. Explanations may be found in ethnic or cultural differences. However, as Shona-speaking people predominate in most of the provinces surveyed, different ethnicities are unlikely to contribute significantly to the observed differences in frequencies. In Harare, in particular, most of the pupils enumerated (26/33 = 78.8%) belonged to the Shona ethnic group, which discourages alliances between members of the same "totem" and there was no evidence of consanguineous marriages in the small sample of eight families interviewed there. It is possible that, instead of invoking ethnic or cultural differences, genetic drift could be the key mechanism underpinning the differences.

III. NAMIBIA: NATIONAL CENSUS

Namibia was the first country in Africa to publish data on albinism from its national census (Namibia Statistics Agency, no date). It is a large country in South West Africa with a small, scattered population of 2.1 million according to the 2011 census. The majority (57%) live in rural areas, with a population density of 2.9 per square kilometer. This section gives an initial analysis, by the current authors, of the albinism data extracted from the census.

Prevalence

Over the census period in August 2011 a total of 1204 people with albinism were recorded in Namibia, giving a frequency of 1 in 1755 (see Table 4.4), the highest national frequency recorded in Africa to date. Assuming Hardy–Weinberg equilibrium, this gives a heterozygote carrier rate of ~1 in 21.

TABLE 4.4   Frequency of Albinism in Namibia[a]

	Total Population	Population With Albinism	Frequency of Albinism (1 in x)
SEX COMPARISON			
Male	1,021,912	585	1747
Female	1,091,165	619	1763
Total	2,113,077	1204	1755
URBAN/RURAL COMPARISON			
Rural	1,209,753	829	1459
Urban	903,324	375	2409
Total	2,113,077	1204	1755

[a] Data on total population and population with albinism for both sexes and for rural and urban locations in Namibia from the 2011 National Census (Namibia Statistics Agency, 2014, Tables 2.1, 2.5, 5.9, and 5.10 in Regional Profiles).

Sex Parity

The total figure disaggregates into 619 females and 585 males with albinism, giving frequencies of 1 in 1763 for females and 1 in 1747 for males. The proportion of females in the total population (51.6%) is therefore very close to the proportion of females among those with albinism (51.4%), indicating that no sex bias is observed in this data set. This sex parity is as expected for a condition with an autosomal recessive pattern of inheritance and is also observed in the Zimbabwe data discussed above.

Urban–Rural Prevalence

Table 4.4 also shows that there were 829 people with albinism in rural areas and 375 in urban areas, giving a rural frequency of 1 in 1459 and an urban frequency of 1 in 2409. Chi-squared analysis confirms that this is a very significant difference ($P < .001$). This is in contrast to the Zimbabwe schools data discussed above for which, across the country as a whole, the urban frequency is significantly greater than the rural frequency. Further work needs to be done to understand the factors that lead to these significant rural–urban differences and why they might vary between countries.

Geographic Distribution

The Namibian census results disaggregated by region (Namibia Statistics Agency, 2014) are shown in Table 4.5. The frequencies of albinism range from 1 in 1077 to 1 in 5679, a noticeably wider variation than in Zimbabwe. This is perhaps a consequence of the low-population density and relative isolation of population groups. The four regions with the highest frequencies, Oshikoto, Ohangwena, Oshana, and Omusati, form a cluster in the north of the country and are the regions in which Oshiwambo is the majority language. It therefore seems very likely that ethnic differences are playing a role in the distribution of albinism in Namibia.

TABLE 4.5 Frequency of Albinism in Regions of Namibia[a]

Region	Total Population	Population With Albinism	Frequency of Albinism (1 in x)
Oshikoto	181,973	169	1077
Ohangwena	245,446	195	1259
Oshana	176,674	129	1370
Omusati	243,166	172	1414
Kavango	223,352	155	1441
Khomas	342,141	157	2179
Otjozondjupa	143,903	62	2321
Zambezi	90,596	37	2449
Erongo	150,809	48	3142
Kunene	86,856	27	3217
!Karas	77,421	21	3687
Omaheke	71,233	18	3957
Hardap	79,507	14	5679
Namibia	2,113,077	1204	1755

[a] Data on total population and population with albinism in Namibia from the 2011 National Census (Namibia Statistics Agency, 2014, Tables 2.5 and 5.9 in Regional Profiles).

IV. TANZANIA: NATIONAL CENSUS

Tanzania is an East African country with a large population (over 44 million people in 2012). The 2012 census asked whether each enumerated person had albinism or not. The published results provide the most extensive national data set on albinism that has been obtained to date. They consist of the numbers of people with albinism in each of the 30 regions of Tanzania, simultaneously disaggregated by sex, 5-year age group, and rural–urban classification (The United Republic of Tanzania, 2014).

Prevalence and Geographic Distribution

The census estimate of the overall frequency of albinism in Tanzania was 1 in 2673 (Table 4.6). Estimates for each region can be obtained by combining data from Table 2.5 and Table 11.1 in the main census report classification (The United Republic of Tanzania, 2014). They range from 1 in 2015 for Dodoma in the center of the country to 1 in 4586 in Kagera in the North West. As in the case of Zimbabwe (discussed above), there are suggestions of spatial correlation in the regional frequencies. For example, the three regions with the highest frequencies after Dodoma, namely Kilimanjaro, Arusha, and Mara, all lie along the northern border of Tanzania with Kenya. Similarly, four regions down the western side of the country, Kagera, Kigoma, Rukwa, and Mbeya, are among the five mainland regions with the lowest frequencies. The frequency of albinism in the islands of Zanzibar was also low, at 1 in 3724. Current research involving the authors is focusing on the application of geospatial statistical methods to the Tanzanian census data to determine whether these correlations are significant and to relate them to factors that might contribute to explanations of the frequency differences between regions.

TABLE 4.6   Frequency of Albinism in Tanzania[a]

	Total Population	Population with Albinism	Frequency of Albinism (1 in x)
RURAL LOCATIONS			
Male	15,184,970	6015	2486
Female	15,915,754	4638	3371
Total	31,100,724	10,653	2871
URBAN LOCATIONS			
Male	6,199,738	2950	2102
Female	6,749,414	2696	2503
Total	12,949,152	5646	2294
TOTAL: RURAL AND URBAN LOCATIONS			
Male	21,384,708	9059	2361
Female	22,665,168	7418	3055
Total	44,049,876	16,477	2673

[a] Data on total population and population with albinism for both sexes and rural and urban locations in Tanzania from the 2012 National Census of 2012 (United Republic of Tanzania, 2014, Tables 11.3 and 11.4).

Sex Disparity and Urban–Rural Prevalence

A summary of the national census data on albinism simultaneously disaggregated by sex and urban–rural location is given in Table 4.6. The overall frequency of 1 in 2673 for albinism in Tanzania disaggregates into a frequency of 1 in 3055 for females and 1 in 2361 for males. Chi-squared analysis confirms that this female frequency is very significantly lower than the male frequency ($P<.001$), setting this Tanzanian data set apart from those from Zimbabwe and Namibia.

The rural frequency of 1 in 2871 is also very significantly lower than the urban frequency of 1 in 2294 ($P<.001$), which agrees with the analogous data from Zimbabwe but is the opposite from that in Namibia. This rural–urban difference persists when the data are further disaggregated by sex: the female rural frequency is very significantly lower than the female urban frequency, and the same is true for the male frequencies (both $P<.001$). In both rural and urban areas, the female frequencies are significantly lower than the male frequencies ($P<.001$). Overall these results suggest that there is a greater tendency for people with albinism in Tanzania to live in urban rather than rural areas.

Age-Structured Prevalence

Table 4.7 gives the prevalence of albinism disaggregated by sex and 20-year age group along with the male frequency divided by the female frequency. The lower frequency of albinism among females is evident in all age groups except 40–59. However, it is particularly prominent in the youngest age group (0–19 years) where the ratio of the male frequency to the female frequency is 1.52.

The age-structured data also show that the frequencies of albinism in both female and male populations increase with age: albinism is more than three times as common in the 80+ age group than it is in the under 20s. This is contrary to the frequently made observation that people with albinism in Africa have shorter life expectancies than their darkly pigmented peers due to life-limiting skin cancers.

Discussion

The significant disparities between the male and female frequencies in Tanzania are very clear but completely unexpected. There are no precedents for these differences in the data sets from Zimbabwe and Namibia. It is possible that an explanation for these disparities in

TABLE 4.7   Age-Structured Frequency of Albinism in Tanzania[a]

Age Group (years)	Frequency of Albinism (as 1 in x)			Male:Female Frequency Ratio
	Male	Female	Total	
0–19	2582	3913	3114	1.52
20–39	2411	2870	2638	1.19
40–59	2356	2365	2361	1.00
60–79	1381	1661	1512	1.20
80+	758	974	865	1.28
Total	2361	3055	2673	1.29

[a] Calculated from age-structured data on the total population and population with albinism in Tanzania from the 2012 National Census (United Republic of Tanzania, 2014, Table 11.2).

Tanzania might be found in the country's recent history of widely reported attacks on people with albinism. This may have resulted in census respondents not reporting that the potentially more vulnerable members of their household had albinism for fear of putting them at risk. In particular, there may have been a significant underreporting of females with albinism, especially in the youngest and oldest age groups. Further research is urgently needed to test this hypothesis and to explore other possible explanations. It is possible that underreporting of both male and female young members of a household with albinism might account for the overall lower prevalence of albinism in the youngest age group. However, it seems unlikely that this is the main cause of the differences between all the age groups. Another possible explanation is that the age gradient is a consequence of increased population mixing.

If there was indeed underreporting of people with albinism in the census, then the estimate of 1 in 2673 for the frequency of albinism in Tanzania is an underestimate. Current research includes work to produce revised estimates, which take into account reasonable hypotheses on how the underreporting occurred.

A much higher figure of 1 in 1400 often quoted for the national prevalence of albinism can be traced back to an informal and unsubstantiated estimate that in the early 1980s "the city of Dar es Salaam [had] one million residents with an estimated population of 700 albinos" (Luande et al., 1985, p. 1824). The paper actually reported on a study of 350 people with albinism living in Dar es Salaam and registered at the Tanzania Tumor Center, providing evidence that at that time the frequency in this city was at least 1 in 2850 (350 in 1 million). The census showed very significant in-country frequency variations and so this local frequency cannot be extrapolated to the country as a whole.

The only reported molecular genotyping to determine the frequency of carriers of the predominant mutant allele in the Tanzanian population, the 2.7 kb deletion mutation, gave a heterozygote frequency of 1 in 24 among 240 individuals from colonies of leprosy patients (Tuli et al., 2012). Assuming Hardy–Weinberg equilibrium, this would correspond to an OCA frequency of 1 in 2300 for this mutation. As this is not the only mutation in *OCA2* that has been found in Tanzania (Spritz et al., 1995), it suggests a higher overall frequency for OCA. However, as the authors themselves recognized, their sample is small and conclusions drawn from it need to be treated with caution.

V. PERSPECTIVES FROM POPULATION GENETICS

The high frequency of OCA in sub-Saharan Africa and analysis of "large sample" albinism data from Zimbabwe, Namibia, and Tanzania discussed above prompt a number of important questions:

1. Why does albinism persist and have such a relatively high frequency in sub-Saharan Africa?
2. What causes the very significant differences between frequencies of albinism observed in different geographic regions, population groups, and urban and rural locations?
3. What are the causes of the sex and age differences seen in the Tanzanian census data?

This section discusses insights into each of these questions that can be obtained from population genetics.

Persistence of Albinism

As described in the next chapter (see Chapter 5), albinism among the Bantu-speaking people of Central, East, and Southern Africa is primarily due to a particular mutation, a 2.7 kb deletion in the *OCA2* gene. The apparent presence of this mutation among all Bantu-speaking groups strongly suggests that it occurred before the beginning of the spread of the Bantu speakers from a region in present-day Cameroon, an event supported by archaeological and linguistic evidence and dated to at least 3000 years ago.

Researchers have argued that the persistence of albinism, despite its deleterious consequences, needs explaining: "historically, some tribes have killed newborns with albinism and those that survive are often discriminated against as marriage partners. Since people with albinism have fewer children, as a result of murders and social customs, there must be a reason that heterozygous frequency for *OCA2* is so prevalent, especially in Tanzania" (Brilliant, 2015, p. 223). Tuli et al. (2012) have commented that selective mechanisms acting on those with albinism are not efficient at removing mutations that result in albinism. Instead they proposed that the high frequency of OCA in Africa is maintained by heterozygote advantage and tested the hypothesis that the 2.7 kb deletion allele confers increased resistance to leprosy, analogous to the increased resistance to malaria conferred by sickle cell trait. This was the motivation for their study to determine the carrier frequency among those with leprosy (Tuli et al., 2012), but no evidence was found that the 2.7 kb deletion carrier rate was higher in the study group than in the population at large.

In fact, simple population genetics models show that it is not necessary to invoke selection to account for the persistence of a mutant allele. In its simplest form the Wright–Fisher model (Fisher, 1922; Wright, 1931) describes the evolution of the frequencies of two alleles of a gene in a population, assuming nonoverlapping generations, random mating, and no selection or mutation. Fisher, Wright, and many subsequent authors treated the case of constant population size for which the model predicts that one allele eventually dies out and the other fixates in the population. The allele that fixates is most likely to be the one that initially has the highest frequency in the population. This might again suggest that albinism will eventually disappear from a population. However, human populations are not constant in size. If the Wright–Fisher model is modified to allow for growing populations, a very different story emerges—an allele that exists even at a very low frequency in a population can persist if the population is growing fast enough. Otto and Whitlock (1997) have shown that in appropriate parameter regions the effect of population growth on the frequency of an allele is similar to that of a positive selection pressure, and that a high-enough growth rate can even compensate for negative selective pressures. Subsequent work has shown that if the size of the population is bounded, then extinction or fixation will eventually occur (Waxman, 2012), but the time to extinction or fixation can become extremely large. In populations that are growing exponentially (a reasonable model for human populations) these times can become infinite. It follows that a growing population is sufficient to account for the persistence of albinism and it is not necessary to invoke any form of selective advantage. This does not imply that selection pressures are absent from the population genetics of albinism, only that they are not necessary to ensure its persistence and high frequency in African populations. Clues about associations that could lead to selection pressures are most likely to come from genomic studies (Sturm and Duffy, 2012).

Variations in Frequency Between Population Groups

The analysis of the data from Zimbabwe and Namibia given above revealed very significant differences between the frequencies of albinism in different regions of the countries, including regions that neighbor each other. Although not discussed in detail here, the same is true for Tanzania. A conventional explanation of this might be that the populations in the two regions are subject to different selection pressures or marriage patterns (see Chapter 3). However, it seems very likely that in many cases the differences can be accounted for by genetic drift, random changes that occur in allele frequencies from generation to generation. The effects of genetic drift are particularly pronounced when populations are small, which may account for the greater variations between regions that are observed in Namibia compared with Zimbabwe and Tanzania.

The Wright–Fisher model can be used to simulate the extent to which the frequencies of a mutation in two initially identical but distinct populations can drift apart "by chance" even in the absence of differential selection pressures. A current research project involving the authors aims to use these simulations as the basis for a statistical test to determine whether the observed differences between the frequencies in two different populations can be entirely accounted for by genetic drift.

Even when there are good reasons for supposing that differences in frequencies between two regions do reflect ethnic or cultural differences, such as the much higher frequencies in the Oshiwambo-speaking regions of Namibia compared with the other regions, genetic drift is still likely to be playing a major role if the population groups tend not to intermarry. It follows that a significant difference in frequencies between two different ethnic/cultural groups does not necessarily have to be explained by different marriage patterns, though these could still be significant factors.

The significant differences observed between rural and urban frequencies in all three countries considered in this chapter are very unlikely to be due to genetic drift because of the relatively recent onset of urbanization and high levels of migration and intermarrying between urban and rural areas. In this case, the authors believe that socioeconomic factors are most likely to provide the best explanations. However, further research is needed to understand what these might be and why they lead to greater *urban* frequencies in Zimbabwe and Tanzania, but greater *rural* frequencies in Namibia, given that all three countries are predominantly rural.

More research is needed to determine the extent to which genetic drift caused by random frequency variations can lead to observed significant differences between subpopulations. In some contexts, it will be the dominant process, whereas in others it may only be a contributing factor.

Sex Disparity and Age Variation in the Tanzanian Census

Finally, we turn to the results from the Tanzanian census. Here it is interesting to speculate that the huge decrease in the frequency of albinism with age group reflects a decrease in the number of children being born with albinism as a result of changing marriage patterns over the lifetime of the oldest members of the community. Over the last century there has been a very significant increase in population mobility, as demonstrated, for example, by the growth of the urban population from under 10% in the 1960s to around 30% at the 2012 census (The

United Republic of Tanzania, 2015, Table 5.1, p. 35), increasing the pool of possible marriage partners. Further research using population genetics models is needed to see whether it is feasible for this mechanism to account for differences on the scale seen in Tanzania.

The fact that albinism is not sex-linked immediately tells us that the much lower observed frequency among females than males in Tanzania, not seen in Zimbabwe or Namibia, is very unlikely to have a genetic origin. Here further research is needed to ascertain whether the disparity is due to the sampling of census data used to obtain the estimates or to underreporting of females during the enumeration process. If so, then lessons need to be learned from the Tanzanian census so that future censuses in Tanzania and elsewhere can be improved. If these explanations are ruled out, then it is necessary to consider the much more disturbing possibility that the data reflect a genuine disparity and research must be conducted to determine the causes of the disparity.

This section has focused on how population genetics might contribute to explanations of the phenomena that have been revealed by analysis of the large sample data from Zimbabwe, Namibia, and Tanzania. Conversely, the study of albinism could provide insights that are useful for the development of theory and methods in this field. It is one of the most easily identifiable autosomal recessive conditions and so, as the studies reported on here show, it is one that lends itself to the collection of large-scale population-wide frequency data that can be used to test ideas and results from population genetics.

VI. SUMMARY AND DISCUSSION

This chapter summarizes the analysis of three large-scale data sets giving nationwide information on the prevalence of OCA in three African countries in east and southern Africa. The Zimbabwean postal survey of schools was conducted in the mid-1990s as an academic study by a researcher at the University of Zimbabwe; the Namibian and Tanzanian data come from the national census in those countries in 2011 and 2012, respectively.

Summary of Key Results

A comparison of the national results is given in Table 4.8. Namibia has the highest national frequency for albinism recorded to date, though the figure for Tanzania is believed to be an underestimate.

The key differences between the countries emerging from the data can be summarized as follows:

- In Zimbabwe (school population) and Namibia (total population) there are no significant differences between the frequencies of albinism in the male and female populations, as predicted by the pattern of inheritance. However, in Tanzania the frequency in the female population is very significantly lower than that in the male population.
- In Zimbabwe both the frequency of albinism in schools and the absolute number of school students with albinism were greater in urban areas (notably the capital Harare) than in rural areas. In Namibia the frequency and the absolute numbers were both higher in rural areas, whereas Tanzania had a higher frequency in urban areas than rural areas but a higher absolute number of people with albinism in rural areas.

TABLE 4.8   Summary of Frequency Data From Three Countries in Africa[a]

Country	Data Set	Frequency of Albinism	Male:Female Frequency Ratio	Urban:Rural Frequency Ratio	% in Urban Areas
Zimbabwe	Postal survey of schoolchildren 1994–5	1 in 4182	1.05	1.39	65.8
Namibia	Population census 2011	1 in 1755	1.01	0.61	31.1
Tanzania	Population census 2012	1 in 2673	1.29	1.25	34.3

[a] The shaded frequency ratios are very significantly different from 1 with P < .001 in all cases.

The analysis from each country also revealed highly significant differences between regions within the country. This highlights the dangers of extrapolating from small epidemiological studies to whole-country estimates. Even larger variations will occur when smaller geographic areas are considered. For example, a study of a small isolated group of Tonga people in the Zambezi valley of Zimbabwe produced a frequency of 1 in 1000 (Lund et al., 1997), which is much higher than any of the provincial frequencies in that country.

Research Priorities

This review of the survey data from Zimbabwe and initial analysis of census data from Namibia and Tanzania raises a number of important research questions, as summarized below:

• What are the causes of the very significant sex disparity and age variations in the albinism frequencies from the Tanzanian census?

Work on this is needed urgently for two reasons. The first is to identify whether census data collection and analysis methodologies need to be improved and to make recommendations that can be incorporated into future censuses in Tanzania and other countries. The second reason is to determine whether there really are significantly fewer females than males with albinism and whether the frequency among younger people is really so much lower than that among the older generations. If these differences are not artifacts of the data collection or analysis, then the reasons why they occur must be found and documented by rigorous research.

• What are the key factors, and their relative importance, underpinning the significant differences in the albinism frequencies of different population groups?

There will be no simple answers to this question. In Section VI above, it was suggested that genetic drift is likely to be a much more important factor than selection when accounting for regional differences, but that urban–rural differences are more likely to be driven by socioeconomic factors, such as sending children with albinism to schools in the city (Harare in Zimbabwe). Marriage patterns may be a significant factor behind differences between ethnic and cultural groups, though genetic drift is also likely to be playing a role in these contexts. Geneticists have frequently speculated on such issues, but there are usually insufficient population data to reach firm conclusions. The feasibility of collecting and analyzing large data sets for albinism provides a significant opportunity to advance research in this field.

Both the research areas described above will require a range of different approaches and benefit from the participation of several disciplines: genetics, social sciences, statistics, and mathematical modeling.

Messages for Advocates and Policy Makers

- Data collection and analysis is needed to reveal problems and to plan and monitor interventions aimed at solving them.

Advocates and policy makers, in both governmental agencies and charitable organizations, need accurate data on people with albinism in their target areas, to enable them to evaluate and plan for the health, educational, security, and welfare needs of families with albinism. They need age-structured data on the numbers with albinism, where they are, and how to contact them. Large data sets such as the national census reveal information on the numbers and locations of people with albinism to facilitate large-scale planning of resources. A register of those with albinism, for example, collected by registration at birth by nurses with basic genetic training, would provide contact details to ensure families benefit from locally available support (see Chapter 10 and Chapter 13).

The importance of having this information is graphically illustrated by the Tanzanian experience of what happened when such data were *not* available. The 2007/8 reports of the mutilation and killing of those with albinism to obtain body parts for use in witchcraft-related charms and rituals led to a political and social outcry in the country and internationally (see Brilliant, 2015). The government responded rapidly with a number of measures including gathering children with albinism into shelters for their immediate protection. Data were not available to assess the scale of the problem, and the shelters were soon overwhelmed as more and more families sent their children there, leading to overcrowding and abuse. The lack of an exit strategy led to the children becoming displaced and stranded away from their own communities (Under the Same Sun, 2012).

- It is important to use accurate and up-to-date data and for data to be reviewed critically and quoted correctly when it is used.

This is essential for problems to be identified early and for planning and monitoring interventions to be convincing and effective. As an example, an often-quoted statistic is that most people with albinism in Tanzania do not survive beyond the age of 40. A recent 2016 example appears in an interview with the director of an organization working with people with albinism in Tanzania: "In Tanzania less than 10% reach age 30, and only about 2% will live to celebrate their 40th birthday because of skin cancer." (Global Disability Watch, 2016)

These unreferenced comments on survival rates probably have their origin in the study in Dar es Salaam by Luande et al. (1985) mentioned in Section IV. The demographic data in Table 4.1 of that paper show that 15% of the study group were aged 31 or older (53 in a group of 350) and 2% were aged 41 or older (8 out of 350), though the authors themselves say "less than 10% of the population were 30 years or older." These figures contrast sharply with those from the 2012 census (The United Republic of Tanzania, 2015), which give a figure of 35%

for the population with albinism aged 30 or over (5714 out of 16,127) and 25% aged 40 and over (4033 out of 16,127). Both sets of figures need to be treated cautiously, but transferring a statistic from a small study at one center in Dar es Salaam in the early 1980s to the whole of Tanzania in 2016 is potentially highly misleading. Hopefully increased advocacy, education, and health provision in the country will have led to very significant increases in survival rates in the intervening decades, though further data need to be collected to confirm this situation.

Acknowledgments

This article reveals initial insights into the analysis of survey and census data on albinism resulting from a cross-disciplinary collaboration between a bioscientist (Lund) and statisticians and mathematicians (including Roberts) based in the United Kingdom and Tanzania. The initial statistical analysis of the Tanzanian census data was conducted by African Institute for Mathematical Sciences (AIMS) Tanzania Masters students, Naomi Kollongei and Laurette Mhlanga, and continued by a Canadian intern at AIMS, Kirsten Mathison. An investigation into the role of genetic drift was conducted by AIMS Tanzania Masters students Alexis Arakaza and Peter Nabutanyi. These students were supervised by Jim Todd (London School of Hygiene and Tropical Medicine), Michelle Stanton (Liverpool School of Tropical Medicine), and Paul Jenkins (Warwick University) in addition to the authors Lund and Roberts.

References

Brilliant, M.H., 2015. Albinism in Africa: a medical and social emergency. Int. Health 7 (4), 223–225.

Fisher, R.A., 1922. On the dominance ratio. Proc. R. Soc. Edinb. 42, 321–341.

Global Disability Watch, 2016. Hunting for Muti: The Chase for People with Albinism. Available from: http://globaldisability.org/2016/05/30/hunting-for-muti-the-chase-for-people-with-albinism.

Kagore, F., Lund, P.M., 1995. Oculocutaneous albinism among schoolchildren in Harare, Zimbabwe. J. Med. Genet. 32 (8), 859–861.

Luande, J., Henschke, M.D., Mohammed, N., 1985. The Tanzanian human albino skin. Natural history. Cancer 55, 1823–1828.

Lund, P., Puri, N., Durham-Pierre, D., King, R.A., Brilliant, M.H., 1997. Oculocutaneous albinism in an isolated Tonga community in Zimbabwe. J. Med. Genet. 34 (9), 733–735.

Lund, P.M., 1996. Distribution of oculocutaneous albinism in Zimbabwe. J. Med. Genet. 33, 641–644.

Namibia Statistics Agency, 2014. Namibia 2011 Regional PHC Profiles. Available from: http://nsa.org.na/page/publications/.

Namibia Statistics Agency, no date. Namibia 2011 Population and Housing Census Main Report. Available from: http://cms.my.na/assets/documents/p19dmn58guram30ttun89rdrp1.pdf.

Otto, S.P., Whitlock, M.C., 1997. The probability of fixation in populations of changing size. Genetics 146, 723–733.

Spritz, R.A., Fukai, K., Holmes, S.A., Luande, J., 1995. Frequent intragenic deletion of the P gene in Tanzanian patients with type II oculocutaneous albinism (OCA2). Am. J. Hum. Genet. 56 (6), 1320–1323.

Sturm, R.A., Duffy, D.L., 2012. Human pigmentation genes under environmental selection. Genome Biol. 13, 248.

The United Republic of Tanzania, 2014. Basic Demographic and Socio-economic Profile Report. (30 volumes, one for each region) Available from: http://www.nbs.go.tz/.

The United Republic of Tanzania, 2015. Migration and Urbanization Report, 2012 Population and Housing Census. Available from: http://www.nbs.go.tz/.

Tuli, A.M., Valenzuela, R.K., Kamugisha, E., Brilliant, M.H., 2012. Albinism and disease causing pathogens in Tanzania: are alleles that are associated with OCA2 being maintained by balancing selection? Med. Hypotheses 79, 875–878.

Under the Same Sun, 2012. Situation Assessment of the Centres of Displaced Persons with Albinism in the Lake Zone and Tanga Regions. Available from: http://www.underthesamesun.com.

Waxman, D., 2012. Population growth enhances the mean fixation time of neutral mutations and the persistence of neutral variation. Genetics 191, 561—577.

Wright, S., 1931. Evolution in Mendelian populations. Genetics 16, 97–159.

Molecular Biology of Albinism

Prashiela Manga[1,2]

[1]New York University School of Medicine, New York, NY, United States; [2]University of the Witwatersrand, Johannesburg, South Africa

I. INTRODUCTION

The term oculocutaneous albinism (OCA) defines a group of hypopigmentation disorders with concomitant loss of visual acuity. Seven forms of OCA that affect the entire body, delineated by the gene that is mutated, have been identified (Table 5.1; Fig. 5.1). Multiple mutations with variable impact on protein function have been identified at some loci, causing phenotype heterogeneity. Although OCAs affect the skin, hair, and eyes, ocular albinism (OA) primarily affects the eyes. Albinism can also be a feature of syndromic disorders, for example, Hermansky–Pudlak syndrome (HPS) and Chediak–Higashi syndrome (CHS). Congenital hypomelanosis disorders, which result in patches or limited sites of hypopigmentation,

Albinism in Africa
http://dx.doi.org/10.1016/B978-0-12-813316-3.00005-2

TABLE 5.1 Albinism Disorders and Associated Genes

Disorder	Subtype	Gene Mutated	Chromosome Location	Gene/Locus MIM#	OMIM#	References
Classical OCA						
OCA1[a]	A	TYR	11q14.3	606933	203100	Tomita et al. (1989)
	B				606952	Giebel et al. (1991)
	TS				606952	Wang et al. (2005b)
OCA2[a]	A	OCA2	15q11.2-q12	611409	203200	Lee et al. (1994)
	AE			611409	203200	Stevens et al. (1995)
	B/BOCA				203200	Manga et al. (2001b)
OCA3[a]		TYRP1	9p23	115501	203290	Boissy et al. (1996) and Manga et al. (1997)
OCA4[a]		SLC45A2	5p13.3	606202	606574	Newton et al. (2001)
OCA5[a]		Unknown	4q24	–	615312	Kausar et al. (2013)
OCA6[a]		SLC24A5	15q21.1	609802	113750	Lamason et al. (2005)
OCA7[a]		C10orf11	10q22.2-q22.3	614537	615179	Gronskov et al. (2013)
Syndromic OCA						
CHS1[a]		LYST	1q42.1-q42.2	606897	214500	Dufourcq-Lagelouse et al. (1999)
HPS1[a]		HPS1	10q23.1-q23.3	604982	203300	Oh et al. (1998)
HPS2[a]		AP3B1	5q14.1	603401	608233	Dell'Angelica et al. (1999)
HPS3[a]		HPS3	3q24	606118	614072	Anikster et al. (2001)
HPS4[a]		HPS4	22cen-q12.3	606682	614073	Suzuki et al. (2002)
HPS5[a]		HPS5	11p14	607521	614074	Zhang et al. (2003)
HPS6[a]		HPS6	10q24.32	607522	614075	Zhang et al. (2003)
HPS7[a]		DTNB1	6p22.3	607145	614076	Li et al. (2003)
HPS8[a]		BLOC123	19q13.32	609762	614077	Morgan et al. (2006)
HPS9[a]		BLOC1S6	15q21.1	604310	614171	Cullinane et al. (2011)
HPS10[a]		AP3D1	19p13.3	607246	617050	Ammann et al. (2016)

Disorder	Subtype	Gene	Locus		OMIM	Reference
Congenital hypomelanosis	Piebaldism[a]	C-KIT/CD117	4q12	164920	172800	Giebel and Spritz (1991)
		SNAI2	8q11.21	602150	172800	Sanchez-Martin et al. (2003)
	Waardenburg Syndrome	WS1[a]	PAX3	2q36.1	606597	193500	Tassabehji et al. (1992)
		WS2A[a]	MITF	3p13	156845	193510	Tassabehji et al. (1994)
		WS2B[a]		1p21-p13.3	600193	600193	Lalwani et al. (1994)
		WS2C		8p23	606662	606662	Selicorni et al. (2002)
		WS2D	SNAI2	8q11.21	608890	602150	Sanchez-Martin et al. (2003)
		WS3[b]	PAX3	2q36.1	606597	148820	(Hoth et al. (1993)
		WS4A	EDNRB	13q22.3	131244	277580	Puffenberger et al. (1994)
		WS4B[b]	EDN3	20q13.32	613265	131242	Edery et al. (1996)
		WS4C[a]	SOX10	22q13.1	613266	602229	Pingault et al. (1998)
Ocular albinism	OA1[c]	GPR143	Xp22.3	300808	300500	Bassi et al. (1995)

CHS, Chediak–Higashi syndrome; HPS, Hermansky–Pudlak syndrome; OCA, oculocutaneous albinism; OMIM, online Mendelian Inheritance in Man (http://www.ncbi.nih.gov/Omim/).

[a] Autosomal recessive.

[b] Autosomal recessive/autosomal dominant.

[c] X-linked recessive.

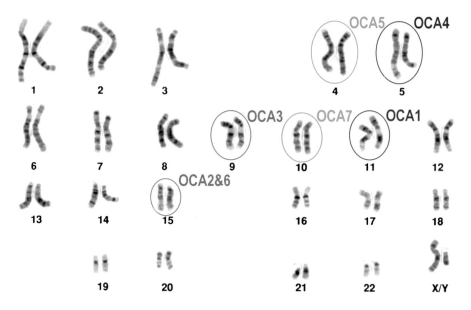

FIGURE 5.1 A normal male karyotype with 22 pairs of autosomal chromosomes and one pair of sex chromosomes showing localization of oculocutaneous albinism (OCA)-related genes. *Modified from AV-9700-4394 obtained from the United States National Cancer Institute (https://visualsonline.cancer.gov/).*

include piebaldism and Waardenburg syndrome. Modes of inheritance of albinism span the spectrum from autosomal recessive disorders such as OCA 1–7 and some forms of OA, to autosomal dominant piebaldism and X-linked OA. In addition, modifier loci can alter the phenotype caused by single gene mutations.

II. BIOLOGY AND BIOCHEMISTRY
OF PIGMENTATION

Melanin is the pigment polymer that colors skin, hair, and eyes. The primary function of melanin is to protect skin from ultraviolet (UV) radiation, and it may also serve as an antioxidant by scavenging reactive oxygen species (Rozanowska et al., 1999). Constitutive skin pigmentation (baseline pigmentation) and the ability to tan (facultative pigmentation) are significant determinants of an individual's risk of developing sun-induced skin cancers (Gilchrest et al., 1999) and susceptibility to premature aging (Hatchome et al., 1987; Porter et al., 1987). Melanin is synthesized in specialized cells, the melanocytes, in the skin, hair, and eyes.

Normal Variation in Skin Pigmentation

Over 400 hundred genes play a role in melanin synthesis and melanocyte function (Montoliu et al., 2014); however only 11 have been confirmed to play a role in determining normal skin pigment variation through genome wide association studies (Nan et al., 2009; Valenzuela et al., 2010), with some genes linked exclusively to either inter- or intrapopulation variation. Many of the other genes are likely to have small effects individually. A similar number of genes have been implicated in determining hair (Branicki et al., 2011) and eye color (Liu et al., 2009; Sturm and Frudakis, 2004) with the melanocortin-1 receptor (*MC1R*) shown to be the "red hair" gene (Rees, 2003) and oculocutaneous albinism type 2 (*OCA2*) the "brown eyes" gene (Sturm and Frudakis, 2004). Skin pigmentation is an adaptive trait (Jablonski, 2012) with dark skin prevalent near the equator preventing folic acid deficiency and protecting against UV damage (Parra, 2007), and lighter skin at higher latitudes to allow for optimal vitamin D synthesis (Jablonski and Chaplin, 2017). Pigmentation has, however, been used as a proxy for both racial classification and determining ancestry. Association studies have shown that in fact skin color correlations with ancestry are not always reliable (Parra et al., 2004).

Single nucleotide polymorphisms (SNPs) at the following loci have been associated with normal pigment variation: *TYR*, *TYRP1*, *OCA2*, *SLC45A2*, *SLC24A5*, *MC1R*, *ASIP*, *KITLG*, *HERC2*, *IRF4*, *TPCN2*, *LYST*, and *BNC2*. A single *SLC24A5* SNP has been reported to account for 25%–38% of skin pigment variation between Europeans and West Africans (Lamason et al., 2005).

Studies have identified a role for *SLC24A5*, *SLC45A2*, and *TYR* in pigment variation among populations in South Asia (Stokowski et al., 2007; Pemberton et al., 2008). Within Europe, skin color variation is associated with *ASIP*, *SLC45A2*, *IRF4*, *HERC2/OCA2*, and *MC1R* SNPs (Liu et al., 2015), whereas *BNC2*, *EGFR*, *LYST*, *MC1R*, *OCA2*, *OPRM1*, *PMEL* (*SILV*), and *TYRP1* play a role in East Asia (Hider et al., 2013). In Africa, genetic variation in major pigment genes, such as *TYR* (Hudjashov et al., 2013) and *MC1R*, is less common (John et al., 2003), although no large-scale association studies have been undertaken.

Differences at the cellular level have also been demonstrated to account for pigment variation. Tyrosinase (an enzyme required to synthesize pigment) protein and activity levels correlate with the amount of melanin synthesized, whereas levels of RNA expression do not. Thus, there is posttranslational regulation of melanin synthesis. Expression of a chaperone protein, HSP701A, which facilitates tyrosinase maturation, has also been shown to correlate with pigment production (Murase et al., 2016). Despite the wide variation in inter- and intrapopulation skin color, there is no difference in basic melanocyte biology.

Melanocyte Differentiation and Function

Melanin is synthesized by pigment cells, which include melanocytes and retinal pigment epithelium. Epidermal melanocytes are dendritic cells derived from the neural crest (reviewed in Sommer, 2011; Bonaventure et al., 2013). During embryogenesis, melanoblasts, the unpigmented melanocyte precursors, populate the interfollicular epidermis, the hair follicle bulge, the uveal tract of the eye, and various structures of the ear. Melanocytes are also found in the substantia nigra of the brain and in the heart (reviewed in Breathnach, 1988; Levin et al., 2009). Melanocyte differentiation is tightly regulated at the genetic level by a number of genes that control proliferation, survival, and migration of precursor cells to the various sites of the body, and their differentiation into active melanocytes.

Melanoblasts migrate to the dermis by 7 weeks of gestation, and markers of differentiation have been noted by 10 weeks (Fujita et al., 1970). Key regulators of this process include the cKIT receptor and the microphthalmia transcription factor (MITF), which is considered the "master regulator" of the melanocyte, capable of modulating expression of several melanocyte specific proteins (Kawakami and Fisher, 2011). *MITF* mutations result in Waardenburg syndrome type II (Tassabehji et al., 1994). MITF also regulates expression of genes in response to UV exposure in the differentiated melanocyte, facilitating survival and tanning response.

Melanocytes synthesize black–brown eumelanin and red–yellow pheomelanin. A third melanin, neuromelanin, is produced by dopaminergic neurons in the midbrain substantia nigra (reviewed in d'Ischia and Prota, 1997). Total melanin amounts and the ratio of eumelanin to pheomelanin determine skin, hair, and eye color (Ito and Wakamatsu, 2011; Wakamatsu et al., 2006). Melanin synthesis takes place in membrane-bound organelles called melanosomes. The toxic intermediates of melanin synthesis (such as reactive oxygen species) are thus confined, limiting the potential for cellular damage. Melanosomes are comprised primarily of proteins synthesized in the endoplasmic reticulum (ER), which are then either modified (for example by glycosylation) in the Golgi or transported to the melanosome directly. Melanosome structure varies depending on whether eu- or pheomelanin is produced. Eumelanosomes are elliptical and more organized, whereas pheomelanosomes are round.

Melanosomes are specialized lysosomes (Orlow, 1995). They share a common lineage with other lysosome-related organelles including alpha- and dense platelet granules, secretory granules, and Weibel–Palade bodies (Marks et al., 2013). Early stage melanosomes are formed from endosomes in the perinuclear area near the Golgi. The Stage I melanosome is defined by the melanosome marker Pmel17/silver, which forms the internal matrix fibrils that the melanin will be deposited onto (Berson et al., 2001). Separation of melanosomes and lysosomes is thought to be regulated by the protein mutated in OA1, GPR143 (Burgoyne et al., 2013). As the proteins and enzymes required for melanogenesis are delivered, the melanosomes matures to Stage II, with increasing melanin deposition marking Stage III. Once the organelle is filled with melanin, synthesis stops following GPR143 signaling (Young et al., 2011). The Stage VI melanosomes are trafficked to the periphery of the cell along microtubules as they mature. The transfer is facilitated by kinesins, which attach melanosomes to the microtubules and transfer them out to the tips of the dendrites. Myosin-Va transfers the melanosomes to actin filaments. In addition, Rab27a and melanophilin are also critical for melanosome movement (reviewed in Aspengren et al., 2009). Following transfer to actin, melanosomes are transferred to keratinocytes where they ultimately form a cap over the nucleus. Melanosomes in Black skin are large and remain as single organelles in keratinocytes, whereas melanosomes in White skin are smaller and incorporated into membrane bound clusters (Minwalla et al., 2001). Epidermal melanocytes have dendrites that intercalate between the keratinocytes allowing one melanocyte to supply melanosomes to up to 30-40 keratinocytes (Hoath and Leahy, 2003). Melanosomes remain intact in Black skin (Thong et al., 2003), while they are degraded in White skin, with visible "melanin dust" in suprabasal layers of the skin (Ebanks et al., 2011). Asian skin contains a combination of the two forms (Thong et al., 2003).

Disruption in delivery of proteins to the melanosome results from mutations in several genes, which cause various forms of HPS (Table 5.1). At least 10 forms, mapping to different loci, have been described. Affected individuals all have OCA, lack platelet dense granules causing a bleeding diathesis, and may experience granulomatous colitis or fatal pulmonary fibrosis (Wei, 2006) because of disruption in biogenesis of lysosome-related organelles.

Several signaling pathways ensure melanocyte survival and regulate the amount of melanin produced by the melanocyte allowing the cell to respond to external factors. The most potent melanocyte stimulator is exposure of the skin to UV. The tanning response is mediated by MC1R, which also promotes melanocyte survival. MC1R, a seven transmembrane domain G-protein-coupled receptor is activated by binding of its ligands, α-melanocyte-stimulating hormone (MSH) and adrenocorticotropic hormone. Both hormones are derived from proopiomelanocortin, which is produced in the pituitary gland, by keratinocytes in the skin and the hair follicle (reviewed in Abdel-Malek et al., 2008). Keratinocytes produce MSH in response to UV radiation, thereby mediating the tanning response (Chakraborty et al., 1999). When bound, MSH activates adenylyl cyclase triggering production of cyclic adenosine monophosphate (cAMP), which in turn causes phosphorylation of the cAMP responsive element-binding protein and binding to its response element, CRE (reviewed in Busca and Ballotti, 2000). This promotes expression of MITF. MC1R activity can be inhibited by binding of its agonist, the agouti/ASIP protein (reviewed in Wolf Horrell et al., 2016). Although the ASIP protein has been shown to regulate hair banding in lower mammals, the function in humans is not known. Variants at the *ASIP* locus have been shown to be associated with normal pigment variation (Liu et al., 2015). Melanocytes also respond to inflammatory factors including Endothelin-1 (also known as steel factor or stem cell factor), inflammatory mediators such as prostaglandins and leukotrienes, as well as basic fibroblast growth factor, which can modulate melanocyte viability and the rate of melanin production (reviewed in Yamaguchi and Hearing, 2009).

Melanin Synthesis

Melanin synthesis begins with the hydroxylation of the amino acid tyrosine to L-3,4-dihydroxyphenylalanine (DOPA) in a reaction catalyzed by tyrosinase. The conversion of DOPA to DOPAquinone is also catalyzed by tyrosinase. Eumelanin is synthesized in the absence of cysteine or glutathione, via a spontaneous reaction or one that requires dopachrome tautomerase (DCT, formerly known as tyrosinase-related protein 2 or TYRP2). DOPAquinone undergoes spontaneous oxidation to produce DOPAchrome, which is in turn targeted by DCT to produce dihydroxycarboxylic acid. Pheomelanin is produced, when DOPAquinone is loaded with cysteines. Other than tyrosinase, the remainder of the reactions are spontaneous and do not require additional catalysts (reviewed in Prota, 2000).

III. MOLECULAR AND BIOCHEMICAL BASIS OF OCULOCUTANEOUS ALBINISM

Garrod first proposed that albinism was an inborn error of metabolism in 1908. He correctly theorized that hypopigmentation was because of lack of an enzyme activity. This enzyme was later shown to be tyrosinase, the rate-limiting enzyme in melanin synthesis (Duliere and Raper, 1930). Synthesis and maturation of tyrosinase is complex and highly regulated. Maturation begins in the ER where proteins known as chaperones facilitate folding, posttranslational modification, and acquisition of tertiary structure. Tyrosinase is further modified in the Golgi before being transported to melanosomes (reviewed in Wang and Hebert, 2006). Mutations that result in OCA1, 2, 3, and 4 all result in retention of tyrosinase in the ER, and/or misrouting that prevents delivery to the melanosome (Toyofuku et al., 2001b; Manga et al., 2001a; Costin et al., 2003). Mutation of proteins involved in ensuring correct transport and delivery of tyrosinase, and other melanosomal proteins lead to the various forms of HPS (reviewed in Sanchez-Guiu et al., 2014).

Tyrosinase maturation can be promoted in some OCA melanocytes in culture by increasing the levels of tyrosine or DOPA. Prior to identification of the genes that cause OCA, a hair bulb test was used to distinguish tyrosinase-positive and tyrosinase-negative OCA. Hair bulbs were incubated in high levels of tyrosine to determine if they produced melanin. In tyrosinase negative, there was no increase in melanin; whereas in tyrosinase positive OCA heavily pigmented hair bulbs were observed postincubation (King and Witkop, 1976).

Oculocutaneous Albinism Type 1

OCA1 is an autosomal recessive disorder caused by *TYR* mutations. The *TYR* gene spans 5 exons and is approximately 65 kb in length encoding a 529 amino acid peptide. The gene encodes an enzyme with tyrosine hydroxylase activity that catalyzes three reactions during melanogenesis. Tyrosinase is a single transmembrane domain protein (Type I membrane protein) with the active site facing the lumen of the melanosome, that is co-translationally modified in the ER (Wang et al., 2005a). Tyrosinase is retained for an unusually long period of time in the ER prior to its exit to the Golgi, compared with similar proteins (Petrescu et al., 1997; Ujvari et al., 2001). This suggests that early processing is highly regulated and/or complex.

Tyrosinase maturation is facilitated by a number of chaperones including heat shock protein 70 (HSP70) family member BiP/GRP78, which is the first chaperone to bind tyrosinase (Wang et al., 2005a). The lectins, calnexin, and calreticulin bind sequentially (Petrescu et al., 2000; Branza-Nichita et al., 1999). Tertiary structure requires disulphide bond formation, which is catalyzed by the oxidoreductase ERp57 (Wang et al., 2005a). ERp57 is thought to be recruited by the calnexin/calreticulin complex. Correctly folded tyrosinase is then transported to the Golgi. In addition to the ubiquitous chaperones, tyrosinase maturation also requires melanocyte-specific factors including TYRP1 (Francis et al., 2003).

Tyrosinase requires copper as a cofactor. Copper is loaded, then removed prior to transport from the Golgi to ensure that the enzyme is not active at any site other than the melanosome. ATP7A is the transport channel that facilitates the availability of copper for tyrosinase.

Over 100 mutations have been described at this locus. The most severe phenotype, OCA1A or "tyrosinase-negative albinism" results from a complete lack of tyrosinase activity and the skin, hair, and eyes remain unpigmented throughout the life of the affected individual. Many of these severe mutations result in misfolding of the mutant protein with attendant inability to exit the ER (Berson et al., 2000; Toyofuku et al., 2001a; Chaki et al., 2011). Milder mutations result in proteins with some activity producing slightly less severe phenotypes. These forms include OCA1B where affected individuals have white skin and hair at birth, but they develop some pigment with age and express less severe ocular findings than in OCA1A; and temperature sensitive OCA1 where tyrosinase is retained in the ER because of misfolding, but it can escape the ER at lower temperatures. The temperature-sensitive enzyme is active in cooler regions of the body resulting in pigmented eyebrows and body hair. The phenotype is homologous to that of the Siamese cat. A third form is platinum OCA where small amounts of pigment accumulate in the hair and eyes in late childhood resulting in a silver tinge (Witkop, 1985). In mice, platinum mutations result in misrouting of tyrosinase due to the loss of the cytoplasmic tail (Beermann et al., 1995).

Tyrosinase polymorphisms have been found to associate with variation in normal skin pigmentation as well as an individual's risk of developing melanoma and vitiligo (Gerstenblith et al., 2007; Jin et al., 2010).

Oculocutaneous Albinism Type 2

OCA2 results from mutations in the human homologue of the mouse pink-eyed dilution gene, *OCA2* (also known as the *P* gene). The gene has 23 coding exons, and one non-coding exon, and spans 345 kb of genomic DNA at 15q11.2-q12 (Lee et al., 1995).

Mutations at the murine *oca2* locus were studied by Russell as far back as the 1940s and shown to cause abnormal melanin granules in the hair of affected mice (Russell and Russell, 1948; Russell, 1949). The murine and human genes were identified (Rinchik et al., 1993; Gardner et al., 1992) and found to encode a 838 amino acid, 110 kDa protein (then referred to as the P protein), with 12 putative transmembrane domains (Rosemblat et al., 1994).

Point mutations and deletions were associated with OCA2 as well as with hypopigmentation in Prader–Willi and Angelman syndromes (Spritz et al., 1997). Over 150 *OCA2* mutations have been reported to date (http://www.hgmd.org/). In addition, duplication of the chromosome 15q region encoding the *OCA2* gene has been associated with diffuse hyperpigmentation (Akahoshi et al., 2001).

OCA2 was mapped to human chromosome 15q11-12 and showed evidence of locus homogeneity in South African Black families (Ramsay et al., 1992). OCA2-associated haplotypes suggested multiple mutations, although there is one common mutation, a 2.7 kb deletion of exon 7, among these families (Stevens et al., 1995). The mutation accounts for about 80% of *OCA2* mutations in Southern Africa chromosomes and has been traced back to a founding mutation in central Africa (Stevens et al., 1995). No other common mutations have been identified among non-2.7 kb alleles in affected individuals from different regions in Africa including South Africa, Lesotho, Zimbabwe, and the Central African Republic (Kerr et al., 2000). Interestingly, a second deletion of 122.5 kb was found to be common among the Navajo of North America. The carrier frequency of the deletion was ~4.5 in 100 suggesting a disease prevalence of 1 in 1500–2000 (Yi et al., 2003).

OCA2 is highly polymorphic and has been shown to be a determinant of normal skin pigmentation as well as eye color. One particular polymorphism located in the gene immediately 5′ to *OCA2*, *HERC-2*, is most informative for eye color (Chaitanya et al., 2014). The polymorphism is thought to result in reduced expression of the *OCA2* gene.

The precise function of the OCA2 protein is not known. Mouse *oca2* mutations have severe effects on the size, number, and pigmentation of melanosomes in ocular tissue (Orlow and Brilliant, 1999). In addition, tyrosinase accumulates in the ER of melanocytes lacking p protein expression. Rather than reaching melanosomes a large proportion of the tyrosinase that exits the ER is instead secreted (Orlow and Brilliant, 1999; Manga et al., 2001a).

No mammalian proteins exhibit homology to OCA2. Several bacterial, archaebacterial, and mycobacterial membrane proteins, some of which are members of the ArsB arsenic transporter family, exhibit homology to both the murine and human OCA2 proteins. Given this transporter-like structure, it was thought that the OCA2 protein was a transporter for tyrosine (Rinchik et al., 1993), protons (Puri et al., 2000; Brilliant, 2001), or thiols (Lamoreux et al., 1995); however, tyrosine transport was no different in wild type and mutant cells (Potterf et al., 1998).

Because tyrosinase maturation and trafficking to the melanosome is disrupted in melanocytes that lack OCA2 protein expression, it may be required for tyrosinase stabilization (Manga et al., 2001a). Alternatively, transport of thiols into the ER may be important for maintaining an environment that enables correct tyrosinase folding (Manga and Orlow, 2011; Fig. 5.2).

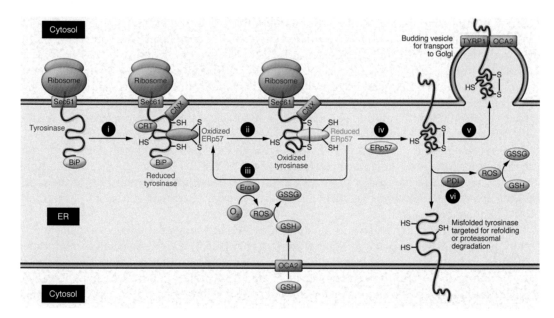

FIGURE 5.2   Tyrosinase is synthesized on ribosomes and translocated into the endoplasmic reticulum (ER) membrane. (i) Tyrosinase is bound by the chaperones binding immunoglobulin protein (BiP), then calnexin (CNX), followed by ERp57 (ii) and calreticulin (CRT). Each chaperone is required to facilitate tyrosinase folding and glycosylation. To fold the peptide, protein disulfide isomerases (PDIs) such as ERp57 catalyze the formation of disulfide bonds. (iii) ERp57 is converted to a reduced state during catalysis. To recycle the protein, the ER oxidoreductin Ero1 oxidizes the PDI, generating reactive oxygen species (ROS) in the process. These ROS are detoxified by reduced glutathione (GSH), which is transported into the ER from the cytosol through a channel. Based on our studies (Staleva et al., 2002), we hypothesize that the OCA2 protein serves as a GSH pump. (iv) ERp57 then catalyzes the isomerization of disulfide bonds to form a folded peptide. Correctly folded tyrosinase is released by the chaperones and then (v) recruited to a vesicle for transport to the Golgi. Recruitment may be facilitated by OCA2 and TYRP1. (vi) In the event that the protein is misfolded, disulfide bonds are removed by PDIs, which generates ROS that are removed by GSH, and tyrosinase is refolded. There are a limited number of protein refolding cycles, after which tyrosinase is transported to and degraded by proteasomes. *GSSG*, oxidized glutathione; *SH/HS*, sulfhydryl group; *S–S*, disulfide bond. *Reproduced from Manga, P., Orlow, S.J., 2011. Informed reasoning: repositioning of nitisinone to treat oculocutaneous albinism. J. Clin. Investig. 121, 3828–3831.*

OCA2 may also play a role in melanosomes. Chaperones involved in protein trafficking facilitate OCA2 delivery to the melanosome (Sitaram et al., 2012), whereas Bellono et al. (2014) demonstrated that heterologously expressed OCA2 could function as a chloride channel that modulates the pH of endolysosomes. Thus, in addition to playing a role in tyrosinase maturation, OCA2 may also regulate tyrosinase enzyme activity by modulating melanosome pH.

There are three OCA2 subtypes: OCA2A (OCA2 without freckles or ephelides), OCA2AE (OCA2 with freckles or ephelides), and OCA2B (brown albinism). It was initially thought that the appearance of ephelides was because of a reversion mutation similar to that observed in the OCA2 mouse model, pink-eyed dilution unstable (Brilliant et al., 1991). The mice carry duplication mutations with a high rate of reversion. In humans, however, individuals homozygous for the 2.7 kb *OCA2* deletion also develop ephelides. The ability to make pigmentation is acquired in sun-exposed areas, suggesting that a UV-induced change is responsible. This change maybe at the DNA level (for example mutations that activate one or more genes that complement OCA2 function), alternatively UV-induced cellular changes in melanosome function may compensate for the absence of OCA2. Because tyrosinase maturation is reduced in OCA2, these changes may increase the amount of functional tyrosinase and promote pigmentation. For example, HSP70 expression correlates with tyrosinase maturation (Murase et al., 2016). HSP70 is also induced by exposure to UV (Semenkov et al., 2015). In some individuals, this increase may result in a sufficient increase in mature tyrosinase and pigment production. The effects would be localized resulting in patchy pigmentation.

OCA2B is much less severe than OCA1 and OCA2A. It was first described in Nigeria (King et al., 1980) and has since been mapped to the *OCA2* locus. All individuals with OCA2B were heterozygous for the common 2.7 kb *OCA2* deletion, although the second pathogenic mutation/s remain unidentified (Manga et al., 2001b).

Oculocutaneous Albinism Type 3

OCA3, also known as rufous OCA, results from autosomal recessive mutations at the tyrosinase-related 1 protein (*TYRP1*) locus (Boissy et al., 1996; Manga et al., 1997). The *TYRP1* gene spans 15–18 kb encoding 8 exons on chromosome 9p23 (Chintamaneni et al., 1991) and encodes a protein of 75 kDa consisting of 537 amino acids (Murty et al., 1992). The protein shares a high degree of homology with tyrosinase. The function of the protein is not known, but it has been shown to stabilize tyrosinase (Manga et al., 2000), contribute to melanosome structure (Johnson and Jackson, 1992), and potentially play a role in determining melanocyte proliferation (Sarangarajan et al., 2000). The protein has been reported to have very low tyrosine hydroxylase activity; however, there is some debate as to whether the protein has enzymatic activity (Zhao et al., 1994).

Tyrp1 has also been shown to be critical for tyrosinase maturation (Toyofuku et al., 2001b). In melanocytes lacking Tyrp1, tyrosinase accumulates in the ER, although to a lesser degree than in *Oca2*-mutant melanocytes. Tyrp1 protein was also found to be required for tyrosinase folding in a cell-free assay system (Francis et al., 2003). Rescue of tyrosinase maturation in this system was achieved with wild-type semipermeabilized melanocytes, but failed when melanocytes mutated at the Tyrp1 locus were used. Kobayashi and Hearing (2007) found evidence of tyrosinase and Tyrp1 complexes in the ER, Golgi, and melanosomes, supporting a role for Tyrp1 in maturation of tyrosinase (see Fig. 5.2).

Sixteen mutations have been reported at this locus. Although the majority of individuals with OCA3 are from southern Africa, where two mutations account for 95% of alleles (Manga et al., 1997), mutations have also been reported in an African American individual (Boissy et al., 1996), several Pakistani families (Jaworek et al., 2012), and a German family (Rooryck et al., 2006).

Oculocutaneous Albinism Type 4

OCA4 is caused by mutations at the *SLC45A2* locus (Newton et al., 2001). The gene encodes the membrane-associated transporter protein (MATP) and spans 7 exons across 40 kb on chromosome 5p13.3. MATP consists of 530 amino acids and is a putative 12 transmembrane domain protein with structural similarity to plant sugar transporters (Fukamachi et al., 2001). MATP function has not been elucidated; however, mutations at this locus also cause disruption of tyrosinase maturation and trafficking, perhaps due to disruption of melanosomal pH and reloading of tyrosinase with copper (Bin et al., 2015). To date, 78 mutations have been reported.

Oculocutaneous Albinism Type 5–7

The gene mutated in OCA5 is yet to be identified. The disorder has been mapped to 4q24 in a consanguineous Pakistani family (Kausar et al., 2013).

SLC24A5 mutations were first identified in a Chinese patient (Wei et al., 2013) and result in OCA6. *SLC24A5*, like *SLC45A2*, encodes a solute transporter, NCKX5. The *SLC24A5* gene is located on chromosome 15q21.1. Alleles of a SNP at this locus, rs1426654, have been shown to segregate differentially in European Americans as compared with African and Asian populations (Lamason et al., 2005; Hernandez-Pacheco et al., 2017). The NCKX5 protein has been found to localize to the trans-Golgi network (Ginger et al., 2008), and knockdown of the protein results in an increase in cholesterol in melanocytes (Wilson et al., 2013). Cholesterol levels have been shown to regulate tyrosinase levels in the melanocyte, suggesting a possible mechanism for the impact on pigmentation (Hall et al., 2004).

OCA7 is caused by *C10orf11* mutations. This rare form has been identified in a family from the Faroe Islands and in a Lithuanian patient (Gronskov et al., 2013). *C10orf11* is localized to chromosome 10q22.2-q22.3 and encodes a 198 amino acid protein. The protein contains three leucine-rich repeats (LRRs) and one C terminal LRR domain. Proteins containing LRRs play diverse roles in the cell, ranging from cell adhesion to RNA processing (Simion et al., 2014).

Syndromic Albinism

OCA is a feature of CHS, an autosomal recessive disorder, which presents with recurrent pyogenic infections, mild-bleeding disorders, and neurologic dysfunction. Hemophagocytic lymphohistiocytosis can develop in a subset of patients and is fatal if it goes untreated. CHS results from mutations in the *CHS1/LYST* gene, which localizes to chromosome 1q42–44 (Barbosa et al., 1996). The function of the CHS1 protein is still unknown; however, it is thought to play a role in lysosome biogenesis (Durchfort et al., 2012). Given the shared lineage between melanosomes and lysosomes, CHS1 may play a role in organelle biogenesis.

HPS is an autosomal recessive disorder that presents with OCA, a bleeding disorder, colitis, and abnormal lysosomal ceroid storage (reviewed in Sanchez-Guiu et al., 2014). To date, 10 genes have been implicated in HPS (Table 5.1). As with CHS, mutations in HPS genes impact organelle biogenesis and protein trafficking.

Interaction Between Proteins and Modifier Loci

Tyrosinase is predicted to interact with the OCA2 and TYRP1 proteins (Orlow et al., 1994). Thus mutations in one protein can affect the function of another. Given that OCA1, 2, and 3 can all result in increased retention of tyrosinase in the ER, complex formation may occur early during protein maturation (Fig. 5.2). *TYRP1* mutations cause a reduction in tyrosinase maturation. When OCA2 is also mutated, the effects are compounded and the phenotypic consequences more severe (Manga et al., 1997).

Mutations at the *MC1R* locus have also been shown to modify the OCA2 phenotype (King et al., 2003). Individuals with OCA2 were found to have red rather than yellow hair when MC1R mutations were present in the same individual. Variation at the *MC1R* locus is rare in the Black population and likely plays a greater role in determining OCA2 phenotype in White populations (John et al., 2003).

IV. MOLECULAR BIOLOGY APPROACHES TO THE TREATMENT OF ALBINISM

Because increased tyrosine levels were found to increase pigmentation in tyrosinase-positive OCA, Summers et al. (2014) performed a clinical trial to test the effects of levodopa (a product of tyrosine metabolism and melanin intermediate) on pigmentation and visual acuity in individuals with albinism They did not, however, identify any therapeutic effects.

Treatment with tyrosine itself is not an option because amino acid levels are tightly regulated in the skin. Nitisinone has, however, been identified as a potential therapeutic for OCA1 forms where tyrosinase has residual activity. Onojafe et al. (2011) noted that nitisinone, which is used to treat hereditary tyrosinemia type 1, caused an increase in serum tyrosine levels. They treated OCA1b mice and saw an improvement in coat color pigmentation. This is the first prospective treatment for OCA which also has potential as a prenatal therapy, to prevent the optic tract misrouting that results in loss of visual acuity. One patient has been reported to have had a successful pregnancy despite nitisinone use during gestation (Vanclooster et al., 2012). However, the effects of nitisinone on a developing fetus are not known and will need to be investigated. Clinical trials are ongoing; however, no reports are available as yet.

More recently, gene editing has become a reality in the treatment of inborn errors of metabolism (Schneller et al., 2017). Because the skin is a large organ, it would be difficult to treat the entire surface of the skin; however, it may be possible to target melanocyte stem cells for gene editing and use a repigmentation approach, similar to that used in the treatment of vitiligo, an acquired depigmentation disorder, to pigment the skin. Increasing pigmentation of the eyes may be more difficult and correction of the optic tract misrouting unlikely; however, even a small improvement in eye pigmentation has been shown to improve visual acuity in adult mice with OA (Surace et al., 2000).

An alternative pharmacologic approach may be the use of a chaperone that can improve tyrosinase maturation. A number of chaperones have been identified for treatment of lysosomal disorders such as Fabry's disease (reviewed in Suzuki, 2014); however, they have not been successful in clinical trials as yet.

Thus, several novel therapies may be of use in the treatment of albinism.

V. CONCLUSION

Pigment synthesis and the impact of melanocytes on various organs, from the skin and hair to the ears and brain, is complex. Significant advances have been made in our understanding of not only the chemistry of melanin synthesis but also the proteins required to produce functional melanocytes. As we develop a broader picture of how these components are regulated and function, the avenues for diagnostics and therapies will be broadened. Identification of the genes involved in each form of albinism is an important step in this process.

References

Abdel-Malek, Z.A., Knittel, J., Kadekaro, A.L., Swope, V.B., Starner, R., 2008. The melanocortin 1 receptor and the UV response of human melanocytes–a shift in paradigm. Photochem. Photobiol. 84, 501–508.

Akahoshi, K., Fukai, K., Kato, A., Kimiya, S., Kubota, T., Spritz, R.A., 2001. Duplication of 15q11.2-q14, including the P gene, in a woman with generalized skin hyperpigmentation. Am. J. Med. Genet. 104, 299–302.

Ammann, S., Schulz, A., Krageloh-Mann, I., Dieckmann, N.M., Niethammer, K., Fuchs, S., Eckl, K.M., Plank, R., Werner, R., Altmuller, J., Thiele, H., Nurnberg, P., Bank, J., Strauss, A., Von Bernuth, H., Zur Stadt, U., Grieve, S., Griffiths, G.M., Lehmberg, K., Hennies, H.C., Ehl, S., 2016. Mutations in AP3D1 associated with immunodeficiency and seizures define a new type of Hermansky-Pudlak syndrome. Blood 127, 997–1006.

Anikster, Y., Huizing, M., White, J., Shevchenko, Y.O., Fitzpatrick, D.L., Touchman, J.W., Compton, J.G., Bale, S.J., Swank, R.T., Gahl, W.A., Toro, J.R., 2001. Mutation of a new gene causes a unique form of Hermansky-Pudlak syndrome in a genetic isolate of central Puerto Rico. Nat. Genet. 28, 376–380.

Aspengren, S., Hedberg, D., Skold, H.N., Wallin, M., 2009. New insights into melanosome transport in vertebrate pigment cells. Int. Rev. Cell Mol. Biol. 272, 245–302.

Barbosa, M.D., Nguyen, Q.A., Tchernev, V.T., Ashley, J.A., Detter, J.C., Blaydes, S.M., Brandt, S.J., Chotai, D., Hodgman, C., Solari, R.C., Lovett, M., Kingsmore, S.F., 1996. Identification of the homologous beige and Chediak-Higashi syndrome genes. Nature 382, 262–265.

Bassi, M.T., Schiaffino, M.V., Renieri, A., De Nigris, F., Galli, L., Bruttini, M., Gebbia, M., Bergen, A.A., Lewis, R.A., Ballabio, A., 1995. Cloning of the gene for ocular albinism type 1 from the distal short arm of the X chromosome. Nat. Genet. 10, 13–19.

Beermann, F., Orlow, S.J., Boissy, R.E., Schmidt, A., Boissy, Y.L., Lamoreux, M.L., 1995. Misrouting of tyrosinase with a truncated cytoplasmic tail as a result of the murine platinum (c^p) mutation. Exp. Eye Res. 61, 599–607.

Bellono, N.W., Escobar, I.E., Lefkovith, A.J., Marks, M.S., Oancea, E., 2014. An intracellular anion channel critical for pigmentation. Elife 3, e04543.

Berson, J.F., Frank, D.W., Calvo, P.A., Bieler, B.M., Marks, M.S., 2000. A common temperature-sensitive allelic form of human tyrosinase is retained in the endoplasmic reticulum at the nonpermissive temperature. J. Biol. Chem. 275, 12281–12289.

Berson, J.F., Harper, D.C., Tenza, D., Raposo, G., Marks, M.S., 2001. Pmel17 initiates premelanosome morphogenesis within multivesicular bodies. Mol. Biol. Cell 12, 3451–3464.

Bin, B.H., Bhin, J., Yang, S.H., Shin, M., Nam, Y.J., Choi, D.H., Shin, D.W., Lee, A.Y., Hwang, D., Cho, E.G., Lee, T.R., 2015. Membrane-associated transporter protein (MATP) regulates melanosomal pH and influences tyrosinase activity. PLoS One 10, e0129273.

Boissy, R.E., Zhao, H., Oetting, W.S., Austin, L.M., Wildenberg, S.C., Boissy, Y.L., Zhao, Y., Sturm, R.A., Hearing, V.J., King, R.A., Nordlund, J.J., 1996. Mutation in and lack of expression of tyrosinase-related protein-1 (TRP-1) in melanocytes from an individual with brown oculocutaneous albinism: a new subtype of albinism classified as "OCA3". Am. J. Hum. Genet. 58, 1145–1156.

Bonaventure, J., Domingues, M.J., Larue, L., 2013. Cellular and molecular mechanisms controlling the migration of melanocytes and melanoma cells. Pigment Cell Melanoma Res. 26, 316–325.

Branicki, W., Liu, F., Van Duijn, K., Draus-Barini, J., Pospiech, E., Walsh, S., Kupiec, T., Wojas-Pelc, A., Kayser, M., 2011. Model-based prediction of human hair color using DNA variants. Hum. Genet. 129, 443–454.

Branza-Nichita, N., Petrescu, A.J., Dwek, R.A., Wormald, M.R., Platt, F.M., Petrescu, S.M., 1999. Tyrosinase folding and copper loading in vivo: a crucial role for calnexin and alpha-glucosidase II. Biochem. Biophys. Res. Commun. 261, 720–725.

Breathnach, A.S., 1988. Extra-cutaneous melanin. Pigment Cell Res. 1, 234–237.

Brilliant, M.H., Gondo, Y., Eicher, E.M., 1991. Direct molecular identification of the mouse pink-eyed unstable mutation by genome scanning. Science 252, 566–569.

Brilliant, M.H., 2001. The mouse p (pink-eyed dilution) and human P genes, oculocutaneous albinism type 2 (OCA2), and melanosomal pH. Pigment Cell Res. 14, 86–93.

Burgoyne, T., Jolly, R., Martin-Martin, B., Seabra, M.C., Piccirillo, R., Schiaffino, M.V., Futter, C.E., 2013. Expression of OA1 limits the fusion of a subset of MVBs with lysosomes - a mechanism potentially involved in the initial biogenesis of melanosomes. J. Cell Sci. 126, 5143–5152.

Busca, R., Ballotti, R., 2000. Cyclic AMP a key messenger in the regulation of skin pigmentation. Pigment Cell Res. 13, 60–69.

Chaitanya, L., Walsh, S., Andersen, J.D., Ansell, R., Ballantyne, K., Ballard, D., Banemann, R., Bauer, C.M., Bento, A.M., Brisighelli, F., Capal, T., Clarisse, L., Gross, T.E., Haas, C., Hoff-Olsen, P., Hollard, C., Keyser, C., Kiesler, K.M., Kohler, P., Kupiec, T., Linacre, A., Minawi, A., Morling, N., Nilsson, H., Noren, L., Ottens, R., Palo, J.U., Parson, W., Pascali, V.L., Phillips, C., Porto, M.J., Sajantila, A., Schneider, P.M., Sijen, T., Sochtig, J., Syndercombe-Court, D., Tillmar, A., Turanska, M., Vallone, P.M., Zatkalikova, L., Zidkova, A., Branicki, W., Kayser, M., 2014. Collaborative EDNAP exercise on the IrisPlex system for DNA-based prediction of human eye colour. Forensic Sci. Int. Genet. 11, 241–251.

Chaki, M., Sengupta, M., Mondal, M., Bhattacharya, A., Mallick, S., Bhadra, R., Ray, K., 2011. Molecular and functional studies of tyrosinase variants among indian oculocutaneous albinism type 1 patients. J. Investig. Dermatol. 131, 260–262.

Chakraborty, A.K., Funasaka, Y., Slominski, A., Bolognia, J., Sodi, S., Ichihashi, M., Pawelek, J.M., 1999. UV light and MSH receptors. Ann. N.Y. Acad. Sci. 885, 100–116.

Chintamaneni, C.D., Ramsay, M., Colman, M.A., Fox, M.F., Pickard, R.T., Kwon, B.S., 1991. Mapping the human CAS2 gene, the homologue of the mouse brown (b) locus, to human chromosome 9p22-pter. Biochem. Biophys. Res. Commun. 178, 227–235.

Costin, G.E., Valencia, J.C., Vieira, W.D., Lamoreux, M.L., Hearing, V.J., 2003. Tyrosinase processing and intracellular trafficking is disrupted in mouse primary melanocytes carrying the underwhite (uw) mutation. A model for oculocutaneous albinism (OCA) type 4. J. Cell Sci. 116, 3203–3212.

Cullinane, A.R., Curry, J.A., Carmona-Rivera, C., Summers, C.G., Ciccone, C., Cardillo, N.D., Dorward, H., Hess, R.A., White, J.G., Adams, D., Huizing, M., Gahl, W.A., 2011. A BLOC-1 mutation screen reveals that PLDN is mutated in Hermansky-Pudlak Syndrome type 9. Am. J. Hum. Genet. 88, 778–787.

Dell'angelica, E.C., Shotelersuk, V., Aguilar, R.C., Gahl, W.A., Bonifacino, J.S., 1999. Altered trafficking of lysosomal proteins in Hermansky-Pudlak syndrome due to mutations in the beta 3A subunit of the AP-3 adaptor. Mol. Cell 3, 11–21.

D'ischia, M., Prota, G., 1997. Biosynthesis, structure, and function of neuromelanin and its relation to Parkinson's disease: a critical update. Pigment Cell Res. 10, 370–376.

Dufourcq-Lagelouse, R., Lambert, N., Duval, M., Viot, G., Vilmer, E., Fischer, A., Prieur, M., De Saint Basile, G., 1999. Chediak-Higashi syndrome associated with maternal uniparental isodisomy of chromosome 1. Eur. J. Hum. Genet. 7, 633–637.

Duliere, W.L., Raper, H.S., 1930. The tyrosinase-tyrosine reaction: the action of tyrosinase on certain substances related to tyrosine. Biochem. J. 24, 239–249.

Durchfort, N., Verhoef, S., Vaughn, M.B., Shrestha, R., Adam, D., Kaplan, J., Ward, D.M., 2012. The enlarged lysosomes in beige j cells result from decreased lysosome fission and not increased lysosome fusion. Traffic 13, 108–119.

Ebanks, J.P., Koshoffer, A., Wickett, R.R., Schwemberger, S., Babcock, G., Hakozaki, T., Boissy, R.E., 2011. Epidermal keratinocytes from light vs. dark skin exhibit differential degradation of melanosomes. J. Investig. Dermatol. 131, 1226–1233.

Edery, P., Attie, T., Amiel, J., Pelet, A., Eng, C., Hofstra, R.M., Martelli, H., Bidaud, C., Munnich, A., Lyonnet, S., 1996. Mutation of the endothelin-3 gene in the Waardenburg-Hirschsprung disease (Shah-Waardenburg syndrome). Nat. Genet. 12, 442–444.

Francis, E., Wang, N., Parag, H., Halaban, R., Hebert, D.N., 2003. Tyrosinase maturation and oligomerization in the endoplasmic reticulum require a melanocyte-specific factor. J. Biol. Chem. 278, 25607–25617.

Fujita, H., Asagami, C., Oda, Y., Yamamoto, K., Uchihira, T., 1970. Electron microscope study on the embryonic differentiation of the epidermis in human skin. Arch. Histol. Jpn. 32, 355–373.

Fukamachi, S., Shimada, A., Shima, A., 2001. Mutations in the gene encoding B, a novel transporter protein, reduce melanin content in medaka. Nat. Genet. 28, 381–385.

Gardner, J.M., Nakatsu, Y., Gondo, Y., Lee, S., Lyon, M.F., King, R.A., Brilliant, M.H., 1992. The mouse pink-eyed dilution gene: association with human Prader-Willi and Angelman syndromes. Science 257, 1121–1124.

Garrod, A.E., 1908. Inborn errors of metabolism. Croonian lectures. Lancet 2, 73.

Gerstenblith, M.R., Goldstein, A.M., Tucker, M.A., Fraser, M.C., 2007. Genetic testing for melanoma predisposition: current challenges. Cancer Nurs. 30, 452–459 quiz 462–463.

Giebel, L.B., Spritz, R.A., 1991. Mutation of the KIT (mast/stem cell growth factor receptor) protooncogene in human piebaldism. Proc. Natl. Acad. Sci. U.S.A. 88, 8696–8699.

Giebel, L.B., Tripathi, R.K., Strunk, K.M., Hanifin, J.M., Jackson, C.E., King, R.A., Spritz, R.A., 1991. Tyrosinase gene mutations associated with type 1B ("yellow") oculocutaneous albinism. Am. J. Hum. Genet. 48, 1159–1167.

Gilchrest, B.A., Eller, M.S., Geller, A.C., Yaar, M., 1999. The pathogenesis of melanoma induced by ultraviolet radiation. N. Engl. J. Med. 340, 1341–1348.

Ginger, R.S., Askew, S.E., Ogborne, R.M., Wilson, S., Ferdinando, D., Dadd, T., Smith, A.M., Kazi, S., Szerencsei, R.T., Winkfein, R.J., Schnetkamp, P.P., Green, M.R., 2008. SLC24A5 encodes a trans-golgi network protein with potassium-dependent sodium-calcium exchange activity that regulates human epidermal melanogenesis. J. Biol. Chem. 283, 5486–5495.

Gronskov, K., Dooley, C.M., Ostergaard, E., Kelsh, R.N., Hansen, L., Levesque, M.P., Vilhelmsen, K., Mollgard, K., Stemple, D.L., Rosenberg, T., 2013. Mutations in c10orf11, a melanocyte-differentiation gene, cause autosomal-recessive albinism. Am. J. Hum. Genet. 92, 415–421.

Hall, A.M., Krishnamoorthy, L., Orlow, S.J., 2004. 25-hydroxycholesterol acts in the Golgi compartment to induce degradation of tyrosinase. Pigment Cell Res. 17, 396–406.

Hatchome, N., Aiba, S., Kato, T., Torinuki, W., Tagami, H., 1987. Possible functional impairment of Langerhans' cells in vitiliginous skin. Reduced ability to elicit dinitrochlorobenzene contact sensitivity reaction and decreased stimulatory effect in the allogeneic mixed skin cell lymphocyte culture reaction. Arch. Dermatol. 123, 51–54.

Hernandez-Pacheco, N., Flores, C., Alonso, S., Eng, C., Mak, A.C., Hunstman, S., Hu, D., White, M.J., Oh, S.S., Meade, K., Farber, H.J., Avila, P.C., Serebrisky, D., Thyne, S.M., Brigino-Buenaventura, E., Rodriguez-Cintron, W., Sen, S., Kumar, R., Lenoir, M., Rodriguez-Santana, J.R., Burchard, E.G., Pino-Yanes, M., 2017. Identification of a novel locus associated with skin colour in African-admixed populations. Sci. Rep. 7, 44548.

Hider, J.L., Gittelman, R.M., Shah, T., Edwards, M., Rosenbloom, A., Akey, J.M., Parra, E.J., 2013. Exploring signatures of positive selection in pigmentation candidate genes in populations of East Asian ancestry. BMC Evol. Biol. 13, 150.

Hoath, S.B., Leahy, D.G., 2003. The organization of human epidermis: functional epidermal units and phi proportionality. J. Investig. Dermatol. 121, 1440–1446.

Hoth, C.F., Milunsky, A., Lipsky, N., Sheffer, R., Clarren, S.K., Baldwin, C.T., 1993. Mutations in the paired domain of the human PAX3 gene cause Klein-Waardenburg syndrome (WS-III) as well as Waardenburg syndrome type I (WS-I). Am. J. Hum. Genet. 52, 455–462.

Hudjashov, G., Villems, R., Kivisild, T., 2013. Global patterns of diversity and selection in human tyrosinase gene. PLoS One 8, e74307.

Ito, S., Wakamatsu, K., 2011. Diversity of human hair pigmentation as studied by chemical analysis of eumelanin and pheomelanin. J. Eur. Acad. Dermatol. Venereol. 25, 1369–1380.

Jablonski, N.G., Chaplin, G., 2017. The colours of humanity: the evolution of pigmentation in the human lineage. Philos. Trans. R. Soc. Lond. B Biol. Sci. 372.

Jablonski, N.G., 2012. Human skin pigmentation as an example of adaptive evolution. Proc. Am. Philos. Soc. 156, 45–57.

Jaworek, T.J., Kausar, T., Bell, S.M., Tariq, N., Maqsood, M.I., Sohail, A., Ali, M., Iqbal, F., Rasool, S., Riazuddin, S., Shaikh, R.S., Ahmed, Z.M., 2012. Molecular genetic studies and delineation of the oculocutaneous albinism phenotype in the Pakistani population. Orphanet J. Rare Dis. 7, 44.

Jin, Y., Birlea, S.A., Fain, P.R., Gowan, K., Riccardi, S.L., Holland, P.J., Mailloux, C.M., Sufit, A.J., Hutton, S.M., Amadi-Myers, A., Bennett, D.C., Wallace, M.R., Mccormack, W.T., Kemp, E.H., Gawkrodger, D.J., Weetman, A.P., Picardo, M., Leone, G., Taieb, A., Jouary, T., Ezzedine, K., Van Geel, N., Lambert, J., Overbeck, A., Spritz, R.A., 2010. Variant of TYR and autoimmunity susceptibility loci in generalized vitiligo. N. Engl. J. Med. 362, 1686–1697.

John, P.R., Makova, K., Li, W.H., Jenkins, T., Ramsay, M., 2003. DNA polymorphism and selection at the melanocortin-1 receptor gene in normally pigmented southern African individuals. Ann. N.Y. Acad. Sci. 994, 299–306.

Johnson, R., Jackson, I.J., 1992. Light is a dominant mouse mutation resulting in premature cell death. Nat. Genet. 1, 226–229.

Kausar, T., Bhatti, M.A., Ali, M., Shaikh, R.S., Ahmed, Z.M., 2013. OCA5, a novel locus for non-syndromic oculocutaneous albinism, maps to chromosome 4q24. Clin. Genet. 84, 91–93.

Kawakami, A., Fisher, D.E., 2011. Key discoveries in melanocyte development. J. Investig. Dermatol. 131, E2–E4.

Kerr, R., Stevens, G., Manga, P., Salm, S., John, P., Haw, T., Ramsay, M., 2000. Identification of P gene mutations in individuals with oculocutaneous albinism in sub-Saharan Africa. Hum. Mutat. 15, 166–172.

King, R.A., Witkop Jr., C.J., 1976. Hairbulb tyrosinase activity in oculocutaneous albinism. Nature 263, 69–71.

King, R.A., Creel, D., Cervenka, J., Okoro, A.N., Witkop, C.J., 1980. Albinism in Nigeria with delineation of new recessive oculocutaneous type. Clin. Genet. 17, 259–270.

King, R.A., Willaert, R.K., Schmidt, R.M., Pietsch, J., Savage, S., Brott, M.J., Fryer, J.P., Summers, C.G., Oetting, W.S., 2003. MC1R mutations modify the classic phenotype of oculocutaneous albinism type 2 (OCA2). Am. J. Hum. Genet. 73, 638–645.

Kobayashi, T., Hearing, V.J., 2007. Direct interaction of tyrosinase with Tyrp1 to form heterodimeric complexes in vivo. J. Cell Sci. 120, 4261–4268.

Lalwani, A.K., Baldwin, C.T., Morell, R., Friedman, T.B., San Agustin, Milunsky, A., Asher, J.H., Wilcox, E.R., Farrer, L.A., 1994. A locus for Waardenburg syndrome type II maps to chromosome 1p13.3-2.1. Am. J. Hum. Genet. 55, A14.

Lamason, R.L., Mohideen, M.A., Mest, J.R., Wong, A.C., Norton, H.L., Aros, M.C., Jurynec, M.J., Mao, X., Humphreville, V.R., Humbert, J.E., Sinha, S., Moore, J.L., Jagadeeswaran, P., Zhao, W., Ning, G., Makalowska, I., Mckeigue, P.M., O'donnell, D., Kittles, R., Parra, E.J., Mangini, N.J., Grunwald, D.J., Shriver, M.D., Canfield, V.A., Cheng, K.C., 2005. SLC24A5, a putative cation exchanger, affects pigmentation in zebrafish and humans. Science 310, 1782–1786.

Lamoreux, M.L., Zhou, B.K., Rosemblat, S., Orlow, S.J., 1995. The pinkeyed-dilution protein and the eumelanin/pheomelanin switch: in support of a unifying hyopthesis. Pigment Cell Res. 8, 263–270.

Lee, S.T., Nicholls, R.D., Bundey, S., Laxova, R., Musarella, M., Spritz, R.A., 1994. Mutations of the P gene in oculocutaneous albinism, ocular albinism, and Prader-Willi syndrome plus albinism. N. Engl. J. Med. 330, 529–534.

Lee, S.T., Nicholls, R.D., Jong, M.T., Fukai, K., Spritz, R.A., 1995. Organization and sequence of the human P gene and identification of a new family of transport proteins. Genomics 26, 354–363.

Levin, M.D., Lu, M.M., Petrenko, N.B., Hawkins, B.J., Gupta, T.H., Lang, D., Buckley, P.T., Jochems, J., Liu, F., Spurney, C.F., Yuan, L.J., Jacobson, J.T., Brown, C.B., Huang, L., Beermann, F., Margulies, K.B., Madesh, M., Eberwine, J.H., Epstein, J.A., Patel, V.V., 2009. Melanocyte-like cells in the heart and pulmonary veins contribute to atrial arrhythmia triggers. J. Clin. Investig. 119, 3420–3436.

Li, W., Zhang, Q., Oiso, N., Novak, E.K., Gautam, R., O'brien, E.P., Tinsley, C.L., Blake, D.J., Spritz, R.A., Copeland, N.G., Jenkins, N.A., Amato, D., Roe, B.A., Starcevic, M., Dell'angelica, E.C., Elliott, R.W., Mishra, V., Kingsmore, S.F., Paylor, R.E., Swank, R.T., 2003. Hermansky-Pudlak syndrome type 7 (HPS-7) results from mutant dysbindin, a member of the biogenesis of lysosome-related organelles complex 1 (BLOC-1). Nat. Genet. 35, 84–89.

Liu, F., Van Duijn, K., Vingerling, J.R., Hofman, A., Uitterlinden, A.G., Janssens, A.C., Kayser, M., 2009. Eye color and the prediction of complex phenotypes from genotypes. Curr. Biol. 19, R192–R193.

Liu, F., Visser, M., Duffy, D.L., Hysi, P.G., Jacobs, L.C., Lao, O., Zhong, K., Walsh, S., Chaitanya, L., Wollstein, A., Zhu, G., Montgomery, G.W., Henders, A.K., Mangino, M., Glass, D., Bataille, V., Sturm, R.A., Rivadeneira, F., Hofman, A., Van, I.W.F., Uitterlinden, A.G., Palstra, R.J., Spector, T.D., Martin, N.G., Nijsten, T.E., Kayser, M., 2015. Genetics of skin color variation in Europeans: genome-wide association studies with functional follow-up. Hum. Genet. 134, 823–835.

Manga, P., Orlow, S.J., 2011. Informed reasoning: repositioning of nitisinone to treat oculocutaneous albinism. J. Clin. Investig. 121, 3828–3831.

Manga, P., Kromberg, J.G., Box, N.F., Sturm, R.A., Jenkins, T., Ramsay, M., 1997. Rufous oculocutaneous albinism in Southern African Blacks is caused by mutations in the TYRP1 gene. Am. J. Hum. Genet. 61, 1095–1101.

Manga, P., Sato, K., Ye, L., Beermann, F., Lamoreux, M.L., Orlow, S.J., 2000. Mutational analysis of the modulation of tyrosinase by tyrosinase- related proteins 1 and 2 in vitro. Pigment Cell Res. 13, 364–374.

Manga, P., Boissy, R.E., Pifko-Hirst, S., Zhou, B.K., Orlow, S.J., 2001a. Mislocalization of melanosomal proteins in melanocytes from mice with oculocutaneous albinism type 2. Exp. Eye Res. 72, 695–710.

Manga, P., Kromberg, J., Turner, A., Jenkins, T., Ramsay, M., 2001b. Southern Africa, brown oculocutaneous albinism (BOCA) maps to the OCA2 locus on chromosome 15q: P-gene mutations identified. Am. J. Hum. Genet. 68, 782–787.

Marks, M.S., Heijnen, H.F., Raposo, G., 2013. Lysosome-related organelles: unusual compartments become mainstream. Curr. Opin. Cell Biol. 25, 495–505.

Minwalla, L., Zhao, Y., Le Poole, I.C., Wickett, R.R., Boissy, R.E., 2001. Keratinocytes play a role in regulating distribution patterns of recipient melanosomes in vitro. J. Investig. Dermatol. 117, 341–347.

Montoliu, L., Gronskov, K., Wei, A.H., Martinez-Garcia, M., Fernandez, A., Arveiler, B., Morice-Picard, F., Riazuddin, S., Suzuki, T., Ahmed, Z.M., Rosenberg, T., Li, W., 2014. Increasing the complexity: new genes and new types of albinism. Pigment Cell Melanoma Res. 27, 11–18.

Morgan, N.V., Pasha, S., Johnson, C.A., Ainsworth, J.R., Eady, R.A., Dawood, B., Mckeown, C., Trembath, R.C., Wilde, J., Watson, S.P., Maher, E.R., 2006. A germline mutation in BLOC1S3/reduced pigmentation causes a novel variant of Hermansky-Pudlak syndrome (HPS8). Am. J. Hum. Genet. 78, 160–166.

Murase, D., Hachiya, A., Fullenkamp, R., Beck, A., Moriwaki, S., Hase, T., Takema, Y., Manga, P., 2016. Variation in Hsp70-1A expression contributes to skin color diversity. J. Investig. Dermatol. 136 (8), 1681–1691.

Murty, V.V., Bouchard, B., Mathew, S., Vijayasaradhi, S., Houghton, A.N., 1992. Assignment of the human TYRP (brown) locus to chromosome region 9p23 by nonradioactive in situ hybridization. Genomics 13, 227–229.

Nan, H., Kraft, P., Qureshi, A.A., Guo, Q., Chen, C., Hankinson, S.E., Hu, F.B., Thomas, G., Hoover, R.N., Chanock, S., Hunter, D.J., Han, J., 2009. Genome-wide association study of tanning phenotype in a population of European ancestry. J. Investig. Dermatol. 129, 2250–2257.

Newton, J.M., Cohen-Barak, O., Hagiwara, N., Gardner, J.M., Davisson, M.T., King, R.A., Brilliant, M.H., 2001. Mutations in the human orthologue of the mouse underwhite gene (uw) underlie a new form of oculocutaneous albinism, OCA4. Am. J. Hum. Genet. 69, 981–988.

Oh, J., Ho, L., Ala-Mello, S., Amato, D., Armstrong, L., Bellucci, S., Carakushansky, G., Ellis, J.P., Fong, C.T., Green, J.S., Heon, E., Legius, E., Levin, A.V., Nieuwenhuis, H.K., Pinckers, A., Tamura, N., Whiteford, M.L., Yamasaki, H., Spritz, R.A., 1998. Mutation analysis of patients with Hermansky-Pudlak syndrome: a frameshift hot spot in the HPS gene and apparent locus heterogeneity. Am. J. Hum. Genet. 62, 593–598.

Onojafe, I.F., Adams, D.R., Simeonov, D.R., Zhang, J., Chan, C.C., Bernardini, I.M., Sergeev, Y.V., Dolinska, M.B., Alur, R.P., Brilliant, M.H., Gahl, W.A., Brooks, B.P., 2011. Nitisinone improves eye and skin pigmentation defects in a mouse model of oculocutaneous albinism. J. Clin. Investig. 121, 3914–3923.

Orlow, S.J., Brilliant, M.H., 1999. The pink-eyed dilution locus controls the biogenesis of melanosomes and levels of melanosomal proteins in the eye. Exp. Eye Res. 68, 147–154.

Orlow, S.J., Zhou, B.K., Chakraborty, A.K., Drucker, M., Pifko-Hirst, S., Pawelek, J.M., 1994. High-molecular weight forms of tyrosinase and the tyrosinase-related proteins: evidence for a melanogenic complex. J. Investig. Dermatol. 103, 196–201.

Orlow, S.J., 1995. Melanosomes are specialized members of the lysosomal lineage of organelles. J. Investig. Dermatol. 105, 3–7.

Parra, E.J., Kittles, R.A., Shriver, M.D., 2004. Implications of correlations between skin color and genetic ancestry for biomedical research. Nat. Genet. 36, S54–S60.

Parra, E.J., 2007. Human pigmentation variation: evolution, genetic basis, and implications for public health. Am. J. Phys. Anthropol. Suppl. 45, 85–105.

Pemberton, T.J., Mehta, N.U., Witonsky, D., Di Rienzo, A., Allayee, H., Conti, D.V., Patel, P.I., 2008. Prevalence of common disease-associated variants in Asian Indians. BMC Genet. 9, 13.

Petrescu, A.J., Butters, T.D., Reinkensmeier, G., Petrescu, S., Platt, F.M., Dwek, R.A., Wormald, M.R., 1997. The solution NMR structure of glucosylated N-glycans involved in the early stages of glycoprotein biosynthesis and folding. Embo J. 16, 4302–4310.

Petrescu, S.M., Branza-Nichita, N., Negroiu, G., Petrescu, A.J., Dwek, R.A., 2000. Tyrosinase and glycoprotein folding: roles of chaperones that recognize glycans. Biochemistry 39, 5229–5237.

Pingault, V., Bondurand, N., Kuhlbrodt, K., Goerich, D.E., Prehu, M.O., Puliti, A., Herbarth, B., Hermans-Borgmeyer, I., Legius, E., Matthijs, G., Amiel, J., Lyonnet, S., Ceccherini, I., Romeo, G., Smith, J.C., Read, A.P., Wegner, M., Goossens, M., 1998. SOX10 mutations in patients with Waardenburg-Hirschsprung disease. Nat. Genet. 18, 171–173.

Porter, J., Beuf, A.H., Lerner, A., Nordlund, J., 1987. Response to cosmetic disfigurement: patients with vitiligo. Cutis 39, 493–494.

Potterf, S.B., Furumura, M., Sviderskaya, E.V., Santis, C., Bennett, D.C., Hearing, V.J., 1998. Normal tyrosine transport and abnormal tyrosinase routing in *pink-eyed dilution* melanocytes. Exp. Cell Res. 244, 319–326.

Prota, G., 2000. Melanins, melanogenesis and melanocytes: looking at their functional significance from the chemist's viewpoint. Pigment Cell Res. 13, 283–293.

Puffenberger, E.G., Hosoda, K., Washington, S.S., Nakao, K., Dewit, D., Yanagisawa, M., Chakravart, A., 1994. A missense mutation of the endothelin-B receptor gene in multigenic Hirschsprung's disease. Cell 79, 1257–1266.

Puri, N., Gardner, J.M., Brilliant, M.H., 2000. Aberrant pH of melanosomes in pink-eyed dilution (p) mutant melanocytes. J. Investig. Dermatol. 115, 607–613.

Ramsay, M., Colman, M.A., Stevens, G., Zwane, E., Kromberg, J., Farrall, M., Jenkins, T., 1992. The tyrosinase-positive oculocutaneous albinism locus maps to chromosome 15q11.2-q12. Am. J. Hum. Genet. 51, 879–884.

Rees, J.L., 2003. Genetics of hair and skin color. Annu. Rev. Genet. 37, 67–90.

Rinchik, E.M., Bultman, S.J., Horsthemke, B., Lee, S.T., Strunk, K.M., Spritz, R.A., Avidano, K.M., Jong, M.T.C., Nicholls, R.D., 1993. A gene for the mouse pink-eyed dilution locus and for human type II oculocutaneous albinism. Nature 361, 72–76.

Rooryck, C., Roudaut, C., Robine, E., Musebeck, J., Arveiler, B., 2006. Oculocutaneous albinism with TYRP1 gene mutations in a Caucasian patient. Pigment Cell Res. 19, 239–242.

Rosemblat, S., Durham-Pierre, D., Gardner, J.M., Nakatsu, Y., Brilliant, M.H., Orlow, S.J., 1994. Identification of a melanosomal membrane protein encoded by the pink eyed dilution (type II oculocutaneous albinism) gene. Proc. Natl. Acad. Sci. U.S.A. 91, 12071–12075.

Rozanowska, M., Sarna, T., Land, E.J., Truscott, T.G., 1999. Free radical scavenging properties of melanin interaction of eu- and pheo-melanin models with reducing and oxidising radicals. Free Radic. Biol. Med. 26, 518–525.

Russell, L.B., Russell, W.L., 1948. A study of the physiological genetics of coat color in the mouse by means of the dopa reaction in frozen sections of skin. Genetics 33, 237–262.

Russell, E.S., 1949. A quantitative histological study of the pigment found in the coat-colour mutants of the housemouse. IV. The nature of the genic effects of five major allelic series. Genetics 34, 146–166.

Sanchez-Guiu, I., Torregrosa, J.M., Velasco, F., Anton, A.I., Lozano, M.L., Vicente, V., Rivera, J., 2014. Hermansky-Pudlak syndrome. Overview of clinical and molecular features and case report of a new HPS-1 variant. Hamostaseologie 34, 301–309.

Sanchez-Martin, M., Perez-Losada, J., Rodriguez-Garcia, A., Gonzalez-Sanchez, B., Korf, B.R., Kuster, W., Moss, C., Spritz, R.A., Sanchez-Garcia, I., 2003. Deletion of the SLUG (SNAI2) gene results in human piebaldism. Am. J. Med. Genet. 122A, 125–132.

Sarangarajan, R., Zhao, Y., Babcock, G., Cornelius, J., Lamoreux, M.L., Boissy, R.E., 2000. Mutant alleles at the brown locus encoding tyrosinase-related protein-1 (TRP-1) affect proliferation of mouse melanocytes in culture. Pigment Cell Res. 13, 337–344.

Schneller, J.L., Lee, C.M., Bao, G., Venditti, C.P., 2017. Genome editing for inborn errors of metabolism: advancing towards the clinic. BMC Med. 15, 43.

Selicorni, A., Guerneri, S., Ratti, A., Pizzuti, A., 2002. Cytogenetic mapping of a novel locus for type II Waardenburg syndrome. Hum. Genet. 110, 64–67.

Semenkov, V.F., Michalski, A.I., Sapozhnikov, A.M., 2015. Heating and ultraviolet light activate anti-stress gene functions in humans. Front. Genet. 6, 245.

Simion, C., Cedano-Prieto, M.E., Sweeney, C., 2014. The LRIG family: enigmatic regulators of growth factor receptor signaling. Endocr. Relat. Cancer 21, R431–R443.

Sitaram, A., Dennis, M.K., Chaudhuri, R., De Jesus-Rojas, W., Tenza, D., Setty, S.R., Wood, C.S., Sviderskaya, E.V., Bennett, D.C., Raposo, G., Bonifacino, J.S., Marks, M.S., 2012. Differential recognition of a dileucine-based sorting signal by AP-1 and AP-3 reveals a requirement for both BLOC-1 and AP-3 in delivery of OCA2 to melanosomes. Mol. Biol. Cell 23 (16), 3178–3192.

Sommer, L., 2011. Generation of melanocytes from neural crest cells. Pigment Cell Melanoma Res. 24, 411–421.

Spritz, R.A., Bailin, T., Nicholls, R.D., Lee, S.T., Park, S.K., Mascari, M.J., Butler, M.G., 1997. Hypopigmentation in the Prader-Willi syndrome correlates with P gene deletion but not with haplotype of the hemizygous P allele. Am. J. Med. Genet. 71, 57–62.

Staleva, L., Manga, P., Orlow, S.J., 2002. The pink-eyed dilution protein modulates arsenic sensitivity and intracellular glutathione metabolism. Mol. Cell. 13, 1406–1420.

Stevens, G., Van Beukering, J., Jenkins, T., Ramsay, M., 1995. An intragenic deletion of the P gene is the common mutation causing tyrosinase-positive oculocutaneous albinism in southern African Negroids. Am. J. Hum. Genet. 56, 586–591.

Stokowski, R.P., Pant, P.V., Dadd, T., Fereday, A., Hinds, D.A., Jarman, C., Filsell, W., Ginger, R.S., Green, M.R., Van Der Ouderaa, F.J., Cox, D.R., 2007. A genomewide association study of skin pigmentation in a South Asian population. Am. J. Hum. Genet. 81, 1119–1132.

Sturm, R.A., Frudakis, T.N., 2004. Eye colour: portals into pigmentation genes and ancestry. Trends Genet. 20, 327–332.

Summers, C.G., Connett, J.E., Holleschau, A.M., Anderson, J.L., De Becker, I., Mckay, B.S., Brilliant, M.H., 2014. Does levodopa improve vision in albinism? Results of a randomized, controlled clinical trial. Clin. Exp. Ophthalmol. 42, 713–721.

Surace, E.M., Angeletti, B., Ballabio, A., Marigo, V., 2000. Expression pattern of the ocular albinism type 1 (Oa1) gene in the murine retinal pigment epithelium. Investig. Ophthalmol. Vis. Sci. 41, 4333–4337.

Suzuki, T., Li, W., Zhang, Q., Karim, A., Novak, E.K., Sviderskaya, E.V., Hill, S.P., Bennett, D.C., Levin, A.V., Nieuwenhuis, H.K., Fong, C.T., Castellan, C., Miterski, B., Swank, R.T., Spritz, R.A., 2002. Hermansky-Pudlak syndrome is caused by mutations in HPS4, the human homolog of the mouse light-ear gene. Nat. Genet. 30, 321–324.

Suzuki, Y., 2014. Emerging novel concept of chaperone therapies for protein misfolding diseases. Proc. Jpn. Acad. Ser. B Phys. Biol. Sci. 90, 145–162.

Tassabehji, M., Read, A.P., Newton, V.E., Harris, R., Balling, R., Gruss, P., Strachan, T., 1992. Waardenburg's syndrome patients have mutations in the human homologue of the Pax-3 paired box gene. Nature 355, 635–636.

Tassabehji, M., Newton, V.E., Read, A.P., 1994. Waardenburg syndrome type 2 caused by mutations in the human microphthalmia (MITF) gene. Nat. Genet. 8, 251–255.

Thong, H.-Y., Jee, S.-H., Sun, C.-C., Boissy, R.E., 2003. The pattern and mechanism of melanosome distribution in keratinocytes of human skin as one determining factor of skin colour. Br. J. Dermatol. 149, 498–505.

Tomita, Y., Takeda, A., Okinaga, S., Tagami, H., Shibahara, S., 1989. Human oculocutaneous albinism caused by single base insertion in the tyrosinase gene. Biochem. Biophys. Res. Commun. 164, 990–996.

Toyofuku, K., Wada, I., Spritz, R.A., Hearing, V.J., 2001a. The molecular basis of oculocutaneous albinism type 1 (OCA1): sorting failure and degradation of mutant tyrosinases results in a lack of pigmentation. Biochem. J. 355, 259–269.

Toyofuku, K., Wada, I., Valencia, J.C., Kushimoto, T., Ferrans, V.J., Hearing, V.J., 2001b. Oculocutaneous albinism types 1 and 3 are ER retention diseases: mutation of tyrosinase or Tyrp1 can affect the processing of both mutant and wild-type proteins. Faseb J. 15, 2149–2161.

Ujvari, A., Aron, R., Eisenhaure, T., Cheng, E., Parag, H.A., Smicun, Y., Halaban, R., Hebert, D.N., 2001. Translation rate of human tyrosinase determines its N-linked glycosylation level. J. Biol. Chem. 276, 5924–5931.

Valenzuela, R.K., Henderson, M.S., Walsh, M.H., Garrison, N.A., Kelch, J.T., Cohen-Barak, O., Erickson, D.T., John Meaney, F., Bruce Walsh, J., Cheng, K.C., Ito, S., Wakamatsu, K., Frudakis, T., Thomas, M., Brilliant, M.H., 2010. Predicting phenotype from genotype: normal pigmentation. J. Forensic Sci. 55, 315–322.

Vanclooster, A., Devlieger, R., Meersseman, W., Spraul, A., Kerckhove, K.V., Vermeersch, P., Meulemans, A., Allegaert, K., Cassiman, D., 2012. Pregnancy during nitisinone treatment for tyrosinaemia type I: first human experience. JIMD Rep. 5, 27–33.

Wakamatsu, K., Kavanagh, R., Kadekaro, A.L., Terzieva, S., Sturm, R.A., Leachman, S., Abdel-Malek, Z., Ito, S., 2006. Diversity of pigmentation in cultured human melanocytes is due to differences in the type as well as quantity of melanin. Pigment Cell Res. 19, 154–162.

Wang, N., Hebert, D.N., 2006. Tyrosinase maturation through the mammalian secretory pathway: bringing color to life. Pigment Cell Res. 19, 3–18.

Wang, N., Daniels, R., Hebert, D.N., 2005a. The cotranslational maturation of the type I membrane glycoprotein tyrosinase: the heat shock protein 70 system hands off to the lectin-based chaperone system. Mol. Biol. Cell 16, 3740–3752.

Wang, T., Waters, C.T., Jakins, T., Yates, J.R., Trump, D., Bradshaw, K., Moore, A.T., 2005b. Temperature sensitive oculocutaneous albinism associated with missense changes in the tyrosinase gene. Br. J. Ophthalmol. 89, 1383–1384.

Wei, A.H., Zang, D.J., Zhang, Z., Liu, X.Z., He, X., Yang, L., Wang, Y., Zhou, Z.Y., Zhang, M.R., Dai, L.L., Yang, X.M., Li, W., 2013. Exome sequencing identifies SLC24A5 as a candidate gene for nonsyndromic oculocutaneous albinism. J. Investig. Dermatol. 133, 1834–1840.

Wei, M.L., 2006. Hermansky-Pudlak syndrome: a disease of protein trafficking and organelle function. Pigment Cell Res. 19, 19–42.

Wilson, S., Ginger, R.S., Dadd, T., Gunn, D., Lim, F.L., Sawicka, M., Sandel, M., Schnetkamp, P.P., Green, M.R., 2013. NCKX5, a natural regulator of human skin colour variation, regulates the expression of key pigment genes MC1R and alpha-MSH and alters cholesterol homeostasis in normal human melanocytes. Adv. Exp. Med. Biol. 961, 95–107.

Witkop Jr., C.J., 1985. Inherited disorders of pigmentation. Clin. Dermatol. 3, 70–134.

Wolf Horrell, E.M., Boulanger, M.C., D'orazio, J.A., 2016. Melanocortin 1 receptor: structure, function, and regulation. Front. Genet. 7, 95.

Yamaguchi, Y., Hearing, V.J., 2009. Physiological factors that regulate skin pigmentation. Biofactors 35, 193–199.

Yi, Z., Garrison, N., Cohen-Barak, O., Karafet, T.M., King, R.A., Erickson, R.P., Hammer, M.F., Brilliant, M.H., 2003. A 122.5-kilobase deletion of the P gene underlies the high prevalence of oculocutaneous albinism type 2 in the Navajo population. Am. J. Hum. Genet. 72, 62–72.

Young, A., Jiang, M., Wang, Y., Ahmedli, N.B., Ramirez, J., Reese, B.E., Birnbaumer, L., Farber, D.B., 2011. Specific interaction of Galphai3 with the Oa1 G-protein coupled receptor controls the size and density of melanosomes in retinal pigment epithelium. PLoS One 6, e24376.

Zhang, Q., Zhao, B., Li, W., Oiso, N., Novak, E.K., Rusiniak, M.E., Gautam, R., Chintala, S., O'brien, E.P., Zhang, Y., Roe, B.A., Elliott, R.W., Eicher, E.M., Liang, P., Kratz, C., Legius, E., Spritz, R.A., O'sullivan, T.N., Copeland, N.G., Jenkins, N.A., Swank, R.T., 2003. Ru2 and Ru encode mouse orthologs of the genes mutated in human Hermansky-Pudlak syndrome types 5 and 6. Nat. Genet. 33, 145–153.

Zhao, H., Zhao, Y., Nordlund, J.J., Boissy, R.E., 1994. Human TRP-1 has tyrosine hydroxylase but no dopa oxidase activity. Pigment Cell Res. 7, 131–140.

Dermatological Aspects of Albinism

Sian Hartshorne[1], Prashiela Manga[2,3]

[1]Mediderm, Plettenberg Bay, South Africa; [2]New York University School of Medicine,
New York, NY, United States; [3]University of the Witwatersrand, Johannesburg, South Africa

OUTLINE

I. INTRODUCTION

This chapter focuses on the skin and skin-related problems that confront people with oculocutaneous albinism (OCA). Hypopigmentation of the skin and hair is a key feature of OCA. The reduction in pigmentation makes the skin vulnerable to solar damage and photoaging and significantly increases the risk of developing skin cancer. These risks vary depending on geographical location, which impacts sun exposure (for example, number of sunny days experienced and the intensity of the sun as determined by latitude and altitude), and access to preventative and therapeutic healthcare (whether there are opportunities to avoid the sun, to use sunscreen creams, and to benefit from early detection and treatment of sun damage if it occurs).

The topics covered in this chapter include: a basic introduction to the structure and function of the skin, a discussion of how sun damage occurs, the clinical features of albinism, and how best to manage skin care. The consequences of too much sun exposure on the skin are described, as well as diagnoses, prognoses, management, and treatment of the damage. The nature and features of the two types of cancer most prevalent in people with albinism (i.e., squamous and basal cell carcinoma), as well as those of other rarer tumors, such as cutaneous melanoma, are specified and treatment options described.

II. BIOLOGY OF THE SKIN

The skin is a large and complex organ, which forms a barrier between the body and the environment. It is a physical barrier that prevents water loss, blocks the entry of toxins and infectious agents, and protects the underlying tissues against environmental assaults such as ultraviolet light (UV) and chemical exposure. Several features of the skin facilitate these functions.

The skin is formed by the epidermis and the underlying dermis, separated by a basement membrane (Fig. 6.1). The epidermis is stratified, with keratinocytes at various stages of differentiation forming each layer.

The stratum basale consists of keratinocytes and melanocytes, specialized cells that produce the pigment melanin that gives skin its color. There is one melanocyte for every 10 basal keratinocytes (Haass and Herlyn, 2005). The single layer of undifferentiated keratinocyte stem

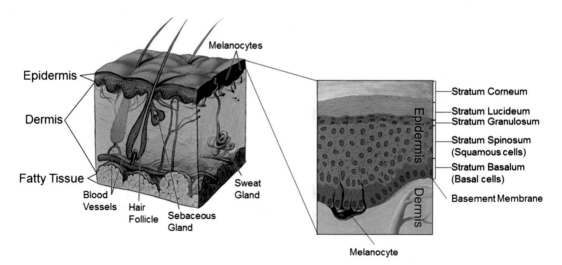

FIGURE 6.1   Anatomy of the skin. *Image from the United States National Cancer Institute Visuals Online, Don Bliss (Illustrator).*

cells that populates the basal layer divides to give rise to the spinous layer. Large polygonal keratinocytes with prominent junctions linking adjacent cells give the stratum its spinous appearance. The primary function of the stratum spinosum is to produce the keratin filaments and protein components that strengthen the skin (Marks et al., 2006).

Melanocytes have long dendrites that intercalate between differentiating keratinocytes and provide portals for the transfer of pigment to the keratinocytes (reviewed in Boissy, 2003). Each melanocyte is associated with 30–40 keratinocytes (Fitzpatrick and Breathnach, 1963). The transferred melanin forms a cap over the nucleus to prevent damage to the genetic material (Boissy, 2003). Two forms of melanin, black/brown eumelanin and red/yellow pheomelanin, are produced. The amount and type of melanin made varies between individuals and is categorized using the Fitzpatrick scale. The scale also takes into account response to the sun (tanning ability and susceptibility to sunburn) (Fitzpatrick, 1988). The pheomelanin/eumelanin ratio is higher in fairer skin types (Fitzpatrick I to III) and lower in darker skin types (Fitzpatrick IV to VI). Eumelanin provides greater protection against UV damage, while pheomelanin may facilitate solar damage (Brash, 2016).

As the cells continue to differentiate, they begin to flatten forming the granular layer, so called because the cells have prominent granules containing a proline-rich, sulfur-containing protein structure called keratohyalin, which is thought to contribute to the stability and reduced permeability of the skin (Matoltsy and Matoltsy, 1970). Keratohyalin associates with filaments in the stratum corneum, which forms the uppermost layer of the skin, where the cells are completely flattened, have lost their nuclei, and are embedded in a lipid matrix (Elias et al., 1981) that further reduces skin permeability (Scheuplein and Blank, 1971). This barrier is critical for preventing water loss and preventing or reducing the absorption of chemicals from the environment. The cells of the stratum corneum are eventually sloughed off allowing for turnover of the skin. The constant renewal provides a mechanism for removing damaged cells, particularly those with genetic damage.

The epidermis is supported by the dermis and a subdermal fat layer. The epidermal layer is not vascularized, thus nutrients are supplied through the dermis. Fibroblasts, the most common dermal cells, produce extracellular components of the dermis including collagen and elastic fibers (Marks et al., 2006). The dermis also supports hair follicles, which form in epidermal invaginations. Like epidermal renewal, hair growth is also cyclical. Hair follicle growth occurs in the anagen phase, with regression during the catagen phase and a resting telogen phase. The hair itself is shed, with cycle lengths varying depending on body site. Hair follicles contain melanocyte stem cells, which divide and populate the growing hair bulb (reviewed in Nishimura, 2011; Myung et al., 2009). As the hair shaft grows, melanin is transferred to the cortical and medulla keratinocytes giving hair its color (Stenn and Paus, 2001).

Mechanisms Underlying Solar Damage

The skin is continually exposed to environmental toxins such as UV that can damage genetic material in the cells. Sunlight is the most common source of UV exposure. There are three types of solar UV radiation, categorized by wavelength: UVA has a wavelength of 315–400 nm, UVB a wavelength of 280–315 nm, and UVC a wavelength of 100–280 nm. UVA and UVB cause the most solar damage to the skin because earth's atmosphere filters out the majority of UVC radiation (D'orazio et al., 2013). Although UVB exposure is important for stimulating production of vitamin D (Rajakumar et al., 2007), UV exposure also causes inflammation, photoaging, and DNA damage that increases skin cancer risk (D'orazio et al., 2013, Elwood and Jopson, 1997). If the solar damage is severe, keratinocytes will activate cell death programs to remove compromised cells (Bayerl et al., 1995). Failure to remove damaged cells could lead to the development of cancers.

The link between sun exposure and DNA damage was made evident in studies of a group of disorders known as xeroderma pigmentosum. Affected individuals have compromised DNA repair systems and are highly sensitive to the sun (Kraemer and Digiovanna, 1993). There are multiple mechanisms through which DNA can be damaged by UV. UVB or UVA can trigger the formation of covalent bonding between two pyrimidine nucleotides (thymine or cytosine) producing dimers that can interfere with DNA replication and result in mutations (Ikehata and Ono, 2011). In addition to direct DNA damage, UV radiation also generates reactive oxygen species that can damage the cell. Free radicals can cause oxidation of DNA nucleotides. Guanine can be converted to 8-hydroxy-2'-deoxyguanine, which causes mispairing with adenine instead of cytosine resulting in a mutation from guanine/cytosine to adenine/thymine (Nishimura, 2002).

Coupled with the immediate effects of sun exposure are longer-term effects including epidermal thickening. Severe sun damage can lead to the formation of precancerous lesions known as actinic keratoses (AKs), which in turn can transform into squamous cell carcinoma (SCC). Malignant melanoma (MM), which is particularly rare among individuals of African descent, including people with OCA (Van Der Westhuizen et al., 2015), is the most aggressive type of skin cancer. Other rare cancers associated with UV light damage, such as Merkel cell carcinomas, have also been described.

In addition to DNA damage, UV exposure can also induce inflammation through the activation of cytokines (Clydesdale et al., 2001). In rare instances, severe sunburn can also trigger postinflammatory hyperpigmentation and/or hypopigmentation (Tomita et al., 1989). Sunburn promotes melanin synthesis resulting in the tanning effect; whether the same mechanisms promote hyperpigmentation is not known.

III. CLINICAL FEATURES OF ALBINISM

OCA is a group of genetic dermatological disorders characterized by hypomelanosis (decreased or absent pigmentation) of the skin, hair, and eyes. OCA is due to defects in the genes responsible for melanocyte differentiation and melanin biosynthesis. The broader group of OCA consists of many different types resulting from mutations in more than 15 genes, with some that are yet to be identified (Wright et al., 2015) (see Chapter 5 for a full discussion of the types of albinism). In a subset of nonsyndromic OCA, which includes OCA1, -2, and -3, the effects of the mutations are limited to melanocyte function. The number and distribution of melanocytes within the epidermis is not affected in these cases (Oetting et al., 2003).

In the most severe type, OCA1A (which is very rare in Africa), there is no active tyrosinase enzyme and therefore no melanin is synthesized throughout the lifetime of the affected person. As no pigment can be produced, the skin is pink, the eyes are blue/grey with a prominent red reflex, and the hair white (King et al., 1995).

OCA2 is the most common type of albinism in the Black population in sub-Saharan Africa. Melanocytes express some tyrosinase activity, and thus there is some pigmentation of the hair, eyes, and skin. The skin can have a yellowish color and patients may also be able to tan slightly. With age, dark brown freckles may develop on sun-exposed areas of the skin (OCA2AE). The iris of the eye may be less translucent and may be brown in color (King et al., 1995).

The OCA3 phenotype varies depending on the ancestry of the affected individual. In African individuals, OCA3 presents with reddish skin and ginger hair (Kromberg et al., 1990). An African American individual with OCA3 had light brown skin and hair (Boissy et al., 1996), while in affected individuals in a Pakistani family, the phenotype was described as a "milder" version of OCA1 (Forshew et al., 2005). The ultrastructure and metabolic defects in melanocytes from people with OCA3 in Africa have been investigated (Kidson et al., 1993). Both eumelanin and pheomelanin are synthesized; however, melanosomes, the membrane-bound structures in which melanin is contained, are not fully matured. In addition, melanosomes were 30% smaller than normal. Furthermore, the reddish skin suggests that the pheomelanin/eumelanin ratio is higher than normal in these individuals.

Ocular defects, including photophobia, are common to all cases of OCA, as are refractory problems of the eye. Horizontal or rotatory nystagmus (repeated eye movement, usually in a horizontal but sometimes in a vertical direction) also occurs in almost all cases (see Chapter 7). In the tyrosinase-positive/OCA2 cases, nystagmus can improve with age.

Ocular albinism (OA) primarily affects the pigment cells of the eyes, although there are some changes seen in epidermal melanocytes. Affected individuals are usually males because the predominant form is OA1, which differs from OCA in that the inheritance is X-linked recessive. OA results in significant visual defects, while the skin and hair color may be fairer than that of relatives.

IV. CLINICAL EFFECTS OF SUNLIGHT EXPOSURE IN THE SKIN IN ALBINISM

As a result of decreased or absent melanin, the skin loses its natural ability to protect cells from the damaging effects of the sun. The immediate observable effect of sunlight on the skin of patients with OCA is redness (erythema) and even occasional blistering. Histological changes in the skin can be seen following UVB damage. Keratinocytes become dyskeratotic (abnormally keratinized) and accumulate vacuoles. The number of Langerhans cells (mobile immune cells that surveil the epidermis below the stratum corneum) is reduced. The dermis is also impacted, with increased blood flow causing the redness. Furthermore, there is a significant change in expression of cytokines that promote inflammation (Gilchrest et al., 1981).

With repeated exposure to sunlight, some individuals develop pigmented freckles (ephelides) and solar lentigines (small, flat, brown spots) with age (Bothwell, 1997). Individuals who develop these pigmented lesions have been shown to have a reduced occurrence of skin cancer (Kromberg et al., 1989). The skin also develops a thickened appearance, which may have a cobblestone look reflecting solar elastoses.

Chronic exposure to the sun may result in photoaging of the skin and heavy wrinkling in the areas most frequently exposed (Wright et al., 2015). In Tanzania chronic skin damage was noted in all infants with OCA by 1 year of age, and by 20 years every subject in the study had subclinical malignant change (Luande et al., 1985). While even in South Africa, further from the Equator and at a higher latitude, 14% of a sample of 111 subjects with albinism, under the age of 20 years, had malignant or premalignant lesions (Kromberg et al., 1989). In another study in rural Tanzania, 91% of affected people, over the age of 20 years, had AKs, while by age 30 years the figure was 100% (Lookingbill et al., 1995).

V. MANAGEMENT AND CARE OF THE SKIN IN ALBINISM

With the loss of the protective function of normal melanin in the skin from the damaging effects of sunlight, it is of extreme importance that affected people take precautions to protect their skin from sunlight. Avoidance of sunlight, protective clothing (made of cotton or other thick fabric) and wide brimmed hats, and the application of high sun protection factor (SFP) sunscreen must be routine practice to decrease the risk of developing skin cancer. Avoiding the damaging effects of sunlight can have the added cosmetic benefit of decreasing the photoaging effects on the skin such as wrinkles, skin thickening, and dyspigmentation.

It has been reported that the incidence of early onset AKs and skin cancer was higher in countries closer to the equator and in these regions there was a more ominous course, possibly due to neglect and late presentation for treatment (Luande et al., 1985). The authors, working in Tanzania, comment that because of the heat of summer most people dress lightly, wearing short-sleeved shirts, and women tended to wear skirts. This increased the risk of developing SCC on the exposed arms in men and women with albinism, as well as the legs in women. In rural areas most work is outdoors and school classrooms, often, tended to be outside, thereby increasing the number of hours of sun exposure to more than 6 h/day. Educating not only those affected with albinism but also the general public, especially teachers and health workers, as to the adverse effect of sun exposure in people with albinism, therefore, is of utmost importance. At the same time, screening programs should be initiated, in schools and communities, so that skin lesions can be detected and treated early, when such treatment is most effective (Lekalakala et al., 2015). It has been suggested that a centralized registry for people with albinism could facilitate free annual skin examinations, increase rates of early detection, and ensure access to treatment (Opara and Jiburum, 2010).

Sunscreens need to be of high SPF to defend against both UVB and UVA. Although not cosmetically appealing, zinc paste can be used as an excellent and inexpensive total block to both. It can be mixed into a paste using castor oil, which can make application easier. The use of zinc paste on the lips can have the dual benefit of sun protection and moisturization.

Dryness of the skin can also occur due to the photodamaging effects of sunlight. To moisturize the skin it is better to use emollients such as ung emulsifications. Creams and watery lotions tend to not be very effective at improving the dryness of the skin. Avoidance of harsh soaps during washing will also decrease the dryness of the skin. In extreme cases it is advisable to use ung emulsifications not only as a moisturizer but also as a cleanser during washing (application of the ung emulsifications to the skin, followed by use of lukewarm water to wash off the skin).

VI. SKIN DISORDERS COMMON IN ALBINISM

The most common skin lesions in OCA are AKs. Individuals may also develop pigmented lesions, particularly on sun-exposed areas. There has been some discrepancy in the reporting of pigmented lesions, which have been referred to as dendritic freckles (Findlay, 1962), ephelides (Tsuji and Saito, 1978), lentigines (King et al., 1980), and goose foot lentigines (Bothwell, 1997) Although skin cancer is generally rare among individuals of African ancestry, individuals with OCA have a much higher risk, particularly those who live in areas where UV exposure risk is high.

Sixty-one individuals with OCA2 and 65 control subjects from the Johannesburg area in South Africa were evaluated for the presence of pigmented skin lesions and malignancies (Bothwell, 1997). Nevi and pigmented macules could be identified on non–sun-exposed areas, a finding confirmed by Van Der Westhuizen et al. (2015) who undertook a prospective study of 16 patients with OCA. Dendritic freckles (present in 43% of OCA subjects in the Bothwell study) and solar keratoses were found to occur on sun-exposed areas, whereas nevi were generally found on non–sun-exposed areas.

While Bothwell reported that 18% of OCA subjects had SCCs, the van der Westhuizen study found 87% of the subjects had a skin malignancy (56% with basal cell and 62% with SCC), although no melanomas were found. Kiprono et al. (2014) studied 134 biopsies from 86 patients. Most biopsies were from the head and neck areas with 46% reported as basal cell and 54% as SCCs. They also reported the occurrence of one acral melanoma.

Actinic Keratoses

Clinical Features

AKs (Fig. 6.2) develop in areas of the skin exposed to sunlight especially the face, hands, arms, and legs. The scalp can be affected, but hair provides partial protection against sunlight and therefore fewer keratoses develop on the scalp.

AKs vary in size ranging from a few millimeters to over 2 cm. They can be flesh colored or pink to red in color and are rough, dry, and scaly. In some areas, such as the dorsum (top) of the hands, AKs can be very thick and hard. They result from abnormal keratinocyte growth.

Treatment

There is a 0.1%–20% risk that an AK will develop into SCC (Costa et al., 2015); however, there is no way of identifying which lesion will undergo transformation. Because of this risk it is important to attempt to clear or destroy as many AKs as possible.

Damage induced by UV radiation can reduce immune responses in the skin, increasing the risk that the cancers can evade the body's surveillance systems (Simon et al., 1991). In addition, mutations in several genes including p16 (INK4a) (Mortier et al., 2002), p14 (ARF), p15 (INK4b), and p53 (Kanellou et al., 2008) have been linked to progression from AK to SCC.

Daily application of sunscreens and reduced sunlight exposure can promote spontaneous resolution of the AKs. However, no studies have been undertaken to determine if this holds true for those with albinism. It is advisable as part of a treatment protocol to inform all patients with albinism to apply sunscreen every day (this includes winter and days with cloud cover) and to avoid the sun.

Therapies used to treat AKs include the following:

Cryotherapy is the use of liquid nitrogen to freeze AKs (Costa et al., 2015). The availability of liquid nitrogen is problematic because it rapidly turns to the gaseous phase of nitrogen as it warms up when exposed to air. It is a painful procedure and patients, especially children, may be fearful. If a blister forms, it is important to inform the patient to keep the skin clean and to use a topical antiseptic cream or ointment if any infection develops.

Topical 5-fluorouracil, known commercially as Efudix (in South Africa and United Kingdom) or Carac (in the United States), is a suitable treatment for use in the treatment of widespread AKs (Dogliotti, 1973). It is usually applied to the affected areas twice daily for 2–3 weeks. To improve absorption, plastic wrap can be used around the hands or arms or legs after application and kept on for 1 day. This decreases application to once a day and can improve clinical efficacy. Patients with albinism occasionally need to use the treatment for periods longer than 3 weeks to improve the clinical outcome.

Imiquimod is a topical therapy, which can similarly be used to treat widespread areas of AK (Stockfleth et al., 2016). The treatment is applied at night on Monday, Wednesday, and Friday for 4 weeks. If after 2 weeks the treated area is very red and inflamed then the treatment can be interrupted for 1 week and then restarted. If there is considerable inflammation, the patient may have flu-like symptoms, which normally disappear within 24 h of interrupting the treatment. Imiquimod has the disadvantage of being expensive as the cream is supplied in small sachets, which can only be used over small areas of affected skin.

Photodynamic therapy may be another treatment option for widespread AKs. This treatment requires application of photosensitizers such as topical 5-aminolevulinic acid or methyl

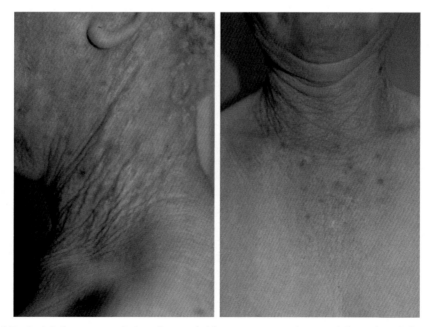

FIGURE 6.2   Actinic keratoses and photodamaged skin on sun-exposed areas of the neck and chest with thickened wrinkled appearance of the neck. *Photographs courtesy of Professor Ncoza Dlova, University of Kwazulu Natal.*

aminolevulinate, which have a high affinity for hyperproliferative cells and thus AKs. Three hours after application, during which time the drugs are converted to protoporphyrin IX, the affected area is exposed to red or blue light. The wavelengths of these lights correspond to the absorption spectrum of protoporphyrin IX, effectively causing cellular damage to the AKs (Serra-Guillen et al., 2012). The treated area is usually very red for about 4–6h and after around 7–10days the skin should heal very well. This approach is less effective for the treatment of thick lesions.

Alternative therapies. AKs are sometimes itchy and the very nature of their dry scaly appearance often makes the patient scratch the lesions continuously, especially on areas such as the top of the hands, the arms, and legs. In response to the scratching the lesions may become very thickened and hard. They also may reach extremely large and irregular sizes. The use of micropore plasters helps to soften the hard keratotic nodules. In addition to inhibiting the scratching, as the plasters are pulled off the lesions become wet and soft and the hard tissue starts to come away. It may take many months before the lesions shrink and even disappear. Another option is to use 10% salicylic acid mixed into aqueous cream and to apply it to the affected areas daily. If the skin becomes too red, 10% lactic acid in aqueous cream can be used as an alternative. In combination with all the other forms of therapy, these techniques will help to soften hard and thick AKs, and thereby should help to improve the clinical outcome.

VII. MALIGNANT SKIN CANCER

Chronic exposure to solar UV radiation results in an increase in the risk of skin cancer, particularly in people with albinism. These skin cancers can be divided into the nonmelanoma skin cancers (squamous and basal cell carcinomas), and cutaneous melanoma. SCC is generally more common than basal cell carcinoma (BCC) among people with albinism in Africa; however, cutaneous melanoma is very rare in this population group.

Squamous Cell Carcinoma

SCC is the most common skin cancer among individuals with albinism (Yakubu and Mabogunje, 1993; Kromberg et al., 1989) (Fig. 6.3). In Nigeria, people with albinism were found to account for 67% of patients in a 2 year study at a skin cancer clinic at Imo State University Teaching Hospital. SCCs were found to be the most common skin cancer (Opara and Jiburum, 2010).

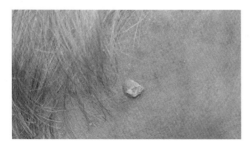 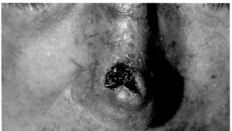

FIGURE 6.3 Squamous cell carcinoma on head (Kelly Nelson, M.D., Photographer) and nose. *Photographs from the United States National Cancer Institute Visuals Online.*

Clinical Features

AKs can give rise to SCC. Inflammation and any pain or extreme itch associated with AKs can indicate transformation. If the cellular changes are limited to the epidermis (uppermost level) of the skin, then this is known as an SCC in situ or Bowens Disease. Clinically these present as a flat, red, and scaly plaque usually greater than 1 cm in diameter. Progression to a deeper SCC is seen as an erythematous-raised keratotic plaque or nodule. These lesions are most common on sun-exposed areas of the body. This includes the lower lip area and eyelids, which are very difficult to treat. Although the risk of malignant metastases is low, it is possible and therefore imperative to treat any suspicious lesion as early as possible. In a 15-year study of albinism in the Cameroon, SCC was the most common cause of death (Aquaron, 1990).

Treatment

Surgery is the best treatment for any suspicious growth or lesion that is growing or changing. Combination with radiotherapy can improve outcome (Burton et al., 2016). For very difficult and poorly positioned lesions, Mohs microsurgery is the ideal surgical choice. This would help reduce excessive surgery and have the benefit of increased certainty of removal of the entire lesion.

Radiotherapy is a therapeutic choice for a difficult lesion. Various options ranging from superficial X-rays to brachytherapy can be used depending on the available facilities (Delishaj et al., 2016; Burton et al., 2016).

For SCC in situ or Bowen's disease **photodynamic therapy** is an alternative treatment option especially for a very large or irregular lesion, which would be difficult to surgically excise.

Basal Cell Carcinoma

Clinical Features

BCC can present in four different clinical types:

1. A *nodular BCC* (Fig. 6.4) presents as a soft, pink nodule, which can start to break down in the middle forming an ulcer, which bleeds spontaneously. For this reason, these lesions have also been known as "rodent ulcers."
2. A *superficial BCC* is often missed as it can be a very subtle slow-growing lesion. It is usually pink to red, flat, irregular, and may have a few small ulcers, which may also bleed.
3. A third type, the *pigmented BCC* is far less likely to occur in patients with albinism because melanin is present in the lesion.
4. The fourth known BCC type is a more aggressive and rare type known as a *morpheaform BCC*. The most consistent clinical feature is a white fibrotic area of growth within the pink to red plaque. This type of BCC is often more widespread in distribution than that which can be seen clinically. It is thus more aggressive and difficult to treat and most likely to recur.

Treatment

Cryotherapy can occasionally work for small superficial BCCs. However, it is generally ineffective in treating BCCs.

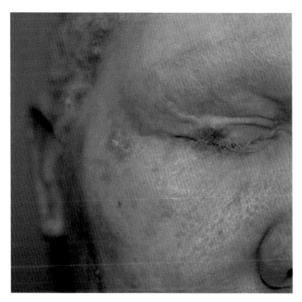

FIGURE 6.4   Basal cell carcinoma on upper eyelid and actinic keratoses on the upper cheek. *Photograph courtesy of Professor Ncoza Dlova, University of Kwazulu Natal.*

Topical therapies such as **5-fluorouracil and imiquimod** may also be effective. In albinism these treatments may have to be used for longer than the usual time periods prescribed. It is best therefore to continue treatment until clinical clearance has been achieved.

There is not much research with regard to **5-fluorouracil** as a treatment for BCCs, but when there are few other options it is worth attempting to use the cream under occlusion and for longer treatment periods.

Imiquimod is well described as a topical treatment for superficial BCCs. It is a cream applied at night from Monday to Friday. No cream is applied on Saturday or Sunday. This helps to calm the reaction that occurs after application. After 2 weeks of use, if there is extreme redness and discomfort, then the treatment should be stopped for 1 week. Thereafter the treatment can be restarted and continued for another 4 weeks; thus the total treatment should be 6 weeks.

Imiquimod should not be used on nodular, pigmented, and morphoeic BCCs. However, imiquimod may be useful when used on a large and irregular nodular BCC to shrink the surrounding affected area, which can then decrease the size of the surgery which may be needed to excise the lesion.

Photodynamic therapy is also useful in the treatment of BCCs. The treatment must be repeated 1 week after the first treatment. If not completely clear, the treatment can be repeated a third time. This treatment is best for superficial BCCs. It is not effective for large nodular, pigmented, or morphoeic BCCs. Nodular BCCs can be treated with photodynamic therapy if the carcinoma is first curetted to remove most of the nodular growth to debulk the lesion and allow the active cream to penetrate as many BCC cells as possible.

Surgery is needed for most nodular, pigmented, and morphoeic BCCs. For difficult areas, such as the face, and for aggressive lesions, such as morphoeic BCCs, Mohs microsurgery yields the best results. This surgery involves excising the involved area and immediately examining the excised lesion with microscopy. This allows the surgeon exact knowledge regarding the edges of the lesion, and extra areas are cut out according to where positive cells are still seen. The result is therefore an exact removal with minimal effect on normal healthy skin.

Radiation therapy can be performed for difficult areas, but it is expensive and time-consuming and may have long-term side effects including fibrosis, scarring, and tightening of the skin.

Malignant Melanoma

Clinical Features

The majority of MMs are flat, irregularly pigmented lesions. They are extremely rare among Africans with albinism; however, amelanotic MM can occur (Van Der Westhuizen et al., 2015). These often look like SCCs, except for a possible small amount of pigmentation. An examination technique known as dermoscopy can facilitate a differential diagnosis.

Treatment

MMs are highly aggressive tumors and are best treated quickly with excision in their early stages. If allowed to grow and to penetrate into the dermal areas, the risk of metastatic spread is significant.

Once a diagnosis is confirmed a reexcision is done based on the thickness of the original tumor. Sentinel lymph node biopsy is performed for more advanced lesions and involves excision of one or several lymph nodes, which drain from the initial tumor site. This procedure can aid diagnosis, disease staging, and in predicting prognosis.

Other Malignant Tumors

Other malignant and rarer tumors of the skin include Merkel cell carcinoma, other adnexal tumors of the skin, and lymphomas of the skin. Any unusual growth of the skin, which has changed or grown over a relatively short period, must either be excised if small enough or biopsied for diagnosis. The only form of treatment for the majority of other malignant tumors of the skin is surgery to excise the lesions. Mohs microsurgery can also be done, especially for unusual lesions in difficult areas such as the face. Radiation therapy or chemotherapy may be more effective for tumors such as lymphomas.

VIII. CONCLUSION

Improved self-esteem is associated with reduced freckling, wrinkling, or other signs of photodamage and results in enhanced quality of life of affected individuals. In all cases it is of utmost importance that prevention strategies are utilized to reduce the risk of premalignant or malignant tumors of the skin.

Education in sun avoidance and management of the skin is the first key to a healthy patient with albinism. Mortality subsequent to malignancies of the skin is avoidable and unacceptable in the modern era of knowledge and research.

References

Aquaron, R., 1990. Oculocutaneous albinism in Cameroon. A 15-year follow-up study. Ophthalmic Paediatr. Genet. 11, 255–263.

Bayerl, C., Taake, S., Moll, I., Jung, E.G., 1995. Characterization of sunburn cells after exposure to ultraviolet light. Photodermatol. Photoimmunol. Photomed. 11, 149–154.

Boissy, R.E., 2003. Melanosome transfer to and translocation in the keratinocyte. Exp. Dermatol. 12, 5–12.

Boissy, R.E., Zhao, H., Oetting, W.S., Austin, L.M., Wildenberg, S.C., Boissy, Y.L., Zhao, Y., Sturm, R.A., Hearing, V.J., King, R.A., Nordlund, J.J., 1996. Mutation in and lack of expression of tyrosinase-related protein-1 (TRP-1) in melanocytes from an individual with brown oculocutaneous albinism: a new subtype of albinism classified as "OCA3". Am. J. Hum. Genet. 58, 1145–1156.

Bothwell, J., 1997. Pigmented skin lesions in tyrosinase-positive oculocutaneous albinos: a study in black South Africans. Int. J. Dermatol. 36, 831–836.

Brash, D.E., 2016. UV-induced melanin chemiexcitation: a new mode of melanoma pathogenesis. Toxicol. Pathol. 44, 552–554.

Burton, K.A., Ashack, K.A., Khachemoune, A., 2016. Cutaneous squamous cell carcinoma: a review of high-risk and metastatic disease. Am. J. Clin. Dermatol. 17, 491–508.

Clydesdale, G.J., Dandie, G.W., Muller, H.K., 2001. Ultraviolet light induced injury: immunological and inflammatory effects. Immunol. Cell Biol. 79, 547–568.

Costa, C., Scalvenzi, M., Ayala, F., Fabbrocini, G., Monfrecola, G., 2015. How to treat actinic keratosis? An update. J. Dermatol. Case Rep. 9, 29–35.

D'orazio, J., Jarrett, S., Amaro-Ortiz, A., Scott, T., 2013. UV radiation and the skin. Int. J. Mol. Sci. 14, 12222–12248.

Delishaj, D., Rembielak, A., Manfredi, B., Ursino, S., Pasqualetti, F., Laliscia, C., Orlandi, F., Morganti, R., Fabrini, M.G., Paiar, F., 2016. Non-melanoma skin cancer treated with high-dose-rate brachytherapy: a review of literature. J. Contemp. Brachyther. 8, 533–540.

Dogliotti, M., 1973. Actinic keratoses in Bantu albinos. Clinical experiences with the topical use of 5-fluoro-uracil. S. Afr. Med. J. 47, 2169–2172.

Elias, P.M., Cooper, E.R., Korc, A., Brown, B.E., 1981. Percutaneous transport in relation to stratum corneum structure and lipid composition. J. Investig. Dermatol. 76, 297–301.

Elwood, J.M., Jopson, J., 1997. Melanoma and sun exposure: an overview of published studies. Int. J. Cancer 73, 198–203.

Findlay, G.H., 1962. On giant dendritic freckles and melanocyte reactions in the skin of the albino Bantu. S. Afr. J. Lab. Clin. Med. 8, 68–72.

Fitzpatrick, T.B., 1988. The validity and practicality of sun-reactive skin types I through VI. Arch. Dermatol. 124, 869–871.

Fitzpatrick, T.B., Breathnach, A.S., 1963. The epidermal melanin unit system. Dermatol Wochenschr 147, 481–489.

Forshew, T., Khaliq, S., Tee, L., Smith, U., Johnson, C.A., Mehdi, S.Q., Maher, E.R., 2005. Identification of novel TYR and TYRP1 mutations in oculocutaneous albinism. Clin. Genet. 68, 182–184.

Gilchrest, B.A., Soter, N.A., Stoff, J.S., Mihm Jr., M.C., 1981. The human sunburn reaction: histologic and biochemical studies. J. Am. Acad. Dermatol. 5, 411–422.

Haass, N.K., Herlyn, M., 2005. Normal human melanocyte homeostasis as a paradigm for understanding melanoma. J. Investig. Dermatol. Symp. Proc. 10, 153–163.

Ikehata, H., Ono, T., 2011. The mechanisms of UV mutagenesis. J. Radiat. Res. 52, 115–125.

Kanellou, P., Zaravinos, A., Zioga, M., Stratigos, A., Baritaki, S., Soufla, G., Zoras, O., Spandidos, D.A., 2008. Genomic instability, mutations and expression analysis of the tumour suppressor genes p14(ARF), p15(INK4b), p16(INK4a) and p53 in actinic keratosis. Cancer Lett. 264, 145–161.

Kidson, S.H., Richards, P.D., Rawoot, F., Kromberg, J.G., 1993. An ultrastructural study of melanocytes and melanosomes in the skin and hair bulbs of rufous albinos. Pigment Cell Res. 6, 209–214.

King, R.A., Creel, D., Cervenka, J., Okoro, A.N., Witkop, C.J., 1980. Albinism in Nigeria with delineation of new recessive oculocutaneous type. Clin. Genet. 17, 259–270.

King, R.A., Mentink, M.M., Oetting, W.S., 1995. Albinism. In: Scriver, C.R., Beaudet, A.L., Sly, W.S., Valle, D.V. (Eds.), The Metabolic and Molecular Basis of Inherited Disease, seventh ed. McGraw-Hill, New York.

Kiprono, S.K., Chaula, B.M., Beltraminelli, H., 2014. Histological review of skin cancers in African Albinos: a 10-year retrospective review. BMC Cancer 14, 157.

Kraemer, K.H., Digiovanna, J.J., 1993. Xeroderma pigmentosum. In: Pagon, R.A., Adam, M.P., Ardinger, H.H., Wallace, S.E., Amemiya, A., Bean, L.J.H., Bird, T.D., Ledbetter, N., Mefford, H.C., Smith, R.J.H., Stephens, K. (Eds.), GeneReviews(R) Seattle (WA).

Kromberg, J.G., Castle, D., Zwane, E.M., Jenkins, T., 1989. Albinism and skin cancer in southern Africa. Clin. Genet. 36, 43–52.

Kromberg, J.G., Castle, D.J., Zwane, E.M., Bothwell, J., Kidson, S., Bartel, P., Phillips, J.I., Jenkins, T., 1990. Red or rufous albinism in southern Africa. Ophthalmic Paediatr. Genet. 11, 229–235.

Lekalakala, P.T., Khammissa, R.A., Kramer, B., Ayo-Yusuf, O.A., Lemmer, J., Feller, L., 2015. Oculocutaneous albinism and squamous cell carcinoma of the skin of the head and neck in sub-Saharan Africa. J. Skin Cancer 2015, 167847.

Lookingbill, D.P., Lookingbill, G.L., Leppard, B., 1995. Actinic damage and skin cancer in albinos in northern Tanzania: findings in 164 patients enrolled in an outreach skin care program. J. Am. Acad. Dermatol. 32, 653–658.

Luande, J., Henschke, C.I., Mohammed, N., 1985. The Tanzanian human albino skin. Natural history. Cancer 55, 1823–1828.

Marks, J.G., Miller, J.J., Lookingbill, D.P., Lookingbill, D.P., 2006. Lookingbill and Marks' Principles of Dermatology. Saunders Elsevier, Philadelphia, PA.

Matoltsy, A.G., Matoltsy, M.N., 1970. The chemical nature of keratohyalin granules of the epidermis. J. Cell Biol. 47, 593–603.

Mortier, L., Marchetti, P., Delaporte, E., Martin De Lassalle, E., Thomas, P., Piette, F., Formstecher, P., Polakowska, R., Danze, P.M., 2002. Progression of actinic keratosis to squamous cell carcinoma of the skin correlates with deletion of the 9p21 region encoding the p16(INK4a) tumor suppressor. Cancer Lett. 176, 205–214.

Myung, P., Andl, T., Ito, M., 2009. Defining the hair follicle stem cell (Part II). J. Cutan. Pathol. 36, 1134–1137.

Nishimura, E.K., 2011. Melanocyte stem cells: a melanocyte reservoir in hair follicles for hair and skin pigmentation. Pigment Cell Melanoma Res. 24, 401–410.

Nishimura, S., 2002. Involvement of mammalian OGG1(MMH) in excision of the 8-hydroxyguanine residue in DNA. Free Radic. Biol. Med. 32, 813–821.

Oetting, W.S., Fryer, J.P., Shriram, S., King, R.A., 2003. Oculocutaneous albinism type 1: the last 100 years. Pigment Cell Res. 16, 307–311.

Opara, K.O., Jiburum, B.C., 2010. Skin cancers in albinos in a teaching Hospital in eastern Nigeria – presentation and challenges of care. World J. Surg. Oncol. 8, 73.

Rajakumar, K., Greenspan, S.L., Thomas, S.B., Holick, M.F., 2007. SOLAR ultraviolet radiation and vitamin D: a historical perspective. Am. J. Public Health 97, 1746–1754.

Scheuplein, R.J., Blank, I.H., 1971. Permeability of the skin. Physiol. Rev. 51, 702–747.

Serra-Guillen, C., Nagore, E., Hueso, L., Traves, V., Messeguer, F., Sanmartin, O., Llombart, B., Requena, C., Botella-Estrada, R., Guillen, C., 2012. A randomized pilot comparative study of topical methyl aminolevulinate photodynamic therapy versus imiquimod 5% versus sequential application of both therapies in immunocompetent patients with actinic keratosis: clinical and histologic outcomes. J. Am. Acad. Dermatol. 66, e131–e137.

Simon, J.C., Tigelaar, R.E., Bergstresser, P.R., Edelbaum, D., Cruz Jr., P.D., 1991. Ultraviolet B radiation converts Langerhans cells from immunogenic to tolerogenic antigen-presenting cells. Induction of specific clonal anergy in CD4+ T helper 1 cells. J. Immunol. 146, 485–491.

Stenn, K.S., Paus, R., 2001. Controls of hair follicle cycling. Physiol. Rev. 81, 449–494.

Stockfleth, E., Sibbring, G.C., Alarcon, I., 2016. New topical treatment options for actinic keratosis: a systematic review. Acta Derm. Venereol. 96, 17–22.

Tomita, Y., Maeda, K., Tagami, H., 1989. Mechanisms for hyperpigmentation in postinflammatory pigmentation, urticaria pigmentosa and sunburn. Dermatologica 179 (Suppl. 1), 49–53.

Tsuji, T., Saito, T., 1978. Multiple naevocellular naevi in brothers with albinism. Br. J. Dermatol. 98, 685–692.

Van Der Westhuizen, G., Beukes, C.A., Green, B., Sinclair, W., Goedhals, J., 2015. A histopathological study of melanocytic and pigmented skin lesions in patients with albinism. J. Cutan. Pathol. 42, 840–846.

Wright, C.Y., Norval, M., Hertle, R.W., 2015. Oculocutaneous albinism in sub-Saharan Africa: adverse sun-associated health effects and photoprotection. Photochem. Photobiol. 91, 27–32.

Yakubu, A., Mabogunje, O.A., 1993. Skin cancer in African albinos. Acta Oncol. 32, 621–622.

Albinism and the Eye

Susan E.I. Williams

University of the Witwatersrand, Johannesburg, South Africa

I. INTRODUCTION

People with albinism all have visual abnormalities because they have hypomelanosis, and melanin and the proteins involved in its synthesis are important in the development and organization of the visual pathway. These abnormalities include poor visual acuity, photophobia, nystagmus, and sometimes strabismus. Both the structure and function of the eye are affected.

In the present chapter, the anatomy of the normal visual system and that of the visual system in albinism will be covered. The parts of the eye will be described, such as the iris, retinal pigment epithelium (RPE), the neurosensory retina, choroid and the visual pathway, and the associated defects found in people with albinism specified. Then, the visual problems experienced by individuals with albinism will be discussed, including nystagmus, strabismus, and refractive errors. Finally, management will be outlined, examining, initially, the general measures that can be taken, such as reducing photophobia and glare, refractive correction, visual aids, and diagnostics. Current treatments for strabismus and nystagmus will be reviewed. The quality of life of individuals with albinism may be reduced by their visual limitations (especially where they remain untreated, as happens in many areas of Africa); therefore lifestyle adjustments will be briefly discussed.

II. THE ANATOMY OF THE NORMAL VISUAL SYSTEM

The visual system consists of the eye and the visual pathway from the eye, which culminates in the visual cortex within the occipital lobe of the brain. Fig. 7.1 represents the normal anatomy of the human eye.

Light passes through the transparent cornea, the pupil, and the lens to stimulate photoreceptors within the retina, the innermost of the three layers of the eye. The choroid and the fibrous sclera form the outermost layers of the eye. The retina is composed of an inner layer, the neurosensory retina, and an outer layer that is a single layer of epithelial cells, the RPE. The neurosensory retina is a transparent layer consisting of several, predominantly neural, cell types. Light generates impulses in the photoreceptors (rods and cones), which relay these impulses to retinal ganglion cells via bipolar cells. The optic nerve is formed by the axons of the retinal ganglion cells as they converge at the optic disc. Fig. 7.2 illustrates the normal visual pathway.

FIGURE 7.1 Normal anatomy of the human eye. *Image courtesy of Dr. S. Williams, University of the Witwatersrand, South Africa.*

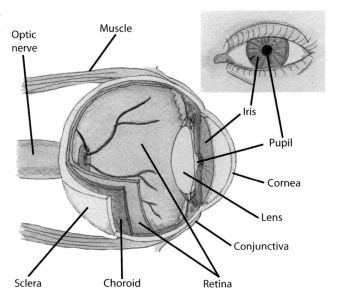

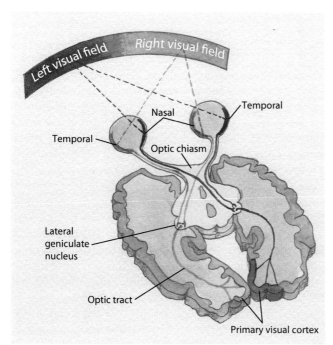

FIGURE 7.2 The normal visual pathway. *Image courtesy of Dr. S. Williams, University of the Witwatersrand, South Africa.*

The optic nerves leave the eye and the orbit and terminate in the optic chiasm in the floor of the third ventricle of the brain. At the optic chiasm, axons that have their origin in retinal ganglion cells in the nasal hemiretina of each eye (receiving visual information from the temporal visual field) decussate (cross the midline) to join the contralateral optic tract. Axons that have their origin in retinal ganglion cells in the temporal hemiretina (receiving visual information from the nasal visual field) do not decussate and become part of the ipsilateral optic tract. This is important for binocular vision as it allows simultaneous cerebral interpretation of the stimuli from any given point within the visual field from both eyes. The optic tracts end in the lateral geniculate nucleus where the retinal ganglion cell axons synapse with the nerve cell bodies whose axons make up the optic radiation. The optic radiation extends from the lateral geniculate nucleus to the primary visual cortex within the occipital lobe of the brain. There are additional connections from here to other parts of the brain.

III. THE ANATOMY OF THE VISUAL SYSTEM IN ALBINISM

Melanin and the proteins involved in its synthesis are an important component of several elements of the eye and visual system. Subsequently a reduction in melanin synthesis will affect both the structure and the function of the eyes and the visual system.

The Iris

The iris, the contractile diaphragm of the eye that restricts the access of light to the retina through the pupillary opening, consists of four layers: the anterior border layer, the stroma, the dilator pupillae muscle, and the posterior pigment epithelium. The connective tissue stroma contains melanocytes that are largely responsible for eye color. More heavily pigmented melanocytes result in the appearance of "brown eyes," whereas less-pigmented melanocytes result in "green eyes" or "blue eyes." The posterior pigment epithelium, however, is heavily pigmented regardless of eye color except in albinism. A reduction or absence of melanin in this layer will allow light to pass through the iris creating the effect of iris transillumination. Because light reflected from the retina is red (the physiological red reflex), individuals with iris transillumination may have irides that appear pink. In albinism, the lack of pigment in the retina increases the saturation of the red reflex, making this effect more prominent. Fig. 7.3 shows anterior segment photographs of a normally pigmented iris in a Black South African (A) compared with that of a 4-year-old Black child with albinism (B) in the same population. Despite the fact that there is pigment in the anterior stroma of the iris of the child with albinism, the paucity of pigment in the posterior pigment epithelium accounts for the peripheral transillumination that gives the peripheral iris a slightly pink appearance (this is particularly evident when one compares it with the normal iris). The photographs, taken on the same day with the same camera, also demonstrate the more prominent red reflex in the pupil of the child with albinism.

The inability of these hypopigmented irides to effectively restrict the amount of light that reaches the retina is a contributory factor to the photophobia experienced by individuals with albinism.

Optical coherence tomography (OCT), an imaging technique that acquires and processes optical signals to produce high-resolution images of ocular tissues, may be used to document the

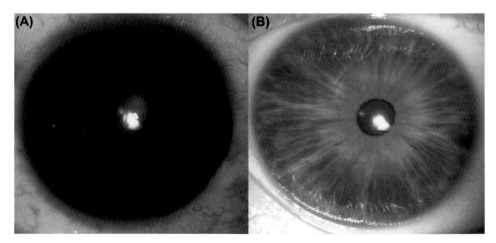

FIGURE 7.3 Iris photographs taken in the same population (Black South African) on the same day with the same camera. (A) Normally pigmented iris. (B) Iris of a child with albinism. *Photographs courtesy of Dr. S. Williams, University of the Witwatersrand, South Africa.*

anatomic changes that occur in the iris in albinism. The thickness of the iris posterior epithelial layer may be measured with OCT and this has diagnostic potential for detecting ocular albinism (Sheth et al., 2013). The grading of iris transillumination has traditionally been used as a diagnostic criterion for albinism, but this is a subjective measure. Sheth et al. (2013) demonstrated that objective iris measurements in albinism acquired with anterior segment OCT correlated better with visual acuity and nystagmus than the grading of iris transillumination, suggesting that iris OCT may be preferable in confirming the diagnosis of albinism where uncertainty exists.

The Retinal Pigment Epithelium

The RPE, that contains melanin granules, is integral to the normal functioning of the neurosensory retina. It is a neuroectoderm-derived, continuous, single layer of epithelial cells that is intimately associated with the neurosensory retina. Both layers together form the retina, the inner layer of the posterior segment of the eye that lines the eye from the margins of the optic disc posteriorly to the ora serrata anteriorly. The visually critical functions of the RPE include providing a selectively permeable blood–retinal barrier; maintaining adhesion of the neurosensory retina; breaking down photoreceptor outer segments; synthesizing the matrix surrounding and supporting the photoreceptors; and transporting and storing vitamins and metabolites. The melanin granules within the RPE absorb light (both visible and ultraviolet light) and reduce free-radical damage as well as light scatter within the eye. Polarization-sensitive OCT has been used to image melanin in the RPE in vivo and has demonstrated depigmentation in individuals with albinism, with the degree of depigmentation possibly associated with visual acuity (Schutze et al., 2014).

The RPE is also important in the development of the neurosensory retina. The different retinal neurons divide and evolve adjacent to the RPE during ocular development. In albinism, defects in melanogenesis in the RPE are accompanied by flawed development of the neuroretina. The relationship between neuroretinal development and alterations in melanin synthesis in the RPE is poorly understood. One hypothesis is that dihydroxyphenylalanine (L-Dopa), an amino acid produced as an intermediate during pigment synthesis, may influence how, and when, neurons, specifically within the neurosensory retina, develop (Roffler-Tarlov et al., 2013). This hypothesis is supported by the finding of absent or low L-Dopa levels in the retinae of albino mice (Roffler-Tarlov et al., 2013). This further suggests a therapeutic possibility for L-Dopa administration in albinism during ocular development, in utero, and potentially via the mother.

The Neurosensory Retina

The neurosensory retina in albinism shows an abnormal organization of neurons, particularly the photoreceptors and ganglion cells that are reduced in number and density in the macula (the visual center of the retina). This is most pronounced in the fovea, the center of the macula, and the region of the retina responsible for central vision, and best visual acuity. Foveal hypoplasia is a characteristic finding in albinism and it is a major contributor to the poor visual acuity observed. It is presumably related to abnormal melanin synthesis within the RPE. Foveal hypoplasia, with reduced cone density and impaired cone specialization, may be identified fundoscopically by a reduced (or absent) foveal light reflex and foveal pit. Fig. 7.4 shows

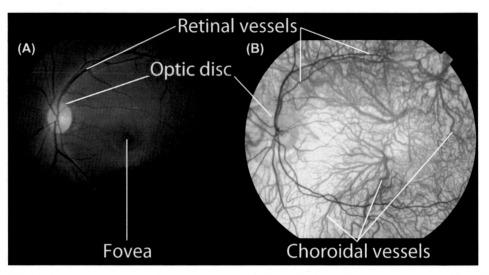

FIGURE 7.4   Fundus photographs taken in the same population (Black South African) on the same day with the same camera. (A) Normally pigmented individual. (B) Fundus of a child with albinism. *Photographs courtesy of Dr. S. Williams, University of the Witwatersrand, South Africa.*

fundus (retinal) photographs of a normally pigmented Black South Africa (A) compared with that of the 4-year-old South African child with albinism (B). The foveal pit is poorly defined in the child with albinism compared with that of the normally pigmented individual.

Foveal hypoplasia may also be visualized with fundus autofluorescence as a lack of foveal attenuation and with infrared reflectance as a concentric macular ring reflex (Cornish et al., 2014). Posterior segment OCT typically demonstrates the absence of a foveal dip, but even when this is not the case, one can expect to find incomplete excavation of the inner retinal layers at the fovea on OCT in subjects with albinism (Wilk et al., 2014). Lee et al. (2015), in a prospective OCT study on retinal development, demonstrated that the normal migration of cone photoreceptors into the fovea, and inner retinal layers away from it, is altered in albinism, but, importantly, it seems that these alterations continue into early childhood. This suggests that early treatment, normalizing the pigment pathway in the RPE, could prevent the progression of foveal hypoplasia and lead to improvements in visual acuity.

The Choroid

The choroid is a thin, richly vascularized, loose connective tissue that lies between the retina and the sclera. The main function of the choroid is to provide a nourishing blood supply to the outer layers of the retina, but it may also have a thermoregulatory role. The tissue is heavily pigmented so that it may absorb the light that has entered the eye and passed through the retina. Without the pigmented choroid, light would reflect off the sclera and pass back through the retina again. Consequently, a paucity of melanin in this tissue increases scattering of light within the posterior segment that may, in turn, increase photophobia and reduce visual acuity. The red reflex is more prominent in albinism because the choroidal vasculature is not obscured by pigment. The dramatically different appearing fundi in Fig. 7.4 (taken using the same camera and lighting on the same day) may be explained by the absent choroidal pigmentation in the child with albinism. The retinal vasculature is visible in both fundi, but the choroidal vasculature is only visible in the fundus photograph of the child with albinism. Choroidal thickness has been shown to be reduced in the subfoveal region in patients with oculocutaneous albinism, indicating a localized choroidal anatomical abnormality (Karabas et al., 2014). It is uncertain how this abnormality relates to foveal hypoplasia, but it provides further evidence for generalized developmental alterations in the macula of individuals with albinism.

The Visual Pathway

The visual pathway consists of the retina, optic nerves, optic chiasm, optic tracts, lateral geniculate bodies, optic radiations, and visual cortex. The pathway is, effectively, part of the central nervous system because the retinae have their embryological origins in extensions of the diencephalon. The portion of this pathway from the retina to the lateral geniculate body (the retinogeniculate projection) is anatomically different in individuals with albinism. In albinism, significantly more nerve fibers (retinal ganglion cell axons) decussate at the optic chiasm (see Fig. 7.5). There is a reduced ipsilateral projection of these fibers arising from fewer, more peripheral, and temporal retinal ganglion cells.

This abnormal crossing leads to abnormalities in the lateral geniculate nucleus that secondarily disrupt normal geniculocortical and interhemispheric visual connections.

An important consequence of abnormal retinogeniculate decussation in albinism is a reduction in binocular, stereoscopic vision. Normal stereopsis can only develop where projections from both eyes are organized together within the lateral geniculate nucleus. A reduced component from the ipsilateral temporal retina in albinism results in poor stereopsis; however, some individuals with albinism have a small degree of stereoscopic vision (Lee et al., 2001). Better stereopsis has been observed in individuals with reduced nystagmus and in individuals with more melanin in their irises and retinae (Lee et al., 2001). It is likely that individuals with albinism who have stereopsis have fewer crossed fibers at the optic chiasm than those who do not have stereopsis. Alternately there may be some degree of compensation for the crossing abnormalities that takes place within the corpus callosum to facilitate binocularity and allow for some stereopsis (Lee et al., 2001).

The retinogeniculate pathway consists of retinal ganglion cell axons (the cell bodies are part of the neurosensory retina), which develop along a preprogrammed pathway from the retina to the optic nerves, chiasm, optic tract, and eventually dorsal lateral geniculate nucleus. The factors responsible for determining the pathway taken by the axons once they reach the chiasm are still not fully understood. Initially this pathfinding was thought to be mediated by a melanin abnormality because the observation was made, in animal models, that there were more decussated fibers when there was less retinal melanin (Guillery, 1996). The tyrosinase enzyme is essential in the control of melanogenesis. Jeffery et al. (1994) demonstrated normal chiasmal anatomy in an albino mouse model where tyrosinase activity was introduced with a tyrosinase transgene. This was thought to represent additional evidence that alterations in melanin synthesis are responsible for the visual pathway abnormalities in albinism.

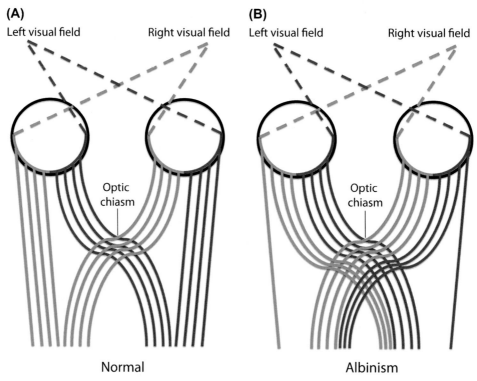

FIGURE 7.5 Decussation of retinal ganglion cell fibers at the optic chiasm. (A) Normally pigmented individual where fibers from the nasal hemiretinae decussate at the chiasm. (B) Abnormal decussation in albinism with more fibers decussating and a reduction in uncrossed fibers. *Image courtesy of Dr. S. Williams, University of the Witwatersrand, South Africa.*

More recently, however, it has been postulated that pathfinding in the visual pathway may be mediated by tyrosinase per se (Cronin et al., 2003). Tyrosinase is thought to initiate a signal in early neuronal development that establishes the path eventually taken by the ganglion cell axons to the lateral geniculate nucleus (Cronin et al., 2003).

Visual electrophysiology, specifically in the form of visual evoked potentials (VEPs) to flash and pattern-onset stimulation, provides evidence of the abnormal decussation of retinal ganglion cell axons at the chiasm and may be used in the diagnosis of albinism (Brecelj, 2014). The higher proportion of crossed fibers in albinism causes hemispheric asymmetry on VEP that may be demonstrated with monocular pattern stimulation (Lee et al., 2001).

VEPs in heterozygotes for a genetic mutation for albinism (OCA2) have been investigated (Castle et al., 1988). VEP testing was performed in 15 obligatory carriers (all with children with albinism) to determine whether they had a similar visual anomaly to that observed in their affected child. All showed symmetry on monocular pattern stimulation and thus no evidence of abnormal decussation at the chiasm. Therefore, VEP testing is not helpful in detecting carriers for OCA2.

Despite the aberrant visual connections in albinism, early reports of the use of magnetic resonance imaging (MRI) demonstrated a normal appearance of both the visual pathways and their interhemispheric connections (Brodsky et al., 1993). These early reports probably reflect the resolution of MRI scans at the time. With higher MRI resolution and with the use of multiple proton density-weighted images, changes in the visual pathway in albinism have been demonstrated (Mcketton et al., 2014). The presence of smaller optic nerves (in diameter), a smaller chiasm and smaller optic tracts, and a reduction in the volume of the chiasm and lateral geniculate nucleus volumes are all features that could, potentially, be used diagnostically. It is likely that more subtle changes may be visualized as the resolution of these scans continues to increase.

IV. NYSTAGMUS

Nystagmus, one of the typical ocular features associated with albinism, consists of involuntary repetitive eye movements. The nystagmus typically consists of a horizontal movement, but the movement may also be vertical or torsional. The cause for the nystagmus associated with albinism is probably the result of a complex interplay between motor and sensory components (Davis et al., 1997).

The motor component of the nystagmus observed in albinism is likely related to the abnormal visual pathway described above. The aberrant decussation of nerve fibers at the chiasm results in abnormalities in the lateral geniculate nucleus. The fibers in the optic tract carry the asymmetrical information from the lateral geniculate nucleus to the visual cortex. There are connections from the visual cortex to other parts of the brain (the cortex and the thalamus) responsible for the control of eye movements. Abnormal innervation along the entire pathway may, therefore, result in abnormal control of eye movements (Guillery, 1996).

Foveal hypoplasia, in conditions other than albinism, is associated with a secondary nystagmus (Perez et al., 2014). This suggests that the nystagmus observed in albinism may also be the consequence of foveal hypoplasia, with an element of sensory nystagmus related to poor central vision and visual scanning or "seeking behavior."

Nystagmus may aggravate the already compromised visual acuity in albinism because of a reduction in foveation time (the time during which the eyes are at the point of fixation with little or no movement). Reducing the nystagmus may therefore improve visual acuity in albinism. This reduction may occur naturally because most individuals affected by nystagmus have what is called a "null" position, a position of the eyes where the nystagmus amplitude and frequency are lowest and therefore visual acuity is best. To utilize the "null" position the head is moved so that the direction of gaze corresponds to that eye position. If the "null" position is eccentric, the result may be an abnormal head position. Surgical treatment for nystagmus, particularly for nystagmus associated with abnormal head turns, has been available for decades in the form of the Kestenbaum–Anderson technique (Davis et al., 1997). The rationale is to adjust the extraocular muscles so that the null point is centralized. For those without an anomalous head position, Dubner et al. (2016) have shown that nystagmus intensity may be decreased and visual acuity improved with four-muscle tenotomy surgery. Both of these approaches have been used with success and some improvement in visual acuity in albinism (Davis et al., 1997; Dubner et al., 2016). Unfortunately, these techniques represent specialized nystagmus surgery that should only be undertaken by surgeons with the requisite expertise. Consequently, the surgery may not be routinely available in most centers. The surgical procedures may be accompanied by complications, particularly restricted postoperative ocular motility, and require appropriate postoperative evaluations and care.

V. STRABISMUS

Strabismus is a common ocular finding associated with albinism. One study on 153 (77 males and 76 females) Black South African children with albinism showed that one-third (34.6%) had strabismus (2–25 prism diopters) (Raliavhegwa et al., 2001). Strabismus surgery, which entails altering the insertion of one or more of the rectus muscles, is most effective in the presence of good stereopsis that encourages fusion postoperatively. In the absence of stereopsis, and therefore fusion, the ocular deviation may recur or persist postoperatively. Furthermore, the abnormalities in stereopsis observed in albinism mean that strabismus surgery in these patients rarely achieves binocularity and depth perception, but this can occasionally be achieved and may be worthwhile in certain situations (Tavakolizadeh and Farahi, 2013; Villegas et al., 2010). Rectus muscle surgery may also improve nystagmus in the patient with albinism. Even when there is no improvement in either visual acuity or nystagmus, the quality of life benefits from improved appearance alone may justify the procedure. The potential benefits of strabismus surgery need to be carefully evaluated against the risks that include postoperative ocular movement restriction and the risks associated with a surgical procedure under general anesthesia. Strabismus surgery may not be available in rural settings.

In addition to strabismus being commonly associated with albinism, it is worth noting that it is also often associated with a positive angle kappa (Merrill et al., 2004). A positive angle kappa reflects a discrepancy between the pupillary axis and the visual axis. In albinism this may reflect eccentric fixation from foveal hypoplasia (Merrill et al., 2004). A positive angle kappa means that exodeviations appear larger than they actually are and esodeviations appear smaller. Strabismus may also be incorrectly diagnosed in the presence of this finding and the measurement of ocular deviations in albinism should, therefore, be undertaken with care.

VI. REFRACTIVE ERRORS

Refractive errors are more common in albinism compared with the healthy population, with high myopia, hyperopia, and astigmatism all commonly reported (Anderson et al., 2004). Schulze Schwering et al. (2015) recently evaluated refractive errors in a large group (120 children) with albinism from Malawi in sub-Saharan Africa. All types of refractive errors were encountered, although myopia was the most common (58%), and the mean refractive error was a spherical equivalent of −4 diopters.

However, visual and structural features in siblings with albinism may differ. Heinmuller et al. (2016) investigated 111 patients in 54 families in Minneapolis. Their findings showed that stereoacuity was present in sibling pairs in 33% of cases, absent in 33%, and found in only one sibling in 33%. Best-corrected visual acuity was equal in only 20% of sibling pairs. They concluded that, although strong concordance in structural features (such as foveal grading and macular melanin) occurred in sibling pairs, there was discordance in visual function. Parents, therefore, should be counseled with care because the visual defects associated with albinism often differ in siblings and they cannot be reassured that if their first affected child has better than expected visual function, the next affected one would be similarly mildly affected.

Nystagmus may cause with-the-rule astigmatism, presumably because the cornea may be molded by the repetitive movements of the eye (Healey et al., 2013). However, it is thought that the most important component accounting for refractive errors in albinism is impaired emmetropization (Karabas et al., 2014). Emmetropization occurs when the eye changes shape to bring light rays into focus on the retina. Emmetropization appears to be controlled by choroidal growth factors that are released in response to visual stimuli and that affect scleral metabolism (Karabas et al., 2014). Foveal hypoplasia alone does not appear to mediate failure of emmetropization (Healey et al., 2013). It may rather be caused by the additional structural abnormalities associated with albinism such as reduced numbers of rod photoreceptors, reduced cone density at the macula, and reduced choroidal thickness (Karabas et al., 2014).

With refractive correction, individuals with albinism typically remain visually impaired because of foveal hypoplasia, intraocular light scatter, and nystagmus. For this reason, historically, glasses were often not prescribed in albinism. There is, however, ample evidence to suggest that, despite still having subnormal best-corrected vision, most individuals with albinism experience significant visual acuity improvement with spectacle correction (Anderson et al., 2004; Schulze Schwering et al., 2015). Furthermore, compliance with spectacle wear has been shown to be good, and refractive correction has been shown to improve stereopsis and ocular alignment (Anderson et al., 2004).

VII. MANAGEMENT

Individuals with albinism have poor vision secondary to the sensory and motor factors described above and this contributes to a poorer quality of life (Maia et al., 2015). They have an abnormal visual pathway, foveal hypoplasia, high refractive errors and astigmatism, nystagmus, and intraocular light scatter, all of which contribute to poor vision. The differing contributions of each of these factors account for wide variations in visual acuity in albinism. Some individuals may have near normal visual acuity, whereas others may be severely visually impaired (Kutzbach et al., 2009). Careful management can, however, enable anyone with albinism to maximize his or her visual potential.

General Measures

Reducing Photophobia and Glare

Glare and photophobia are among the most common ocular complaints in albinism. Avoiding high-ambient light situations is the first step for increasing comfort. For children with albinism, seating them away from windows and light sources may be beneficial. Shading the eyes with the use of a hat or cap, especially outdoors, may also help.

Glare and photophobia may both be symptomatically improved with tinted lenses.

Contact lenses with peripheral special iris tinting and a clear pupillary center may be particularly helpful because the reduction of light entering the eye reduces light scatter and foveal blur and may consequently improve nystagmus and visual acuity as well as comfort (Omar et al., 2012).

Refractive Correction

Routine cycloplegic refraction and refractive correction are obligatory in anyone with albinism. For many individuals with albinism, the provision of the correct spectacles is sufficient to increase the visual acuity and consequently quality of life.

Visual Aids

Most children with albinism should be able to manage in mainstream schools provided that their specific visual challenges are recognized and managed with appropriate visual aids. The same is true of adults with albinism in the workplace. A common complaint in albinism is difficulty with distance acuity, so seating children with albinism near the front of the class is recommended. Furthermore, supplying these children with copies of projected material may be beneficial. Individuals with albinism may struggle with light- and low-contrast material. Providing black on white material that is high-contrast is helpful.

Reading is a complex process that requires both eye movement control and visual acuity. It is impaired in albinism because of the nystagmus, the foveal hypoplasia, and the impaired binocularity. Increasing the print size improves the reading ability in individuals with albinism (Merrill et al., 2011). Also, they should be introduced to computers early because the contrast, illumination, and print size can all be controlled on an individual basis.

Low-vision aids, such as magnifiers, may be required. These may be handheld devices or may be incorporated into prescription spectacles that enhance visual acuity and quality of life (Omar et al., 2012). Referral to a low-vision specialist is recommended.

Diagnostics

The diagnosis of albinism, in the absence of typical clinical findings, may be made using electrophysiology (VEPs) that demonstrates the excessive decussation of retinal nerve fibers. OCT may also be helpful in evaluating foveal hypoplasia. Anterior segment OCT can demonstrate iris thinning that is more objective than iris transillumination grading. Where imaging and electrophysiology are not available, iris transillumination grading may be helpful as may identification of the fundoscopic features of foveal hypoplasia and the presence of nystagmus.

Specific Treatments

Strabismus

It is useful to start treating the individual with albinism and a manifest ocular deviation as early as 6 months of age. Initially the treatment may take the form of eye patching to prevent amblyopia. Occasionally strabismus may be corrected with refractive correction. Poor binocularity is a limiting factor in the success of strabismus surgery but even so, the cosmetic improvement alone can enhance quality of life. In centers where strabismus surgery is available and where postoperative follow-up can be ensured, it is recommended even though there may be no improvement in visual acuity.

Nystagmus

Nystagmus may be dampened by convergence, so prism glasses may be used to induce convergence (Thurtell and Leigh, 2012).

Pharmacotherapy has been used for the nystagmus associated with albinism. The purpose is to reduce the nystagmus without affecting normal eye movements. Baclofen may improve periodic alternating nystagmus and has been used to reduce the periodic component in the nystagmus associated with albinism (Strupp et al., 2013). Other options that may be considered include gabapentin, memantine, acetazolamide, and brinzolamide eye drops (Thurtell and Leigh, 2012). All of these options have potential side effects and the benefits should be weighed against these.

Surgery for nystagmus may be recommended in specific cases. Patients have reported a better quality of life after surgery for the nystagmus, with greater confidence from improved appearance and better work performance (Davis et al., 1997). Where there is a head turn and a null point may be demonstrated, the outcome may be more predictable. Nystagmus surgery may be combined with strabismus surgery but should be undertaken by someone with appropriate clinical experience. It may be better to perform staged surgery, correcting the head turn first and then correcting the strabismus. Nystagmus surgery may only be available in specialized centers.

Lifestyle Adjustments

As a result of visual difficulties, individuals with albinism may be limited in their ability to drive (although many may have driving vision) and in their vocational choice. With the appropriate management, however, they may maximize their visual potential.

VIII. CONCLUSION

The normal visual pathway has been described here together with the pathway in people with albinism, so that they can be compared and the many abnormalities found in affected people can be better understood. The reduction in melanin affects not only the visual system but also the structure and function of the eyes. Various defects can be identified in the iris, the RPE, neurosensory retina, and the choroid, as well as in the visual pathway. These defects result in poor visual acuity, myopia, astigmatism, reduced binocular and stereoscopic vision, photophobia, refractive errors, nystagmus, and occasionally strabismus.

Some of the visual defects found in people with albinism can be managed in various ways, which include surgery, pharmacotherapy, prescription spectacles, tinted lenses, and hats. Furthermore, visual aids such as magnifiers may be useful and practical hints, such as the provision of large-print texts and computers and seating close to the chalkboard and away from any glare, will help with functioning at school and lead to greater confidence in the schoolchild. Although quality of life can be improved in this way for people with albinism, some lifestyle adjustments have to be made to the level of visual impairment that remains.

References

Anderson, J., Lavoie, J., Merrill, K., King, R.A., Summers, C.G., 2004. Efficacy of spectacles in persons with albinism. J. Am. Assoc. Pediatr. Ophthalmol. Strabismus 8, 515–520.

Brecelj, J., 2014. Visual electrophysiology in the clinical evaluation of optic neuritis, chiasmal tumours, achiasmia, and ocular albinism: an overview. Doc. Ophthalmol. 129, 71–84.

Brodsky, M.C., Glasier, C.M., Creel, D.J., 1993. Magnetic resonance imaging of the visual pathways in human albinos. J. Pediatr. Ophthalmol. Strabismus 30, 382–385.

Castle, D., Kromberg, J., Kowalsky, R., Moosa, R., Gillman, N., Zwane, E., Fritz, V., 1988. Visual evoked potentials in Negro carriers of the gene for tyrosinase positive oculocutaneous albinism. J. Med. Genet. 25, 835–837.

Cornish, K.S., Reddy, A.R., Mcbain, V.A., 2014. Concentric macular rings sign in patients with foveal hypoplasia. JAMA Ophthalmol. 132, 1084–1088.

Cronin, C.A., Ryan, A.B., Talley, E.M., Scrable, H., 2003. Tyrosinase expression during neuroblast divisions affects later pathfinding by retinal ganglion cells. J. Neurosci. 23, 11692–11697.

Davis, P.L., Baker, R.S., Piccione, R.J., Kraft, S.P., 1997. Large recession nystagmus surgery in albinos: effect on acuity/comment. J. Pediatr. Ophthalmol. Strabismus 34, 279–285.

Dubner, M., Nelson, L.B., Gunton, K.B., Lavrich, J., Schnall, B., Wasserman, B.N., 2016. Clinical evaluation of four-muscle tenotomy surgery for nystagmus. J. Pediatr. Ophthalmol. Strabismus 53, 16–21.

Guillery, R., 1996. Why do albinos and other hypopigmented mutants lack normal binocular vision, and what else is abnormal in their central visual pathways? Eye 10, 217–221.

Healey, N., Mcloone, E., Mahon, G., Jackson, A.J., Saunders, K.J., Mcclelland, J.F., 2013. Investigating the relationship between foveal morphology and refractive error in a population with infantile nystagmus syndrome foveal morphology and refractive error. Investig. Ophthalmol. Vis. Sci. 54, 2934–2939.

Heinmuller, L.J., Holleschau, A., Summers, C.G., 2016. Concordance of visual and structural features between siblings with albinism. J. AAPOS 20, 34–36.

Jeffery, G., Schütz, G., Montoliu, L.S., 1994. Correction of abnormal retinal pathways found with albinism by introduction of a functional tyrosinase gene in transgenic mice. Dev. Biol. 166, 460–464.

Karabas, L., Esen, F., Celiker, H., Elcioglu, N., Cerman, E., Eraslan, M., Kazokoglu, H., Sahin, O., 2014. Decreased subfoveal choroidal thickness and failure of emmetropisation in patients with oculocutaneous albinism. Br. J. Ophthalmol. 98, 1087–1090.

Kutzbach, B.R., Merrill, K.S., Hogue, K.M., Downes, S.J., Holleschau, A.M., Macdonald, J.T., Summers, C.G., 2009. Evaluation of vision-specific quality-of-life in albinism. J. AAPOS 13, 191–195.

Lee, H., Purohit, R., Sheth, V., Papageorgiou, E., Maconachie, G., Mclean, R.J., Patel, A., Pilat, A., Anwar, S., Sarvananthan, N., 2015. Retinal development in albinism: a prospective study using optical coherence tomography in infants and young children. Lancet 385, S14.

Lee, K.A., King, R.A., Summers, C.G., 2001. Stereopsis in patients with albinism: clinical correlates. J. Am. Assoc. Pediatr. Ophthalmol. Strabismus 5, 98–104.

Maia, M., Volpini, B.M., Santos, D., Rujula, M.J., 2015. Quality of life in patients with oculocutaneous albinism. An. Bras. Dermatol. 90, 513–517.

Mcketton, L., Kelly, K.R., Schneider, K.A., 2014. Abnormal lateral geniculate nucleus and optic chiasm in human albinism. J. Comp. Neurol. 522, 2680–2687.

Merrill, K., Hogue, K., Downes, S., Holleschau, A.M., Kutzbach, B.R., Macdonald, J.T., Summers, C.G., 2011. Reading acuity in albinism: evaluation with MNREAD charts. J. AAPOS 15, 29–32.

Merrill, K.S., Lavoie, J.D., King, R.A., Summers, C.G., 2004. Positive angle kappa in albinism. J. Am. Assoc. Pediatr. Ophthalmol. Strabismus 8, 237–239.

Omar, R., Idris, S.S., Meng, C.K., Knight, V.F., 2012. Management of visual disturbances in albinism: a case report. J. Med. Case Rep. 6, 1.

Perez, Y., Gradstein, L., Flusser, H., Markus, B., Cohen, I., Langer, Y., Marcus, M., Lifshitz, T., Kadir, R., Birk, O.S., 2014. Isolated foveal hypoplasia with secondary nystagmus and low vision is associated with a homozygous SLC38A8 mutation. Eur. J. Hum. Genet. 22, 703–706.

Raliavhegwa, M., Oduntan, A.O., Sheni, D.D.D., Lund, P.M., 2001. Visual performance of children with oculocutaneous albinism in South Africa. J. Med. Genet. 38(Suppl. 1), S35.

Roffler-Tarlov, S., Liu, J.H., Naumova, E.N., Bernal-Ayala, M.M., Mason, C.A., 2013. L-Dopa and the albino riddle: content of L-Dopa in the developing retina of pigmented and albino mice. PLoS One 8, e57184.

Schulze Schwering, M., Kumar, N., Bohrmann, D., Msukwa, G., Kalua, K., Kayange, P., Spitzer, M., 2015. Refractive errors, visual impairment, and the use of low-vision devices in albinism in Malawi. Graefe's Arch. Clin. Exp. Ophthalmol. 253, 655–661.

Schutze, C., Ritter, M., Blum, R., Zotter, S., Baumann, B., Pircher, M., Hitzenberger, C.K., Schmidt-Erfurth, U., 2014. Retinal pigment epithelium findings in patients with albinism using wide-field polarization-sensitive optical coherence tomography. Retina 34, 2208–2217.

Sheth, V., Gottlob, I., Mohammad, S., Mclean, R.J., Maconachie, G.D., Kumar, A., Degg, C., Proudlock, F.A., 2013. Diagnostic potential of iris cross-sectional imaging in albinism using optical coherence tomography. Ophthalmology 120, 2082–2090.

Strupp, M., Kremmyda, O., Brandt, T., 2013. Pharmacotherapy of vestibular disorders and nystagmus. Semin. Neurol. 33, 286–296.

Tavakolizadeh, S., Farahi, A., 2013. Presence of fusion in albinism after strabismus surgery augmented with botulinum toxin (type a) injection. Korean J. Ophthalmol. 27, 308–310.

Thurtell, M.J., Leigh, R.J., 2012. Treatment of nystagmus. Curr. Treat. Options Neurol. 14, 60–72.

Villegas, V.M., Díaz, L., Emanuelli, A., Izquierdo, N.J., 2010. Visual acuity and nystagmus following strabismus surgery in patients with oculocutaneous albinism. P. R. Health Sci. J. 29.

Wilk, M.A., Mcallister, J.T., Cooper, R.F., Dubis, A.M., Patitucci, T.N., Summerfelt, P., Anderson, J.L., Stepien, K.E., Costakos, D.M., Connor Jr., T.B., Wirostko, W.J., Chiang, P.W., Dubra, A., Curcio, C.A., Brilliant, M.H., Summers, C.G., Carroll, J., 2014. Relationship between foveal cone specialization and pit morphology in albinism. Investig. Ophthalmol. Vis. Sci. 55, 4186–4198.

CHAPTER

8

Low-Vision Rehabilitation and Albinism

Rebecca L. Kammer[1,2]

[1]Standing Voice, London, United Kingdom; [2]Kammer Consulting, Anaheim, CA, United States

I. INTRODUCTION

In the prior chapter (see Chapter 7), attention was given to each part of the eye and the associated visual pathway impacted by albinism. As such, we will pay special attention to certain areas as they may be approached in a low-vision evaluation for the purposes of enhancing functional vision.

When someone with loss of function of an organ attends a doctor visit, the emphasis is placed on repairing or treating that organ or associated systems. However, in the case of vision rehabilitation, the emphasis is not on treating or repairing the eyes. Although a health professional's understanding of the eyes and vision are critical, the emphasis is on how the deficits in vision impact the whole person and how the low-vision doctor can ultimately improve an individual's task achievement and, more broadly, their societal context and quality of life.

In 2010, based on work with the International Council of Ophthalmology and the International Society for Vision Research and Rehabilitation, Colenbrander (2010) described the distinctions between vision function (how the eye functions), functional vision (how the person functions with his/her vision), and vision-related quality of life. Vision rehabilitation is aimed at assessing various aspects of vision function and functional vision and, indirectly, enhancing overall quality of life or supporting a person in achievement of his/her visual and functional life goals (see Fig. 8.1).

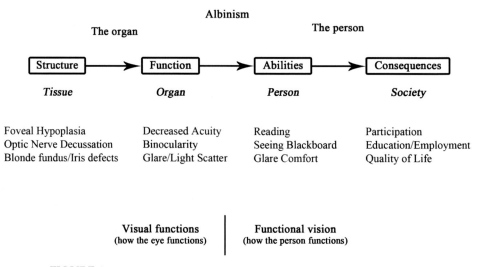

FIGURE 8.1   Comparison of visual function and functional vision in albinism.

The barriers faced by persons with albinism regarding visual impairment are both obvious and hidden. When considering albinism as the cause of impaired vision, certain aspects of vision function are impacted and cause reduced functional vision. The focus of this chapter is on how individuals with albinism are specifically impacted in daily life and how functional vision can be improved or enhanced through devices, magnification, and rehabilitation strategies.

II. LOW VISION AND REHABILITATION

Understanding the areas of vision function impacted most in the condition of albinism is important for addressing task-specific goals. Three primary areas of impaired visual function include reduced visual acuity (or detail discrimination), impaired binocular vision, and glare sensitivity. Also associated with vision function impairment is the high rate of refractive error (e.g., the need for spectacle correction or glasses). Prior to addressing deficits in function or problems with eye anatomy, it is an essential first step to determine the best options for correcting refractive error. Studies consistently emphasize that gains from spectacle correction can be made in individuals with albinism, and in this chapter, specific clinical skills and strategies are noted to make assessment and prescribing for refractive error as effective as possible.

The World Health Organization (WHO) divides low vision into moderate to severe visual impairment, which includes worse than 20/60 or 6/18 vision or impaired field of vision (e.g., constriction, hemianopsia, central scotoma). Usually this definition includes best-corrected vision with glasses or contact lenses, but in many countries where access to care is limited, uncorrected refractive error is considered a cause of impairment. Naidoo et al. (2016) reviewed 243 studies and found that uncorrected refractive error is the leading cause of vision impairment worldwide. In 2002, sub-Saharan Africa was identified as holding 11% of the world's population but represented 20% of the world's blindness (Lewallen and Kello, 2009).

At the most straightforward level, low-vision rehabilitation can be defined as one or more interventions that impact individuals with vision impairment by reducing the level of difficulty in vision-related tasks or goal performance. Although, when more comprehensive services are available, such as occupational therapy, orientation and mobility training, and other rehabilitation training, interventions are aimed more broadly at improvement of quality of life.

Visual Function: Refractive Status and Visual Acuity

The type of refractive error in persons with albinism has been reported to vary in both hyperopic (e.g., farsighted) and myopic (e.g., nearsighted) directions, but this may be due to the type of albinism and region of the world where the study was performed. For example, in Israel, in a sample of 132 people with albinism, a researcher found that the oculocutaneous albinism (OCA) 1A type resulted in a higher incidence of high hyperopia (43% of individuals) than other types (Yahalom et al., 2011). Many small studies do not specify the type of albinism but in several studies in the United States there seems to be a trend toward hyperopia. For example, in a sample of 19 individuals with albinism, hyperopic refractive error was more common; however, the range of spherical equivalent refractive errors extended from −13.75 to +7.30 D (Sampath and Bedell, 2002). Likewise, a regularly cited study in the United States includes only 25 individuals, with the data demonstrating a greater tendency toward hyperopia but also with a wide range (−10.50 to +9.13 D) and a mean occurrence of 2 D of astigmatism (Wildsoet et al., 2000). Astigmatism direction in these studies is usually consistent at an axis of 180 and corneal in nature or, in other words, the curvature of the front of the eye is steeper in the vertical meridian or direction and flatter in the horizontal representing an ellipsoid type of shape rather than spherical. Studies have proposed that this shape or curvature of the cornea is related to the nystagmus direction and the pressure of the eyelids essentially flattening the upper and lower portions of the cornea as the eyes move from side to side (Wildsoet et al., 2000).

Refractive error among individuals with albinism in Africa has been reported to be more myopic with, in some cases, extremely high myopia (−20 to −30 D) (Schulze and Kumar, 2015). Following full prescription of refractive error, reports indicate improvement in best-corrected visual acuity by 1–2 lines, with this difference making a significant functional improvement. In Malawi in a study of 120 individuals with albinism, mean improvement was two lines of acuity (Schulze and Kumar, 2015). In Tanzania, improvement depended on the magnitude and direction of the refractive error with moderate to high myopes or myopic astigmatism gaining at least two lines of acuity (Kammer and Grant, 2014). Kammer and Grant (2014) explored the distribution of uncorrected and corrected visual acuity of 237 patients and established 0.8 logMar (20/125 or 6/38) as the most frequent uncorrected acuity with 0.74 logMar (20/110 or 6/33) as the mean best-corrected acuity (See Table 8.1).

TABLE 8.1   Synthesis of Refractive and Visual Acuity Results From Population Reports From Five African Countries

Country	N	Mean or Median Age and Range	Uncorrected Acuity (logMar)	Best-Corrected Acuity (logMar)	Acuity Gain (>1 Line) (%)	Refractive Error Distribution	
						Myopic and/or Myopic Astigmatism	Hyperopic and/or Hyperopic Astigmatism
Cameroon[a]	35	12.3 (5–37yo) (71% <15yo)	0.96 (20/180) (6/54)	0.82 (20/132) (6/40)		68.6%	14.3% With-the-rule astigmatism, mean −3.62D
Tanzania[b]	237	18 (5–40) (63% <18yo)	0.84 (20/140) (6/40) −4 to <−8D 0.89 >8D 1.12	0.74 (20/110) (6/33) −4 to <−8D 0.74 >8D 0.86	68.4	68.4%	26.2%
Malawi[c]	120	12 (4–25yo)	0.98 (20/200) (6/60)	0.77 (20/125) (6/38)	78	70% (Up to −30)	41% (Up to +15)
South Africa[d]	40	50% <18yo (3–58)*			60	46.3% (Up to −20)	22.5% (Up to +5.5)
Nigeria[e]	138	23 (6–60yo)	2.15–1.0 (71%) (20/60–20/200) (6/18–6/60)			49.7% (Up to −14)	13.4% (Up to +8D) (Mean −3.00)

[a] Mean acuity published in decimal notation but converted to logMar. Study population from a hospital clinic from 2003 to 2011 (Eballe et al., 2013).

[b] Mean calculated indirectly. Mean acuity published categorically by low (<4D), moderate (4 to <8D), and high (>8D). Study population from school-based clinics from 2011 to 2014 (Kammer and Grant, 2014).

[c] Study population from children in government schools from 2010 to 2012 (Schulze and Kumar, 2015).

[d] Exact ages not provided, range estimated. Study population retrospective from a university low-vision clinic from 1993 to 1998. Sacharowitz, H., 1999. An overview of oculocutaneous albinism in South Africa. S. Afr. Optometry 58 (4), 105–109.

[e] Uncertainty if reported findings are from uncorrected or corrected acuities (categorized as % participants with moderate visual impairment) (Udeh et al., 2014).

Considering these findings of best-corrected vision ability and with refractive error trends toward high myopia, strategies for refracting in Africa in both low- and high-resource settings can be formulated.

Prescribing for near for a person with hyperopia (farsighted) and visual impairment can be completed in a few steps. First, refract the patient for distance correction in a trial frame. Next, the patient can hold his or her reading material (schoolbook) where it is easiest to see the text size (see Fig. 8.2). Then, measure the distance and take the reciprocal in meters to determine the accommodative demand (e.g., when holding material at 10 cm, 1/.10 m, or 10 D), and finally, place lenses in a trial frame and test comfort and visual clarity. Some clinicians ultimately prescribe half of the calculated demand as a near prescription (e.g., +5 D). Older students (age 12 or more) and those at university level can demonstrate care and retention for spectacles as low as +2.00 D, and case-by-case cost versus benefit can be determined by the patient/student.

In cases of high myopia or hyperopia, spectacle corrections can induce additional magnification or minification. In hyperopic cases, this additional magnification (up to 0.3× for aphakic prescriptions) can be beneficial, although the individual's eyes can appear larger by 1/3, which may be a cosmetic barrier to full-time wear (see Fig. 8.3). In myopic cases, the minification effect can further degrade the image so that although it may be more in focus, it appears smaller than wearing no correction. This minification may initially negatively impact the user and make adaptation to full-time wear also a barrier. The cosmetic appearance is also impacted in that the eyes appear smaller.

In one case in a remote village, this appearance was perceived by the community as a curse and a belief that wearing spectacles would physically "shrink the eyes" (Personal communication between the author and local community leaders). In these cases, much patient education and demonstration, with and without the glasses, is necessary. Over time, reinforced with education and family support, these myths can be overcome. If financial resources and access to care make contact lenses a treatment option, other benefits can include reduction of spectacle magnification or minification, improvement of spectacle adaptation, and reducing glare with tinted contact lenses (Omar et al., 2012).

Visual Function: Binocularity

In the few studies of individuals with albinism, reports of percentage of patients with strabismus vary widely from 8% (Udeh et al., 2014) to 53% (Hertle, 2013) or reporting is nonexistent. Agreement on the presence of strabismus is undoubtedly related to the skill of the examiners and the manner in which strabismus is evaluated in each study. It is a difficult skill to master in clinically determining the magnitude (size of deviation in prism diopters), direction (eso or exo), and constancy (intermittent vs. constant; alternating or R or L) of strabismus especially in individuals with poor fixation and nystagmus. Clinically, individuals with albinism either have a strongly preferred or better eye or they may alternate; less often, a few individuals with more pigmentation may have rudimentary binocularity. Because of the more common finding of no binocularity, low-vision prescribing and magnification strategies can often be directed to the better-seeing eye especially when a large difference in refractive error (e.g., 2 D) exists between the two eyes. Individuals may appreciate a binocular strategy (e.g., binocular telescopes) at times, but often they perform just as

FIGURE 8.2   Youth with albinism demonstrating the close viewing distance needed to see newspaper print size text. *Photograph courtesy of Dr. R. Kammer.*

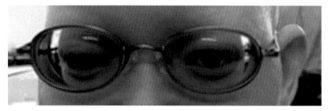

FIGURE 8.3   Teenager with albinism and spectacle correction for high myopia. *Photograph courtesy of H. Freeland.*

well with a monocular approach for the better-seeing eye. A monocular approach is helpful programmatically when considering expense and availability of low-vision devices in rural or low-economic resource areas. In addition, refractive error in the better eye can be the primary emphasis in the binocular prescription of the spectacles. For example, if the worse eye has an additional 2 D of myopia, it is likely that the worse eye has a constant strabismus and is ignored or suppressed. Therefore if that eye is underprescribed, the individual does not notice the discrepancy and the better eye is maintained as the dominant viewing eye. If the individual has an alternating strabismus pattern or seems to be aligned or possibly binocular, then it is likely that the refractive error is similar in each eye and spectacles that are made to correct each eye will be most effective. Considering the principle of just noticeable difference, an individual with 20/120 or 3/36 vision may not be sensitive to about 1.25 D of under- or overcorrected refractive error. This allows for a prescribing strategy that can be utilized in low-resource settings or programmatically when training vision specialists to serve individuals with albinism.

In a population with high-myopic refractive error, it can be difficult to impossible to obtain spectacles because of cost and import availability. For example, for an individual with a −12 D in the better eye and −16 D in the worse-seeing eye, a simple approach would be to prescribe −12 D for each eye. Although accuracy in correcting each eye maximally may be compromised, it reduces the risk of spectacle rejection due to imbalance of image sizes from each eye (aniseikonia) of potential disuse or complaints of glasses being "too strong or uncomfortable" to wear. One strategy to consider when prescribing for the better eye is to ask the individual if they are aware of which eye is better or if they believe they use both eyes equally. Next, evaluating strabismus with cover testing can confirm alignment and potential binocularity versus constant strabismus. Finally, refraction and visual acuity testing can help the clinician decide on monocular versus binocular prescribing strategies.

Visual Function: Glare and Contrast Sensitivity

For individuals with albinism, reduced pigment in the eye causes light to not be absorbed or restricted compared with other individuals, and bright light can cause glare sensitivity and extreme discomfort (Van den Berg, 1986). Glare sources (e.g., light from a window directly on the classroom blackboard) can also cause reduced contrast because of the veiling glare. Contrast is the difference between an object and the background of that object. In daily life, contrast is critical for vision; for example, for a student to see white chalk writing on a blackboard is not typically challenging, but for an individual with vision impairment, high contrast may be critical to reading the information.

An individual's contrast sensitivity (CS) function is a scientific description of his/her ability to detect different size objects when also varying levels of contrast. In albinism, some studies suggest that the CS function is decreased in persons with albinism. A reduction in CS would mean that for a child with albinism, the highest levels of contrast would help him/her perform daily vision tasks. However, just placing more light on the task to enhance contrast, although logical, would be a mistake due to his/her light sensitivity and potential for veiling glare. So, for a classroom that uses orange chalk on a dirty green chalkboard, a student with albinism may not be able to read the print at all. A solution would be to keep the chalkboard clean and use white chalk to create the most contrast possible. Also, avoiding seating the child near a window would also help reduce the chance of compounded disability glare.

Although albinism has several areas of decreased visual function, it is the interaction or combination of several impairments that reduce individual ability to perform visual tasks. The next section explores strategies to enhance vision function either in one area or multiple areas by using magnification and glare control. The low-vision specialist must have a good understanding of how to manage refractive error (prescribing of glasses or contact lenses) and binocularity (how the eyes work together or separately) along with how to enhance vision through magnification and glare control. In fact, enhancing vision with one strategy can often impact multiple vision functions. For example, Loshin and Browning (1983) enhanced CS when using telescopic magnification (due to retinal image size increase), but they were surprised to find an additional increase in the peak sensitivity of the curve. This means that magnification could dually help enhance CS and resolution of small objects (e.g., reading small print on yellowed paper) simultaneously. Another example includes the use of tinted or opaque contact lenses to successfully correct refractive error and reduce glare and discomfort in individuals with albinism and improving vision by significantly more than just with tinted spectacles (Phillips, 1989). Low-vision specialists are trained to consider the complexities of how a visual task requires multiple visual functions and how the condition itself impairs vision function, while also considering the most optimal solutions given an individual's unique life and goals.

Functional Vision: Magnification

To improve functional vision, the goal is to reduce task difficulty (e.g., reading notes on the blackboard), so it is critical to first provide appropriate refractive correction to obtain best possible vision. The second step is to provide magnification, enhancing the visual image or optimizing the visual environment. Other common goals may be to see small print in books, to recognize faces, or to distinguish facial expressions. Seeing road signs and small landmarks when driving are other common goals that may not be identified initially due to uncertainty as to whether they can be attained.

Typically, individuals with albinism have a positive response to magnification. The three types of magnification commonly utilized include optical magnification (e.g., telescopes, near magnifiers), relative distance magnification (bringing objects closer), and relative size magnification (enlarging size of object, e.g., large print notes). Optical magnification can be utilized at near as in the case of a dome or hand-held magnifier, or it can be at distance as in the case of the use of a monocular telescope. For distance, the optics is straightforward with the magnification printed on the device representing the angular change seen by the individual who is wearing his or her spectacle correction. For example, a 4× telescope represents an object that appears 4× larger in size.

For near, the optics are more complicated and the number printed on the side of the magnifier may represent magnification as calculated in specific ways (e.g., conventional magnification or rated magnification). Representation of the device in diopters is the most appropriate for a near device and this allows low-vision specialists who are familiar with these meanings to also consider the spectacle correction (correction of refractive error or lack thereof) before prescribing the appropriate power device. For children, accommodation or the ability to change the power of the eye to create a simulated type of reading lens is more abundant than for an adult of 40 years of age. A visually impaired child naturally holds objects closer to the eye to create relative distance magnification (objects that are closer subtend a larger angular size on the retina and appear larger), but how close and how long they can sustain that distance is questionable. The younger the child, the more flexible is the accommodative system, but in visually impaired children there is some question of how accurate and sustainable is their ability to accommodate or focus their vision for near work.

For individuals with albinism, it is common to see students get closer to the textbook or walk up to the blackboard. If the student is wearing corrective lenses (for myopia, hyperopia, and/or some form of astigmatism) and gets closer to an object, they are purely creating magnification. However, this issue gets more complicated for students not wearing their spectacle correction. A child with myopia naturally has more positive power in the ocular system and a reading aid is naturally simulated without wearing any spectacle correction. This makes it easier to see objects at near as less-accommodative demand is created, but the magnitude of the myopia needs to be considered. For myopia greater than 10 D, for example, the point of clarity at near will be within 10 cm of the eye, but this means that any object further from the eye than that will be blurred on the retina, and this blur is much worse at distance (e.g., blackboard). Some children with high myopia are often mistaken as blind or severely visually impaired and taught Braille. If the student is hyperopic, the eye has less positive power and must accommodate on an everyday basis just to see distant objects in focus. For lower powers of hyperopia (e.g., +2 D), this is not difficult for younger individuals to maintain, but as the age and demand of near work increases (e.g., reading for school), sustaining ongoing accommodation even at distance can cause visual fatigue, headaches, and blur. Although refractive correction for a low to moderate hyperope (e.g., <5 D) may not improve distance visual acuity in an ocular examination, many patients prefer to wear their spectacle lenses for distance for visual comfort. Hyperopes, however, are in more need of near correction so that they can hold print or objects closer and sustain that distance and focus when reading. They may be able to see small print size at near during vision testing at school or in a clinic, but to sustain reading that size of print during the school day can be extremely challenging. Often the best option may be to forego the distance spectacles for hyperopia and prescribe a near reading pair that is higher than usual to reduce some of the eyestrain or accommodation needed.

For reading spectacles or magnification without associated hyperopia or myopia, one common approach to prescribing is to offer half of the near focusing demand in the form of spectacles. Based on experience in running these low-vision clinics in Tanzania, the author recommends evaluating near addition lenses individually and in consideration of symptoms such as headache and visual fatigue. This visual fatigue is much greater in the case of a child with hyperopia and much less in the case of myopia. With myopia, a near point of focus is naturally induced but in high myopia that near point may be too close to the spectacle plane and limit the student's range of reading ability. For example, a child with extremely high myopia, e.g., −20 D distance prescription, would have a near position of best clarity at 5 cm. This may be helpful for some reading but may be too close for any computer or book reading based on posture and neck angle. In these cases, a reduced prescription that provides a near boost such as −16 D might be useful—it provides clarity with some blur at distance but also allows a near clarity point at 1/4 D or 25 cm.

One study found that 60% of the sample demonstrated reduced focusing ability (accommodation) and benefit from a reading pair of spectacles for comfort and clarity and reading performance (Merrill et al., 2011). Reading speed was also found to be similar to speeds of nonimpaired young adults if the print size was large enough. Another study demonstrated that children and adults with albinism have normal to superior reading skills with some loss in skills with greater impairment ranges (MacDonald et al., 2012). The authors suggest increased time to assist reading task completion in those situations.

The use of reading lenses may not be readily adopted because of immaturity, loss, theft, breakage, and confusion about when to wear the glasses and so all these considerations may be considered before a prescription is recommended.

Functional Vision: Telescopes

Telescopes are optical systems that utilize multiple lenses in combination to create angular magnification when viewing distant objects. This type of magnification is for viewing of objects ~2 m and beyond. In telescopes designed specifically for visually impaired individuals, the amount of magnification can vary from 2 to 10× but with increasing magnification in each device, the field of vision becomes smaller.

Telescopes can be modified for near viewing by changing the distance between the lenses (twisting the telescope), but only certain designs enable this much change in lens settings. Another method is to add a lens on the outside of the telescope (reading cap).

The majority of telescopes are used for distance viewing and can be clipped onto the outside of spectacles or drilled into spectacle lenses for viewing either straight ahead (e.g., for watching TV or sporting events) or in the upper portion of the lens as a bioptic and mounted at an angle for spot viewing. One unique type of biocular telescope, the BITA, has been used successfully in individuals with albinism as a lightweight option compared with other bioptics and more cosmetically acceptable because of its small design. It is also commonly prescribed with a gradient tint on the spectacle lens in which it is mounted to help with image awareness, but this tint is also helpful to reduce glare for individuals with albinism (Omar et al., 2012). Other, more common bioptic designs used for albinism typically include the widest possible opening for light to travel through the telescope into the eye, so that when viewing with nystagmus, as much light and image viewing time as possible is available. These custom designs in spectacles and telescopes can be quite expensive and require special labs and companies that are only located in certain countries with highly populated areas. A simple and effective option is the design of a hand-held monocular telescope. Monocular telescopes can be obtained in various powers, but for classroom use to see the instructors face or see notes on the blackboard, most telescopes range from 2 to 6×. With training and practice, most children with albinism can easily adapt to using a monocular in the classroom (see Fig. 8.4). In Tanzania, Kammer and Grant (2014) found that 3 and 4× monocular telescopes were the most useful for OCA2 school-age children with albinism.

FIGURE 8.4   Teenager with albinism using monocular telescope for magnification of distant objects. *Photograph courtesy of H. Freeland.*

Functional Vision: Driving

Driving with visual impairment can be a controversial topic, but it is of high importance for individuals who seek independence and full participation in society. For persons with albinism and specifically those with type II albinism who tend to have better visual acuity than individuals with type I albinism, it may be possible to drive pending local regulations. Three aspects of albinism, which are more favorable for potential driving when compared with other causes of low vision, include stable and nonprogressive vision loss, no visual field loss (blind spots), and adequate color vision. In locations where licensure may be granted, it is usually when vision is better than 6/30 best-corrected vision (e.g., with spectacles if needed). Although driving with low vision (and spectacles when necessary) may be possible in certain situations (e.g., driving in familiar areas, driving during the day), in some states in the United States and in several other countries, bioptic telescopes are often used to assist with magnifying small objects such as print on road signs or for the identification of landmarks (Alabdulmunem et al., 2007). Although the use of bioptics for driving are controversial or even illegal in some countries, there have been studies that demonstrate effective driving competency with the use of bioptics including those in persons with albinism (Owsley et al., 2014). Training in the use of bioptics is an essential aspect of driving and can include up to 30 hours. In one study, bioptic drivers with acuity better than 6/60 (20/200) performed as well as non-impaired drivers but had more difficulty in poor visibility conditions or in unfamiliar areas (Owsley et al., 2014). They drove less often than normally sighted drivers, but spacing and timing of driving reactions were similar to those of normally sighted drivers.

Functional Vision: Photophobia and Glare

Yellow filters are often used in patients with low vision to reduce or absorb short wavelength light scatter that can reduce image quality and to indirectly increase contrast of viewed objects (Wolffsohn et al., 2000). In albinism, this strategy has been tested by Provines et al. (1997) who explored the impact of using two different yellow filters on CS function and found no conclusive change. Nonetheless, the authors argued for the potential of subtle, real world viewing vision improvement. This potential benefit should be weighed with a filter option that reduces the most prominent visual symptom, photophobia. Hoeft and Hughes (1981) found that all eight patients in their study with albinism preferred a 14% or 7% transmission amber-colored filter over other colors (gray green, dark gray green, dark green). Eperjesi et al. (2002) conducted a review of literature to determine if a certain evidence-based prescribing protocol could be found for prescribing filters in low vision. They determined that according to only a few, small uncontrolled studies on the topic the selection of filters would still be based on the clinician making subjective decisions as per the patient.

In Africa, especially in rural areas where outdoor activities are more common than in urban settings, reduction of discomfort from glare and photophobia is an urgent and straightforward problem to solve. Although, some sunglasses may be expensive for resource poor individuals, large donations from nongovernmental organizations can make brown or gray sunglasses available sporadically throughout Africa. Although the quality of lenses vary greatly in these donations, the tint itself can be helpful in reducing light sensitivity. The primary problem is that these sunglasses are plano or without refractive correction, so students or adults must remove their indoor spectacles to wear tinted sun lenses. There are, however, different types of sunglasses available that can help with the problem of combining with thick spectacle lenses such as a fit-over design that can be placed in combination or over spectacles. This type of design is most valuable for individuals with highly myopic prescriptions because if they remove the spectacles to replace with nonprescription sunglasses, the individual must choose between reduction of photophobia and brightness as a trade-off to major blur without their correction. Many people with albinism choose to reduce brightness/glare but this can create adjustment issues for an individual who regularly switches between spectacles. Another option is to have a clip-on sun filter for the high prescription distance spectacles.

Other choices in modern lenses that are generally costly and less accessible include transition or photochromic lenses and polarized lenses. These lenses and other specialized coatings or trademark-tinted colors may be helpful, but expense must be balanced with the trade-off in visual gain.

III. ACCESS TO VISION CARE

A general eye examination usually includes assessing any visual deficits or problems. In the case of a person with albinism, several areas of the eye are impacted resulting in nystagmus, foveal hypoplasia, and often strabismus. These defects may cause confusion or uncertainty in examination strategies of a general eye care provider. It is recommended that general eye health and visual system is evaluated (e.g., best-corrected vision, binocular vision, intraocular pressure, anterior and posterior segment health, and visual fields) and then referral is made to a low-vision specialist for a more detailed assessment of refractive error, binocular vision, and enhancement of visual function.

As albinism is more common in sub-Saharan Africa than in most of the world, local low-vision specialists should all have a special segment in their training dedicated to learning specific strategies for treating albinism and the most common types of albinism in Africa.

Health care programs specific to addressing skin care and vision needs of individuals with albinism are limited or nonexistent depending on the country. Hong et al. (2006) reported the results of a WHO survey of 12 countries regarding national resources related to albinism. Study findings indicated a lack of trained health professionals and other barriers to care including lack of finances and education, as well as high cost of medical treatment and visual devices. The survey respondents also believed individuals with albinism did not understand their own condition and health needs and that most people with albinism were in the lowest economic status group. For vision, if any assistance was provided, it was in the form of sending children with albinism to special schools for the blind.

Within Africa, eye care provision by optometry and ophthalmology is among the most poorly distributed specialties among the health care areas (Oduntan et al., 2015). Optometry is a primary eye care profession and can meet the needs of rural areas of Africa, especially when countries have few trained surgical ophthalmologists who are clustered in urban hospitals centralizing care. Although ophthalmic nurses and refractionists often perform basic services to supplement ophthalmology in many African countries, low-vision services require more advanced skills and training to manage the complex problems associated with vision impairment. Optometry is the primary profession providing low-vision services for visually impaired patients in many continents, but in Africa only seven optometry schools exist among the 54 countries.

Access to care in African countries varies by nation, but, in general, low–gross domestic product countries provide limited to no specific vision or low-vision national programs for individuals with albinism. If centers for low-vision rehabilitation exist and provide access to remote locations, there may not be a need for specific programs for this population. However, specialists would need to be educated on current low-vision approaches that include the very specific needs of persons with albinism.

One example of a unique collaborative low-vision care program in Tanzania is through a nongovernmental organization, Standing Voice. In Tanzania, albinism is poorly understood and associated with many myths and beliefs that have resulted in 76 murders and 72 mutilations and attacks since 2006. Most victims have been children and those not directly victimized struggle for access to education, health care, and employment. In early 2014, Standing Voice and the present author (Dr. Rebecca Kammer, an American low-vision and albinism expert) collaborated to design and deliver an in-country vision program. The in-country team includes local staff and Tanzanian public sector optometrists. Dr. Kammer visits up to twice a year to provide training and support to Tanzanian optometrists who regularly provide the vision care within the program. Key partners include Kilimanjaro Christian Medical Center School of Optometry, the Tanzania Optometric Association, and the Tanzania Albino Association. Children with albinism are identified by Standing Voice through local schools, local media, and albinism associations. Materials for the program (e.g., spectacles and monoculars) are donated or purchased at low cost through key ophthalmic partners and manufacturers. Monocular telescopes (usually 4×) have been sourced from a global manufacturer at a cost of approximately US$12, including shipping, which compares favorably to a typical cost in the United States of approximately $80.

Standing Voice has also trained a network of 64 teachers as "Vision Ambassadors" to track children's progress and offer support in schools. Partnerships with the Tanzanian government occur at the local level with provision of vision event space, assignment of Special Education Officers to support the Vision Ambassadors, and provision of the base optometry salaries. Data are coordinated and collected using an online database specifically designed for both internet access and remote offline use.

Three to four major clinics are conducted annually to provide each patient with an efficient low-vision evaluation; assistive devices including monocular telescopes, sunglasses, and prescription glasses; education in the management of low vision and use of these devices; and registration in an online cloud database. Dr. Kammer included specific prescribing and clinical strategies based on knowledge gained from volunteering with another nonprofit organization, Under The Same Sun, from 2011 to 2014 (Kammer and Grant, 2014). One of the keys to the success of the program to date includes the use of a premade kit of several hundred spectacles that range in lens powers, size of frames, and pupillary distances. The range in lens powers was selected based on the initial hundreds of examinations and prescribing patterns with high rates of myopia and astigmatism (Kammer and Grant, 2014). Considering the strategy of prescribing for the better eye, both lenses in each spectacle are identical in power to each other. Astigmatism can make it very complicated for prescribing and ordering spectacles, but using the knowledge that most individuals with albinism have a specific direction of astigmatism, most patients have been matched within 1 D of their subjective refraction. This premade kit allows for immediate dispense and individualized education about one's vision. If more than 1 D of difference is found in the kit from what the patient needs, a custom pair is ordered from local partner organizations across Tanzania for very low cost and a trained program optometrist dispenses all the spectacles within 2 months of the low-vision event.

By late 2016, the program had served 1000 people in 11 locations across 6 regions of Tanzania with goals of serving an additional 1000 by late 2017 (http://www.standingvoice. org/programmes/vision). The vision program is unique in Tanzania because it builds the capacity of local stakeholders (optometrists, teachers, government) to conduct specialized clinics and incorporate program elements within existing vision clinics nationwide. Standing Voice and Dr. Kammer regularly share program pearls regionally and globally to facilitate replication of the program elsewhere. The program has innovated a scalable model that reaches record numbers of an underserved population while building global knowledge of visual impairment.

IV. SCHOOL AND CLASSROOM RECOMMENDATIONS

Reading (near task) and seeing instructor notes on a board (distance task) are important steps to gaining visual information and ultimately achieving academic success. Although alternative methods can be used (e.g., Braille, auditory), generally, it is preferable for students to read in a similar fashion to student peers and to be able to retrieve input from all types of materials for maximum information gain. There are also potential learning gains from incidental learning, which includes learning outside of the classroom in more informal ways (e.g., reading signs, package labels, internet searching). Many children with albinism, in Africa, have been placed in schools for the blind (Gaigher et al., 2002), and although they may have special attention for vision impairment, the attention is not always beneficial. Profoundly impaired or totally blind children require instruction using Braille, and other differences, in the classroom, can be detrimental to children with albinism. High levels of lighting can be helpful for severely visually impaired children, for example, but for children with albinism who are sensitive to bright lights, these levels are disturbing and visually compromising.

Although outdated, a study of 230 students from several schools for the blind in Kenya and in Uganda reminds us of the realities of rural education and education from a blindness perspective (Tumwesigye et al., 2009). The study explored how many students within the schools for the blind were actually visually impaired and not considered "blind" once they received an eye examination, spectacles, and magnifiers. Of the 230 students, 140 were determined to be low vision but not blind and most of those (110) were able to read enlarged print. In spite of this, all students were taught using Braille only. The study authors proposed that this was because of several factors including teachers hired in the schools being totally blind and most comfortable with blind approaches, high cost of magnifiers or enlarging print, or lack of current knowledge about low-vision approaches in the classroom. Learning with Braille or in blindness strategies is highly limiting for a person with vision in obtaining both higher education and employment opportunities.

In South Africa, there is an example of a school for children with visual impairment where the staff use sighted teaching strategies but with adaptations (Lund and Gaigher, 2002). The school also put specific measures in place for students with albinism. The majority of the students had albinism (112 of 131) and classroom modifications included glare reducing measures, moveable blackboards, and verbal encouragement for students to approach the blackboard. Covered walkways and trees were also part of the outdoor environment to reduce sun exposure. Unfortunately, only a certain percentage of students had spectacles and no specification was provided as to whether they were for full-time or part-time, near or distance. No low-vision devices were mentioned as being available. In this adaptive environment, 38 students (ages 11–16) completed a survey indicating that they were well cared for and that they preferred this type of special accommodating school during their primary school years but felt that these accommodations were less necessary in secondary school. The risk, however, in isolating children with visual impairment from mainstream schools includes the great risk of isolating them from society, which is one of the core issues limiting daily life in general.

Integrating children with albinism and visual impairment in public schools can be beneficial for social integration and visual learning exposure, but instructors would need to allow or support modifications to enable the students to thrive. Children with albinism often have reduced focusing ability or accommodation and reading skills (speed, comprehension), which ultimately impact their speed compared with their normally sighted peers. Therefore, in schools with both normally sighted and visually impaired students with albinism, extra time to complete tasks would be beneficial (e.g., 1.5 times longer). The size of the print does make a difference in speed and comprehension as well. Most children with albinism can achieve a normal reading speed through the use of appropriate low-vision management magnification devices or moving the print closer by 2–3 times (Lovie- Kitchin and Whittaker, 1999). The magnification strategy must be considered after their distance refractive error (glasses prescription) is corrected. A student's classroom or homework reading situation should be considered in concert with the low-vision optometrist or ophthalmologist recommendation. For example, if a child with significant hyperopia and astigmatism is not wearing glasses at all, then the near demand of reading is much greater and it may not be possible for them to maintain the extra strain of near viewing or the clarity of focus without glasses at a regular reading distance much less bringing the text 2–3 times closer for relative distance magnification.

In reviewing the small studies of refractive error in East Africa and the higher prevalence of myopia in these OCA2 populations, it is possible to recommend some guidelines for classrooms and inclusive education strategies (see Table 8.2).

Based on Table 8.1, uncorrected mean visual acuity ranged from 6/40 to 6/60 and best-corrected mean acuity had a tighter range of approximately 6/33–6/40 (or 20/120–20/132). Assuming then an approximate best acuity of 6/38 or 20/125 for distance vision, with glasses correction, magnification needs become ~3×. Three times magnification is based on the goal of improving vision from 6/38 (20/125) to ~6/12 (20/40), which is helpful in reading most small print on blackboard. In a vision program, Kammer and Grant (2014) utilized 3 and 4× monocular telescopes to meet the majority of needs in students and adults with albinism. Another useful accommodation in the classroom to aid in distance viewing (seeing the blackboard or facial expressions of the instructor) is appropriate seating. Moving twice as close to the board is equivalent to using a 2× telescope, so a simple vision support strategy is to allow students with albinism to sit in or near the front of the classroom. This also enables the student to use their telescope (or monocular) to see the board with unobstructed views.

TABLE 8.2 Classroom Guide

Vision Task	Vision Function	Classroom Modification/Low-Vision Devices (Functional Vision)
Reading notes/books in class	Resolution/low vision Contrast sensitivity	• Allow leaning over or close viewing of paper/books • Allow head turn to use best eye if one eye preferred • Wear prescribed spectacles for near if applicable • Use hand or stand magnifier if available and preferred • Allow or provide tilted reading stand on the desk to move print closer to eye (reduces neck strain) • Provide large-print books (sized 2–3 times larger than usual) • Provide copy of teacher lecture notes ahead of time
Seeing notes on blackboard	Resolution/low vision Contrast sensitivity	• Wear spectacles prescribed for distance viewing • Clean blackboard regularly, use white chalk • If white board, shade window to reduce reflections/glare. Use dark markers • Sit in front row • Use 3 or 4× monocular telescope • Compare notes with a peer/classmate to check for accuracy • Instructor writing in block letters, test the size for student visibility
Lighting/squinting	Photophobia Glare from windows	• Sit away from bright windows (let student select seat) • Overhead lighting should be evenly distributed (lights placed in the ceiling or around edges of room) • If available, allow a flexible neck floor or desk lamp if room is extremely dim. • Allow wearing of sunglasses or tinted spectacles in class if student is especially light sensitive (should be worn over distance spectacles and coordinated with vision specialist)
Best seating position	Nystagmus, null point Strabismus, better eye use	• Allow student to turn head to any position to see the board • Allow student to sit on the side or center of the room to improve viewing angle for a head turn
Test taking	All the above	• Allow 1.5–2× time for completion, if written test questions or instructions on board, create a printed paper copy for students with albinism to view at near. • Ensure seating in class is away from glare/windows • Allow use of monocular telescope if transcribing from the blackboard
Computer/tablet use	Resolution/low vision Contrast sensitivity Glare sensitivity	• Wear spectacles for near, allow leaning toward the monitor • Allow student to change the monitor settings to reduce background glare • Allow student or change settings to enlarge text (accessibility modifications in the computer "settings") • Seat the computer away from a window/light so that glare is not reflecting off the monitor or coming from behind the monitor • When using internet, often using the mouse by holding down the left button while using the scroll knob enlarges print on websites • Place enlarged print and reverse contrast keyboard stickers on the keyboard keys

Another strategy involves allowing the student to select a seat that produces the best lighting and visual comfort. As students with albinism are highly sensitive to glare and bright light from windows, a seat away from a window or facing away from a window is beneficial. In addition, vision tends to be worse with higher amplitudes of nystagmus, but there is often a null position in which a head turn or facial posture can slow the amplitude and allow better visual fixation and more visual information gathering. For many individuals, this position is related to convergence, but for some individuals, the null position can be in a dramatic position that requires an angled head position or head tilt in a particular direction. Instructors should allow the child to utilize their best eye and head posture and the corresponding classroom seating position that may assist in best potential vision.

For the near-vision needs of children with best visual acuity in the 6/38 (20/125) range, 3–4× magnification is adequate to enable seeing a small print size and to help students achieve maximal reading speed. This magnification can be provided by holding material 3–4× closer than the standard viewing distance (e.g., 10–13 cm from the spectacle plane), enlarging print optically with magnifiers, and enlarging print itself (relative size magnification). If holding material closer, a student would need about 10 D of accommodation to hold this posture for reading and this magnitude can be fatiguing depending on the age and visual ability of the student. Several strategies exist for low-vision specialists to prescribe reading spectacles for a child, so it is important that the teacher or the school communicate with the eye doctor/low-vision specialist for how spectacles are to be used in the classroom. Older students (e.g., 12 years and older) may be more knowledgeable about how and when to use various spectacles, but children often need reminding or reinforcement from the parent and the teacher to effectively utilize low-vision spectacles or magnifiers in school.

Other magnifiers for near include hand-held magnifiers or stand magnifiers. Both options require some knowledge of the individual's refractive error and best-corrected visual acuity to select the proper power and viewing distance; however, a small hand-held magnifier can be an inexpensive tool for spot reading. For the task of reading books or notes for more than a few minutes, hand-held magnifiers are not appropriate as they require steady hands and a small viewing window. Stand magnifiers, a lens mounted above the reading material on some type of legs or stand, eliminate the need for one hand to hold the device but there is still need to move the print under the stand or move the device itself once the print is out of the field of vision. Stand magnifiers can be costly, unavailable in certain areas of Africa, and they must be well cared for to reduce scratches and preserve their use for several years. These factors make their regular uptake in classroom use challenging.

Video magnifiers or closed-circuit televisions can be another useful category of magnification devices, but these devices require electrical and/or battery power for full use of their variable magnification, contrast, and brightness settings.

Early integration of computers as a tool in the classroom is a powerful aid to the visually impaired student. The multiple benefits include: built-in magnification features; the convenient ability to enlarge text as one is writing or reading; and the ability to control contrast or colors of the monitor background to reduce glare and enhance contrast. Early computer adoption also allows for increased speed in access to learning material and communication with others in preparing for essential higher education and employment skills.

V. CONCLUSION

This chapter synthesizes small studies across Africa along with the author's personal experience of conducting low-vision clinics in East Africa. The goal of this synthesis is to provide practical strategies for assisting individuals with albinism in vision rehabilitation. By examining and understanding the deficits in visual function produced by albinism and the specific characterization of most forms of albinism present in Africa, rehabilitation can be directed at improving functional vision and indirectly supporting social inclusion and improving quality of life.

Individuals with albinism in Africa face many barriers including limited access to modern low-vision medical expertise and to low-vision devices, but with practical approaches, there are solutions to these access gaps. Solutions include initiating or expanding training programs for eye care professionals who can understand the structure of the eye, the intricacies of vision, the principles of optics, and magnification and who apply that knowledge to enhancing functional vision in daily life. This workforce can start in optometry or ophthalmology and could be supplemented with teamwork from orthoptists, low-vision therapists, or other individuals from the fields of medicine or education. Another step in bridging the access gap includes conducting low-vision examinations and simultaneously providing patient education that supports individuals' knowledge about their own condition and how to advocate for themselves in all areas of life. Access to spectacles and magnification devices is another potential barrier, but there are strategies that can reduce the challenges. Strategies include prescribing for the better eye, collaborating with private organizations to provide an ongoing supply of inexpensive spectacles and sunglasses, supplying one or two inexpensive magnifiers (e.g., 4× monocular telescope) per individual, and prescribing reading glasses when appropriate.

For education attainment, integrated schools offer many benefits that may translate to social inclusion and employment attainment. This attainment, however, is completely dependent on incorporation of a low-vision rehabilitation plan that includes spectacles or corrected refractive error, especially when the rates of high myopia are astounding, simple magnification and glare control. Successful integration also depends on classroom modifications and the training of teachers about these modifications.

References

Alabdulmunem, M., Kollbaum, E., Rone, A., 2007. Characteristics of drivers fitted with telescopes for driving. Investig. Ophthalmol. Vis. Sci. 48 (13), 3568 Abstract, ARVO, Ft. Lauderdale, Fl, USA.

Colenbrander, A., March 2010. Assessment of functional vision and its rehabilitation. Acta Ophthalmol. 88 (2), 163–173.

Eballe, A., Mvogo, C., Noche, C., Zoua, M., Dohvoma, A., 2013. Refractive errors in Cameroonians diagnosed with complete oculocutaneous albinism. Clin. Ophthalmol. 7, 1491.

Eperjesi, F., Fowler, C., Evans, B., 2002. Do tinted lenses or filters improve visual performance in low vision? A review of the literature. Ophthalmic Physiol. Opt. 22 (1), 68–77.

Gaigher, R., Lund, P., Makuya, E., 2002. A sociological study of children with albinism at a special school in the Limpopo province. Curationis 25 (4), 4–11.

Hertle, R., 2013. Albinism: particular attention to the ocular motor system. Middle East Afr. J. Ophthalmol. 20 (3), 248–255.

Hoeft, W., Hughes, M., 1981. A comparative study of low-vision patients: their ocular disease and preference for one specific series of light transmission filters. Am. J. Optom. Physiol. Opt. 58 (10), 841–845.

Hong, E., Zeeb, H., Repacholi, M., 2006. Albinism in Africa as a public health issue. BMC Public Health 6, 212.

Kammer, R., Grant, R., 2014. Albinism and Tanzania: development of a national low vision program. Visibility 8 (2), 2–9.

Lewallen, S., Kello, A., 2009. The need for management capacity to achieve VISION 2020 in Sub-Saharan Africa. PLoS Med. 6 (12), e1000184.

Loshin, D.S., Browning, R.A., 1983. Contrast sensitivity in albinotic patients. Am. J. Optom. Physiol. Opt. 60 (3), 158–166.

Lovie-Kitchin, J., Whittaker, S., 1999. Prescribing near magnification for low vision patients. Clin. Exp. Optom. 82 (6), 214–224.

Lund, P., Gaigher, R., 2002. A health intervention programme for children with albinism at a special school in South Africa. Health Educ. Res. 17 (3), 365–372.

MacDonald, J., Kutzbach, B., Holleschau, A., Wyckoff, S., Summers, C., 2012. Reading skills in children and adults with albinism: the role of visual impairment. J. Pediatr. Ophthalmol. Strabismus 49 (3), 184–188.

Merrill, K., Hogue, K., Downes, S., Holleschau, A., Kutzbach, B., MacDonald, J., Summers, C., 2011. Reading acuity in albinism: evaluation with MNREAD charts. J. AAPOS 15 (1), 29–32.

Naidoo, K.S., Leasher, J., Bourne, R.R., Flaxman, S.R., Jonas, J.B., Keeffe, J., Limburg, H., Pesudovs, K., Price, H., White, R.A., Wong, T.Y., Taylor, H.R., Resnikoff, S., Vision Loss Expert Group of the Global Burden of Disease Study, 2016. Global vision impairment and blindness due to uncorrected refractive error, 1990–2010. Optom. Vis. Sci. 93 (3), 227–234.

Oduntan, O., Mashige, R., Ovenseri-Ogbomo, G., 2015. Strategies for reducing visual impairment and blindness in rural and remote areas of Africa. Afr. Vis. Eye Health 74 (1).

Omar, R., Idris, S., Meng, C., Knight, V., 2012. Management of visual disturbances in albinism: a case report. J. Med. Case Rep. 6, 316.

Owsley, C., McGwin, G., Elgin, J., Wood, J., 2014. Visually impaired drivers who use bioptic telescopes: self-assessed driving skills and agreement with on-road driving evaluation. Investig. Ophthalmol. Vis. Sci. 55 (1), 330–336.

Phillips, A., 1989. A prosthetic contact lens in the treatment of ocular manifestations of albinism. Clin. Exp. Optom.

Provines, W., Harville, B., Block, M., 1997. Effects of yellow optical filters on contrast sensitivity function of albino patients. J. Am. Optom. Assoc. 68 (6), 353–359.

Sampath, V., Bedell, H., 2002. Distribution of refractive errors in albinos and persons with idiopathic congenital nystagmus. Optom. Vis. Sci. 79 (5), 292–299.

Schulze, S., Kumar, D., 2015. Refractive errors, visual impairment, and the use of low-vision devices in albinism in Malawi. Graefe's Arch. Clin. Exp. Ophthalmol. 253 (4), 655–661.

Tumwesigye, C., Musukwa, G., Njuguna, M., Shilio, B., Courtwright, P., Lewellan, S., 2009. Inappropriate enrolment of children in schools for the visually impaired in east Africa. Ann. Trop. Paediatr. 29 (2), 135–139.

Udeh, N., Eze, B., Onwubiko, O., Arinze, E., Onwasigwe, R., Umeh, R., 2014. Prevalence and profile of ophthalmic disordersin oculocutaneous albinism: a field report from South-Eastern Nigeria. J. Community Health 39 (6), 1193–1199.

Van den Berg, T., 1986. Importance of pathological intraocular light scatter for visual disability. Doc. Ophthalmol. 61 (3–4), 327–333 PMID:3948666.

Wildsoet, C., Oswald, P., Clark, S., 2000. Albinism: its implications for refractive development. Investig. Ophthalmol. Vis. Sci. 41, 1–7.

Wolffsohn, J., Cochrane, A., Khoo, H., Yoshimitsu, Y., Wu, S., 2000. Contrast is enhanced by yellow lenses because of selective reduction of short-wavelength light. Optom. Vis. Sci. 77 (2), 73–81.

Yahalom, C., Tzur, V., Blumenfeld, A., Greifner, G., Eli, D., Rosenmann, A., Glanzer, S., Anteby, I., 2011. Clinical science: refractive profile in oculocutaneous albinism and its correlation with final visual outcome. Br. J. Ophthalmol.

Further Reading

Brilliant, M., 2015. Albinism in Africa: a medical and social emergency. Int. Health 7 (4), 223–225.

DeCarlo, D., McGwin, G., Searcey, K., et al., 2013. Trial frame refraction versus autorefraction among new patients in a low-vision clinic. Investig. Ophthalmol. Vis. Sci. 54 (1), 19–24.

Karabas, L., Esen, F., Celiker, H., Elcioglu, N., Cerman, E., Eraslan, M., Kasokoglu, H., Sahin, O., 2014. Decreased subfoveal choroidal thickness and failure of emmetropisation in patients with oculocutaneous albinism. Br. J. Ophthalmol. 98, 1087–1090.

Resnikoff, S., Pascolini, D., Etya'ale, D., Kocur, I., Pararajasegaram, R., Pokharel, G., Mariotti, S., 2004. Global data on visual impairment in the year 2002. Bull. World Health Organ. 82 (11), 844–851.

Spritz, R., Fukai, K., Holmes, S., Luande, J., 1995. Frequent intragenic deletion of the P gene in Tanzanian patients with type II oculocutaneous albinism (OCA2). Am. J. Hum. Genet. 56, 1320–1323 PMID: PMC1801108.

Stevens, G., Ramsay, M., Jenkins, T., 1997. Oculocutaneous albinism (OCA2) in Sub-Saharan Africa: distribution of the common 2.7-kb P gene deletion mutation. Hum. Genet. 99, 523–527 PMID: 9099845.

Tuli, A., Valenzuela, R., Kamugisha, E., Brilliant, M., 2012. Albinism and disease causing pathogens in Tanzania: are alleles that are associated with OCA2 being maintained by balancing selection? Med. Hypotheses 79, 875–878.

Psychosocial and Cultural Aspects of Albinism

Jennifer G.R. Kromberg

University of the Witwatersrand and National Health Laboratory Service,
Johannesburg, South Africa

I. INTRODUCTION

Two specific aspects of albinism underlie the discussions covered in this chapter. The first is the relationship between the affected person and his physique, or the somatopsychological problems, and how factors within the person ameliorate or negatively affect adjustment. The second involves the social factors that encompass the affected individual and how they impact on that individual, either aiding adjustment or contributing to maladjustment. Together these issues, which are inextricably intertwined, are termed psychosocial aspects.

Wright (1960, p. xviii), while discussing the search for body–mind interrelationships, states "the fact that man is to some extent influenced by impersonal factors does not keep him from influencing the course of events as well. He is as much a determining organism as one that is determined." Furthermore, in discussing priorities in research, Childs (1978, p. 29) wrote: "…learning about how to harmonize one's endowment with one's environment and how to accept people whose appearance and behavior is 'different' are all as necessary to the solutions posed by birth defects as learning about the molecular mechanisms."

Although albinism is a genetic disorder and the cause and implications of the condition are well known, it is surrounded by psychosocial issues. Some of these have been researched and others become evident when families are interviewed for research projects and/or in genetic counseling sessions or when support group discussions expose them.

In this chapter selected psychosocial issues will be discussed, as well as relevant research findings and some of those with cultural and anthropological implications. Most of these projects have been based in Africa, but a few, completed in other parts of the world, where they contribute to a better understanding of the condition, will be covered briefly.

The topics to be discussed will include the psychosocial impact on the family of the birth of a child with albinism and the associated maternal- and paternal-infant bonding. Then the adjustment of affected children and adults in their home and community settings will be investigated, and the extent of the anxiety and the stigmatization they experience will be examined. The nature of their body image differentiation, their intellectual maturity and identity issues, and their educational problems and quality of life (QOL) will also be explored. Finally, aspects regarding their acceptance or otherwise in the community, the cultural issues and mythology that has surrounded the condition for many centuries will be discussed.

II. IMPACT OF THE BIRTH OF A CHILD WITH ALBINISM ON THE FAMILY

In this section, the impact of the birth on the members of the family will be discussed according to the amount of relevant research on each. Because most of the available research focuses on the mother, this will be the starting point. The little research on fathers will then be presented, followed by a more general discussion on the impact of having an affected child in the family on the parents, siblings, and grandmothers (there appears to be no available relevant research on grandfathers).

Mothers

When a stressful or negatively perceived event occurs, such as the birth of an infant with an observable birth defect or an unusual presentation (such as albinism), a sequence of psychological reactions is initiated in the mother (Weil, 2000). The first response is shock and the rapid mobilization of the defense system; disbelief, numbness, withdrawal, and feelings of unreality are also experienced. These are defense mechanisms and psychophysiological reactions that act as a buffer to prevent more serious psychological disorganization. The ability of the parents to understand and utilize the information offered is severely compromised, but this phase is usually quite short. However, the expected "normal" outcome of the pregnancy is irretrievably lost, and anger, guilt, shame, and depression may result. The feeling of loss may be exacerbated by the anticipation of future losses, such as the limitations the condition will exert on the child. With time, some acceptance of the situation occurs and, usually, appropriate coping patterns develop.

Against the background of both positive and negative community beliefs regarding albinism, the birth of an infant with the condition is a shocking and startling event. This is so particularly for the mother but also for her partner, the siblings, grandparents, and other members of the family. If there is a family history of the condition, or if some family members have a lighter skin color, then it might be partly understandable, but with no history the parents may attribute the condition to maternal impression (something bad that the mother did during the pregnancy, which affected her fetus), divine punishment or blessing, or a sorcerer's curse (witchcraft) (Kromberg et al., 1987). A few fathers of affected children may consider the mother's unfaithfulness to be the cause and the child to be evidence of her infidelity and may desert both mother and baby as a result.

The findings of Kromberg et al. (1987) suggest that even though infants with albinism have no physical malformation but only a light skin, hair, and eye color at birth, this may arouse the same reactions in mothers as would a child with a physical handicap. In another study on the significance of skin color in the newborn infant (Fuller and Geiss, 1985), one brown-skinned mother perceived her white-skinned baby as "ugly and alien." Drotar et al. (1975) interviewed parents of infants with various malformations. They reported that common themes emerged and five stages of parental reactions were observed: shock, denial, sadness and anger, adaptation, and reorganization. In the Kromberg et al. (1987) study a similar pattern was reported with the mother initially showing emotional upset, reluctance to hold and feed the baby, and delayed attachment, followed, slowly over time, by adaptation and increased maternal-infant interaction.

Defects that are visible to everyone tend to be the cause of more maternal anxiety than those that are not so obvious, according to Johns (1971), who studied family reactions to infants with a variety of congenital abnormalities. Because the skin color of infants with albinism in a black community cannot be hidden from the casual observer, the mothers are likely to be affected more deeply than mothers of children with less obvious conditions, such as Down syndrome. However, a condition such as cleft lip and palate, which is an obvious facial malformation, may produce similar strong feelings in the mother (Slutzky, 1969). Mothers cannot look directly at the affected child (which is required for maternal-infant bonding) without seeing the defect. Mothers have been known to abscond from the maternity ward leaving their infant with this condition behind. Similarly, occasionally, mothers who have given birth to a child with albinism have deserted their affected child in the newborn period (Sr. M. Glass, personal communication, 2016).

In the Kromberg et al. (1987) study the black mothers of newborn infants with albinism initially appeared to have difficulty identifying with their child whose skin color differed so much from their own. Usually the birth was so unexpected that, for several mothers, it resulted in emotional detachment, withdrawal, and indifference. Positive changes in the mothers' behavior and attitudes toward the infant, however, were observed over time in this longitudinal study. By 3 months after the delivery, most mothers showed increased interaction with their child and by 9 months most expressed decreased feelings of unhappiness. Acceptance at the behavioral level therefore might occur sooner than that at the cognitive or affective and emotional level. Slutzky (1969) suggests that, when considering rehabilitation, the psychosocial problems of the mother might take priority over the physical disability itself. However, it is recognized that recovery from the effects of having a child with a disability is gradual and takes time (Kennedy, 1970).

Data on depression and other reactions in mothers interviewed in the first 3 months after delivery showed that mothers were significantly more depressed about their affected child than the control mothers with unaffected babies (Kromberg et al., 1987). There was also a significant difference regarding headaches, disapproval of the infant's appearance, wanting the next child to be different (i.e., not affected), more complaints about both the infant and the father (the mother stating that the father was unhappy about the child), and claiming that the child would need more education than other children. By 9 months these initial strong reactions had lessened and the only significant differences were related to concerns about health problems in the infant (mostly realistic) and wanting the next child to be different, which might have indicated that there was still a residue of rejection and a longing for an unaffected child.

At 12 months the infants' achieved milestones were reported by the mothers, and the children with albinism were significantly slower to sit and crawl than the control children. Milestones have been observed in other infants with visual disabilities and found to be slightly delayed (Fraiberg, 1977), but this delay does not reflect on the intelligence of the infant. However, the delay in these infants with albinism could have been influenced both by the partial maternal rejection and the infant's visual defects.

How the initial poor maternal-infant attachment reported in the Kromberg et al. (1987) study affected the individuals involved in later life is not clear. However, attachment style (the ways in which emotional bonds are formed and emotions in close relationships are regulated) is a very basic aspect of the social side of life (Baron and Branscombe, 2012). Furthermore, Bowlby's attachment theory (Bowlby, 1977) focuses on the making and breaking of bonds of affection (e.g., the bond between mother and infant), the character of that bond, and the implications for the later psychological health and development of relationships (Watkins, 1995). The key point of Bowlby's theory is that "there is a strong causal relationship between an individual's experience with his parents and his later capacity to make affectional bonds" (Bowlby, 1977, p 206). Resulting from the expanding research base on attachment theory there is some empirical justification to support the statement that early life experiences can strongly influence subsequent adult functioning and can contribute to vulnerability and psychopathology (Watkins, 1995). Long-term follow-up of affected infants, whose mothers showed delayed attachment, would be worthwhile to gain insights into this aspect of the psychology of albinism. Any counseling offered to them could then be more informed, in future.

Whether the unusual maternal reactions reported above are entirely due to the skin and hair color of the infant or whether the infant's reduced visual acuity and nystagmus contributed to the reduced maternal-infant interaction is difficult to establish. Possibly, because the mothers were reluctant to look at the infant and the infant could not focus on the mother, the development of normal bonding (which requires these functions in mother and child) was difficult and delayed (see Fig. 9.1). Furthermore, the part played by the belief in local myths associated with albinism, which could have contributed to the shock many mothers described when faced with their affected newborn child for the first time, has not been fully investigated. The situation is likely very complex and all these factors probably play a part in determining the mothers' reactions to their affected infant.

Research has shown that parents, after the birth of an infant with a birth defect, have to face intrapsychic, self-esteem, and reality-based challenges (Mintzer et al., 1992). This study showed that the parents can experience assaults on their sense of self, which affect their equilibrium, and interfere with the parenting process and appropriate handling of the infant. An unaffected infant is an active partner in the maternal-infant interaction, but an infant with a disability, particularly a visual disability, may be less able to elicit parenting. If the mother is in a state of disequilibrium then she cannot be an adequate ally to her child and communication between mother and child becomes problematic. One of the infants described in this article (Mintzer et al., 1992) was born with bilateral cataracts and could neither focus on her parents nor make eye contact (infants with albinism have similar, though milder, visual problems, due to their nystagmus). For months the mother felt depressed, defective, and incompetent and was afraid to handle her baby. Self-esteem was restored only when the parents recognized their infant as a separate and individual person,

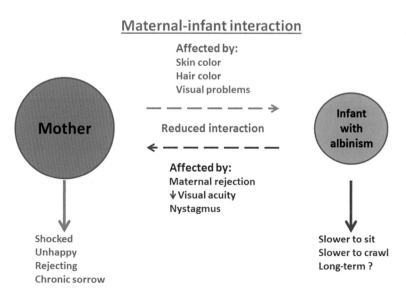

FIGURE 9.1   Factors influencing maternal-infant interaction after the birth of an infant with albinism.

instead of a negative extension of the self, and they could then respond to the special as well as the normal developmental needs of the child.

Fathers

Father–infant relationships have only been studied in a small sample of fathers with an infant with albinism (Kromberg, 1985). In general, these fathers followed the same course as mothers but often not in synchrony. They were upset and shocked at the sight of their newborn child, initially, but gradually they came to a state of acceptance and adapted to the situation. Soon after the birth some fathers admitted to having a history of albinism and therefore were able to accept that the child was theirs, whereas others, especially those with no family history, accused the wife of infidelity and some then deserted the wife and child. In one unusual case, the mother deserted and the father took the infant; later he and his second wife accepted and brought up the child. When unhappiness was investigated, both mothers and fathers were initially very unhappy; however, at 9 months after delivery mothers were much happier, whereas fathers were still at a higher level of discontent (Kromberg, 1985). Parenting attitudes of fathers, in general, appear to be changing in developed countries, as they take a more hands-on, as opposed to a hands-off, approach and as roles of mothers and fathers become more fluid; this may also be happening in developing countries.

Munyere (2004), in a personal experience and autobiographical article, states that it was difficult, in his Maasai community in Kenya, for his mother to persuade his father that he was his biological child. However, although there was initially great resistance, his father eventually accepted him. Munyere writes "my birth was very traumatizing to my entire family, because it was hard to convince the entire community that I was one of them" (Munyere, 2004, p 31). The community looked on children with disabilities as bad omens, so if the father felt the child was not his biological child he could lay the child on the ground where the cattle trod, and if they walked around the child and he survived, the father would recognize the child as his. Also, when an affected child was born, it might be denied breast milk and exposed to harsh conditions so that it would die, before the father saw it. However, for unknown reasons, Munyere was not treated in this way. He survived many years during which his skin, visual, and relationship problems were ignored and remained untreated; however, he did well academically and learned how to cope. He now works in special needs education in Kenya, is married and has three children, and is an inspiring role model for other people with albinism.

Parents

Parents of children with albinism, in Africa, may experience many different negative views from members of their communities (e.g., in Malawi, Lynch et al., 2014). These experiences impact on their social and mental well-being, as well as on their parenting strategies, and may cause their detachment from the community. Parents make choices for and about their affected children that may be influenced by poverty, fears for the safety of the children, or, in the case of education, the idea that one type of school is better, when considering their visual impairment and well-being, than another. These decisions can affect the QOL of affected children, their future enjoyment of life, and whether or not they are able to take control of their lives and gain meaningful employment. However, regardless of their limited resources, parents, in this Malawian study, showed great resilience in dealing with the various barriers that arose in innovative and courageous ways. Among the group of parents interviewed some had wider experience and knowledge of the condition than others. The group, therefore, encouraged these parents to take up an advocacy role and visit villages to talk to the elders about the condition and try to dispel the negative myths and misconceptions that could harm affected people (Lynch et al., 2014).

Other genetic conditions may also affect the family and result in a burden on the parents in various ways. For example, one study on the psychosocial burden of sickle cell disease (SCD) in Cameroon families has been undertaken (Wonkam et al., 2014). This condition is a chronic recessively inherited one, which is more common than albinism in Central African countries. The findings from this study, on 130 parents each with at least one affected child, showed that they experienced moderate to severe difficulty in coping with SCD and, particularly, with the painful episodes (often more than three a year) that the children experience. Another major stressor was the fear of having another affected child. As a result, the majority of parents (90%), who have prenatal diagnosis for SCD, request medical termination when the fetus is affected. The main factor influencing their decision, in this study, was the severity of the condition in the child. However, the decision was also impacted by other sociodemographic factors such as unemployment, single marital status (particularly single mothers), low education, and other affected children.

In the same way the major factor, for the parents, in coping with albinism, also a life-long condition, is the impaired vision in the child. As one mother, commenting on her management and financial problems, explained "Our main problem has been coping with the poor eyesight of my three children with albinism and always having to consider protective glasses, visual aids and sometimes special schooling" (Kromberg, personal communication, 2015). A couple who requested prenatal diagnosis for albinism found the condition in their affected son a heavy burden and difficult to manage; they spoke about skin care, the child not wanting to cooperate in protecting himself from the sun, and their fear of skin cancer (Wessels, personal communication, 2017). Again, the ability to cope with the problems associated with both skin care and poor vision is also impacted by unemployment, single mother status, low education levels, and having many affected children.

Parents with a child with an inherited condition often have strategies for sharing information, particularly the genetic information, with both their affected and their unaffected children, according to the findings of a one study (Gallo et al., 2005). The information sharing may be motivated by the goal of encouraging the unaffected siblings' adaptation to the condition. Parents reported that they assessed the developmental stage and readiness of the sibling, before sharing information. They also saw the sharing process as a developing process that would continue throughout childhood. However, nondisclosure may also occur and sometimes parents consider this to be in the best interests of the unaffected child because disclosure may lead to stigmatization, discrimination, or ostracizing of the child. Studies in children with cancers or HIV/AIDS have shown that they have better coping skills and fewer psychosocial problems when they are well informed about their condition and its consequences. In addition, such knowledge may result in a better integrated self-concept in the affected person (Wiener et al., 1996).

Not much is known about how parents explain genetic risks to their at-risk children. A study conducted by Metcalfe et al. (2011) showed that parents, in the United Kingdom, found discussing risk difficult and emotionally challenging. The children preferred to learn about the risks gradually throughout childhood rather than during a major family event when unexpected discussions might arise. Such sharing of information, with open communication, appeared to help parents and children to cope better. Observation suggests that these findings apply to the situation in South Africa too and local family members and siblings seldom fully understand the genetics of albinism and risks of recurrence, even when they live with an affected member.

Siblings

Siblings are particularly vulnerable in situations where there is a brother or sister with a genetic disorder requiring special and time-consuming attention from the parents, expert care, medication, and/or hospital visits. As a result, it has been suggested that the siblings of disabled children, potentially, face difficulties (Houtzager et al., 2004). Although albinism, in general, is not viewed as very disabling, the stigmatization (as well as the medical problems associated with visual defects and susceptibility to cancer) attached to the condition may affect the siblings and other members of the family. Most studies on siblings have focused on those with a brother or sister with cystic fibrosis, and few, if any, have examined the effects of having a sibling with albinism. One study suggests that the impact of a disabled child on the siblings can affect their behavior (Cowen et al., 1986), another that it can lead to worry and jealousy or resentment in the siblings toward the affected child, as well as confusion about the risk to themselves and their offspring (Hutson and Alter, 2007), and a third that it can result in lowered self-esteem (Metcalfe et al., 2011).

In a study in Zimbabwe 138 school pupils with albinism completed a self-report questionnaire (Lund, 2001). Altogether 65 (47%) of these pupils had at least one normally pigmented sibling. Most (80%) felt that they were not treated any differently by their siblings, due to their albinism ("they treat me like a normal guy"). A further 8% made positive comments stating that extra help or protection was offered by siblings. However, 6% of the pupils described their siblings' negative attitudes, including scolding, mocking, and/or avoidance (e.g., neither sharing clothing or food, nor allowing them to visit their workplace).

Plumridge et al. (2011) investigated British parents' communication with siblings of children affected by various genetic conditions. The aim of the study was to explore parents' communication about risk with the siblings and the level of support they might require. They found that the siblings' needs for information and support are mostly not recognized by health professionals and sometimes by parents. As one parent stated:

> "...and [sibling] would need a lot of psychological support...the siblings do need it. Because I think the person with the disease automatically gets everything in a way... It's the siblings that are left all on their own and there's no help for them at all..." (Plumridge et al., 2011, p. 380. Although this quote was from the parent of a child with Duchenne muscular dystrophy, it could apply, but perhaps to a lesser extent, to the siblings of children with albinism).

In this study some siblings were informed about the genetic nature of the condition, but others were left to find out for themselves. Siblings might be provided with information about symptoms and management of the condition but were less likely to know about their risk of being a carrier of the condition and/or of recurrence for their own offspring. Although Plumridge et al. (2011) included families with six different genetic conditions in their sample, their findings showed that there were many similarities across conditions. They concluded that when siblings were fully informed about the genetic condition in the family and included in relevant discussions, they developed a better understanding of their role in the family and the relationships between family members were more harmonious. However, parents often required help in responding to and supplying the support needs for the siblings and health professionals should be aware of this need.

When the adaptation of a child to his/her disorder (cystic fibrosis, in this case) was investigated, results suggested that adequate family functioning is more important than the condition itself in terms of psychological effects on the affected child (Cowen et al., 1986). However, findings about the personality development of healthy siblings are discrepant and can range from positive self-image and absence of any psychopathology to high anxiety levels, sleep disturbance, and behavior problems. The majority of parents, in the Cowen et al. (1986) study, reported that they did not receive any information from professionals about how to share genetic information with their children. Health professionals need to assist parents to identify ways to promote healthy interaction with their children in this regard.

Grandmothers

Grandmothers play a particularly important part in communities where group decision-making is usually preferred to individual decision-making, reflecting the collectivist rather than the individualistic nature of African culture, and where important decisions may be discussed in the family prior to being finalized (Kromberg and Jenkins, 1997). Grandmothers, who may be the matriarch in the family, therefore, often have an integral and respected role in this cultural system.

Grandmothers' understanding of birth defects has been investigated by one South African research group (Penn et al., 2010). One of the issues discussed in focus groups with these grandmothers was the traditional role they played regarding disabled infants and infanticide. Although this topic was considered somewhat taboo, infanticide was one of the management strategies suggested by the grandmothers. However, although they admitted it was not a strategy used much in recent times, they stated that albinism was one of the conditions subjected to this practice. The term used as a euphemism, by the grandmothers, was "letting go" and a specific case of a child without legs, who was "let go," was discussed in one focus group. In previous times, when most infants were delivered at home in the village, they (probably together with the traditional midwife) could make the decision to end the life of the child born disabled, where necessary. The grandmother would then make up a reason for the child's death to the mother, e.g., he/she was too weak to survive. The grandmothers accepted that this role was no longer possible (at birth) because, recently, most births occurred in hospital In general, however, the role they played was a caring one, which included child care, cooking, fetching water, checking on the children's proper eating habits, teaching life skills, giving advice, and managing health issues.

III. ADJUSTMENT OF THE INDIVIDUAL WITH ALBINISM

Adjustment has been defined as "a condition of harmonious relation to the environment wherein one is able to obtain satisfaction for most of one's needs and to meet fairly well the demands physical and social put upon one" (English and English, 1970, p. 13). How people with albinism adjust and relate to their world, satisfy their needs, cope with society's demands, and manage their lives are questions that have not been fully investigated.

Probably one of the first significant psychological studies to be carried out on people with albinism was undertaken by Beckham (1946) in Chicago, United States. He was interested in the question "Does the albino present with a psychology peculiar to himself?" He interviewed and conducted psychological examinations on 42 affected people; 10 years later he reinvestigated them in a follow-up study. Social histories were collected from all the subjects by a social worker. From these documents, Beckham concluded that people with albinism have some specific problems particularly related to family and personal adjustment. Based on his observations, Beckham stated that many affected people from the black population, even at an early age, develop feelings of insecurity in their homes, and in society generally, because of discrimination against them and the feeling of "being different," and that psychological and/or psychiatric treatment is often necessary to help them face these problems.

The adjustment of young people with albinism has been studied (Kromberg and Jenkins, 1984) in a group of black South Africans. Altogether 35 subjects (mean age 17.3 years) and 35 unaffected matched controls (mean age 16.8 years), with, on average, 8 years of education, participated. The results showed that, in general, the affected subjects were as well adjusted as the controls and no significant differences on total scores for primary, high school, or nonschool-going subgroups were identified. However, as might be expected, although the subjects with albinism did not admit to specific social problems, they had more adjustment problems of a physical and psychosomatic nature than the controls. These problems included blinking, headaches, sweaty palms, and general poor health. On the other hand, perhaps unexpectedly, this group of subjects was less ashamed of their physical appearance, possibly because they were so unique. There were 21 males and 14 females with albinism in the sample, but there were no significant differences on the total adjustment scale scores for these two groups. However, when two clinical psychologists assessed the additional data (collected where the subjects had qualified their responses and added comments), the affected females, specifically, were found to be significantly better adjusted, in general, than the control females. No differences were detected between the two male groups.

Several of the items on the adjustment scale used in this study (Kromberg and Jenkins, 1984) suggested that the subjects with albinism were showing some psychosomatic symptoms (e.g., headaches, sweaty palms, general health complaints). It may be that these subjects were expressing their anxiety, through these symptoms, at living in a society that has partially rejected them (indicated by the frequent use of the word "nkau" or monkey, for the person with albinism). Although the subjects in this sample did not admit to having problems in the area of interpersonal relations or psychosocial adjustment, it is known that depressive, anxious, and neurotic trends can be measured by psychosomatic or health complaints as effectively as by introspective items (Vernon, 1964). Furthermore, it may be that the subjects were drawing their responses from their "public" or "collective" selves, rather than their "private" selves (Triandis, 1989), in this situation. However, further research is required, with more in-depth measuring instruments, to clarify this issue.

The finding in this study (Kromberg and Jenkins, 1984) that significantly more controls, than affected subjects, expressed feelings of dissatisfaction with their physical appearance is intriguing. Why both groups were somewhat concerned in this regard may be associated with their age because teenagers are known to experience some internal conflict during puberty and the rapid development of their body size and shape. However, teenagers with albinism, because of their uniqueness, may be better able to establish an identity for themselves than their unaffected contemporaries. Also, their self-concept may be more independent, because of their being unique, and less interdependent, as defined by their culture (Markus and Kitayama, 2010). Woolf and Dukepoo (1969) suggested that among Hopi Indians the admiration of the whiteness of the community members with albinism was not a matter of identifying them with White Americans but rather represented an "association of whiteness with cleanliness, goodness, and purity." Similarly, in the writer's experience, African persons with albinism do not appear, consciously, to identify themselves with White people; they might, however, instead associate themselves with the characteristics attributed to the color white, such as pureness and virtuousness, which are also held in high regard in local cultures.

Stewart and Keeler (1965) investigated six Cuna Indian people with albinism and six normally pigmented controls, in Central America. Their conclusions were that affected people showed a passive, dependent, feminine orientation with little interest in the opposite sex. They were anxious, insecure, and relied on defense mechanisms, such as intellectualization, religious preoccupation, and denial, to allay anxiety. The subjects tended to deny any difference between themselves and the controls, but this denial collapsed under moderate stress with subsequent hostile, angry, and resentful responses and even withdrawal. The subjects also appeared to show a less accurate perception of reality. Their emotional immaturity was attributed to early parental overprotection or the inheritance of a poorer basic personality or to the interaction of both these factors. However, the authors remarked that their findings should be confirmed with a larger sample.

Research on psychological and personal aspects of albinism has been reviewed by Estrada-Hernandez and Harper (2007), with the aim of investigating the factors associated with adaptation to albinism. They discussed 15 articles dating from 1929 to 2003. Reports state that people with albinism, who have average intellectual abilities and achievement skills, face many challenges that are physical, mental, emotional, and social. They tend to have difficulties with the social impact of the visible differences associated with having albinism; they may develop personality problems, low self-esteem, and poor social and coping skills. The authors conclude that affected people, therefore, may benefit from rehabilitation and counseling interventions aimed at encouraging positive self-esteem and assertiveness.

Biesecker and Erby (2008) investigated adaptation to living with a genetic condition or a genetic risk, in a minireview. They stated that adaptation is a multidimensional concept, which can change over time. The dimensions of the process include intrapersonal and interpersonal psychosocial factors, as well as physical functioning, personality characteristics, and economic resources, which all influence adaptation. The writers draw several broad conclusions from their review:

1. the severity of the disorder does not predict adaptation
2. the majority of individuals adapt well but some require interventions
3. a child's adaptation parallels that of the parents, especially the mother's.

They add that human beings are resilient and can withstand and rebound from disruptive life challenges, becoming stronger and more resourceful in the process. Adaptation requires "struggling effectively, working through and learning from adversity, and integrating the experience into one's life" (Biesecker and Erby, 2008, p. 402).

IV. STIGMATIZATION

A stigmatized person has been broadly defined as one who has an attribute that makes him different from others and characteristics that are less desirable and not seen as "normal" in the group to which he or she belongs (Goffman, 2006). The person is therefore "reduced in our minds from a whole and usual person to a tainted and discounted one" (Goffman, 2006, p. 131). Furthermore, Goffman (2006) suggests that society establishes categories, in which certain characteristics are considered ordinary and natural, resulting in the development of social identity in the acceptable members. In social interactions, social identity is based, initially, on first appearances and it develops further in social situations where the stigmatized person is in contact with unaffected people. Parker and Aggleton (2003) suggest that stigma is a changing social process that is understood through interaction. Both interaction and stigma vary in different situations; also, stigmatization operates in relation to difference and social and structural inequalities.

Stigmatization is well recognized in situations where disabled people, or those who differ in some way from the unaffected, come into contact with members of the community. Especially where the condition is rare and most people have not had any experience of dealing with affected people and do not know how to behave, stigmatization may occur. Wan (2003) suggests that much of the prejudice toward people with albinism is associated with a fear of the unknown. This fear gives stigma its power and reality. However, the greatest stumbling block to affected people playing their full part in society is not the presence of their differences but the myths, fears, and misunderstandings that surround them (Murphy, 1995).

There is little sociological research on how people with albinism cope with the stigmatization they experience. Wan (2003) investigated this issue by obtaining reports from affected individuals themselves. Her international, but small, sample of respondents (12) came from Canada, United States, and Australia (it was assumed that affected individuals experienced stigmatization and marginalization wherever they lived in the world). One respondent stated:

> My mother always told me 'when you are inside this home, everyone here loves you, but when you step over the threshold there's a world out there that does not and may not' *Wan (2003, p. 284)*

As these respondents enrolled in primary schools, in their countries, they had to endure name-calling (e.g., whitey, ghost), humiliation, and physical victimization from both teachers and students and were left with feelings of unhappiness and isolation. However, at high school the levels of discrimination decreased, teachers and pupils were more understanding, and fitting in was easier, although dating was more difficult. As adults they still found they were sometimes discriminated against, marginalized, and rejected by superiors, colleagues, and employers because of both their visual impairment and appearance. Many of them stated that they had found doctors and nurses to be ignorant and insensitive (Wan, 2003).

The participants also reported on stares and some unpleasant encounters with strangers in public areas. To cope with these adversities, people with albinism develop coping mechanisms (Wan, 2003), and eight have been documented: being defiant; being active; cultivating serenity; internalizing; talking; hiding; being flamboyant; and being positive (accepting albinism as part of their identity). Respondents might engage in one or more of these approaches in different circumstances and their preferred strategy may change over time. Such strategies of resistance to being stigmatized are essential to the well-being of people with albinism and contribute to their personal change and adaptation to coping with their condition.

In 1989, Ezeilo, working in Nigeria, published a research note on her exploratory study on the psychological aspects of albinism. She studied two affected males and one affected female who were university students and asked them to write an essay on the merits and demerits of having albinism. She found that one male subject saw no merit at all in having the condition, whereas the other male thought affected people were highly intelligent and the female subject felt having a beautiful skin was a merit. However, all three felt the demerits outweighed the merits and stated that their skin color attracted stigmatization including name-calling, ridicule, and even stone throwing. One subject stated that "one reference to his physical state will shatter all the man and fight within him" (Ezeilo, 1989, p. 1130). They all admitted to withdrawing from social situations to avoid being noticed. They also noted that there were sexual problems and one male felt that no girl would accept him, whereas the female stated she had had boy friends who she rejected because rather than relating to her affectionately they wanted to exploit her sexually, to see what it would be like to sleep with a White (Caucasian) woman. All the subjects preferred the company of unaffected to affected people in their socializing and interpersonal relationships and perhaps this was another means by which they could uphold an identity of "normality." Also, all of them considered society to be unkind and rejecting, although all three had close unaffected friends who helped them with their work. The author (Ezelio, 1989) makes an important point when discussing coping mechanisms and states that learning to cope depends on both family environment and socialization processes. If an affected child is nurtured and regarded as an individual rather than a stereotype, if he is raised to be assertive and self-confident, then he can strive toward self-actualization and developing an independent self, rather than succumbing to society's negative attitudes.

Curiosity, whether associated with stigmatization or not, also affects the interactions between people with albinism and others. This curiosity is often inappropriate and can be felt as an invasion of privacy by the stigmatized person (Goffman, 2006). Similarly, sympathy is sometimes associated with stigmatization and may be used by the stigmatizing person as a response to the feeling that stigmatizing is unacceptable and therefore needs to be managed in this way. Sympathy is, however, also usually inappropriate, unnecessary, and often unappreciated by affected people, who basically want to be treated like any other person.

In a recent study on the experience of felt stigma in 154 people with albinism in the United States, six domains of stigma were investigated (Young, 2012). These included self-esteem, concealment, avoidance versus group affiliation, salience, endorsement of stereotypes, and experience of prejudice and discrimination. The findings confirm that people with albinism are commonly stigmatized with negative stereotypes. Implications for policy and interventions directed at addressing such stigmatization need to be considered and implemented.

The fight against stigma has been comprehensively covered in a review by Heijnders and van der Meij (2006). Although albinism is not mentioned specifically, the authors acknowledge that the stigma associated with the conditions they included (e.g., epilepsy, tuberculosis, intellectual disability, HIV/AIDS) had many aspects in common, especially regarding the strategies used in fighting such stigma. These strategies could be developed at five different levels:

1. the intrapersonal level (treatment, individual and group counseling, cognitive-behavior therapy, empowerment, self-help, advocacy, and support groups),
2. interpersonal level,

3. organizational/institutional level,
4. community level,
5. government level.

Examples of programs on all these levels had been found by the authors, but very few had been evaluated in terms of effectiveness. The effective strategies they identified were focused mainly on the individual and community levels. The authors concluded that a patient-centered approach to fighting stigma was essential, beginning with interventions targeting the intrapersonal level and empowering the affected individual to support the implementing of multilevel stigma-reduction programs at all the other levels.

V. DEVELOPMENT OF SELF-CONCEPT AND SOCIAL IDENTITY

The development of the self has been studied by psychologists, sociologists, and anthropologists for many decades. The self is an "active agent that promotes differential sampling processing and evaluation of information from the environment and thus leads to differences in social behavior" (Triandis, 1989, p. 506). It is dynamic and changes as the environment changes. Three main aspects of the self have been defined (after Triandis, 1989):

1. The private self that includes cognitions (or thinking), which involve personal traits, states or behaviors, or assessment of the self by the self (e.g., I am introverted).
2. The public self, cognitions (or thinking), which concern the generalized others' view of the self or assessment of the self by others (e.g., people think I am introverted).
3. The collective self, cognitions (or thinking), which concern the way in which the self is viewed by a collective or assessment of the self by a specific group (e.g., the family).

It has been suggested that people experience these three selves differently in different cultures and situations and the way they sample them affects their behavior. How people with albinism in Africa sample these selves is not yet apparent. However, the collective self partly determines social identity, which is that part of the individual's self-concept that stems from being a member of a social group, together with the values and emotional significance attached to that membership. In the case of people with albinism this might arouse conflict and confusion because it might be difficult to gain acceptance and knowledge of membership in a social group, and/or to determine to which group he/she belongs, and consequently, to strengthen social identity. Such conflict is shown in the findings of research, which suggest that on the one hand affected people prefer to socialize with unaffected people (e.g., Ezeilo, 1989), prefer to marry unaffected people (Kromberg and Jenkins, 1984), and be seen as any other person, and on the other hand they might wish to be seen as special and deserving of special treatment, because of their condition, and classified as disabled (see Chapter 12).

There appear to be two types of social relations that can be linked to senses of self or modes of being (Markus and Kitayama, 2010). The first type assumes that social relations are developed based on goals and interests of the individuals, whereas the second type assumes that individuals are connected and made meaningful through relationships with others. The first type is termed the "individualist" type and the second "collectivist." The prime example of the first type is found in North American communities, whereas examples of the second type are found in Africa and in other developing countries (but also in some developed countries, such as China and Japan). However, any individual's culture and self are constantly interacting, developing, and forming an environmental awareness that guides action. Because of the initially somewhat marginal existence of the person with albinism it might be suggested that the individualistic or independent approach to developing social relations might be preferred. It could be more difficult to develop relations in an interdependent way, when the affected person is not fully integrated into the community, and many of the other members of the community are discriminating against and reluctant to interact with him/her.

People with albinism are born into a family, who although they initially experience shock, usually, in time, accept the child as their own. This becomes the affected person's first experience of a group in which he/she has membership and through which he/she can start to develop social identity. However, this family group (the in-group) may then have conflict with their community group (the out-group), when they seek to remain part of that group, even though they have a family member with albinism. This conflict may become serious and the village community may expel the whole family or the dyad of mother and child (as happened in the Botswana case described by Livingstone, 1857), if they believe that the family has become a threat, bringing bad luck, to the village community (See Chapter 1). Because the small group usually has difficulty in surviving in isolation, the leader has to try to mitigate the conflict and find a solution to the situation. The parents, or mother or father, may take on the role of ambassador for their affected child, trying to help the village elders to understand the nature of albinism, to demystify the condition, and to accept the family back into the village. Intergroup relations were focused on individual prejudice and discrimination in the case of Livingstone (1857) and this caused the intergroup conflict. Such conflict has been described by Tajfel and Turner (1997). The more intense the conflict the more likely that the individuals in the out-group will behave toward the other group in terms of their group membership, rather than their individual characteristics or interindividual relationships. This behavior can then result in extreme measures being taken, such as excommunication of the in-group from the out-group (Livingstone, 1857). Sometimes, the mere perception of individuals belonging to two distinct groups can trigger intergroup discrimination favoring the out-group or the dominant group, and discriminatory responses against the in-group (the family with a member with albinism, in this case) may result.

Social identity is often partly defined by membership in a community group. However, the essential criteria for such membership are that the individuals concerned define themselves, and are defined by others, as members of the group. These criteria may be difficult for people with albinism to meet because they do not look like the unaffected people in their communities, are surrounded by myths and superstition, and may be unacceptable as members. Fitting into an acceptable social category is essential for the development of social identity. Social categorizations represent "cognitive tools that segment, classify and order the social environment and thus enable the individual to undertake many forms of social action" (Tajfel and Turner, 1997, p. 40). If the people with albinism are denied the opportunity to join a community group, it will be difficult for them to find a system for self-reference, create a place in society, and build a social identity. Social groups provide their members with an identification of themselves in social terms and every person needs such group interaction to establish a satisfying self-image. Albinism support groups and associations (see Chapter 13) therefore, could play an important part in the lives of those affected people who are looking for interaction, for a place in society, and for the clarification of their social identity.

VI. INTELLIGENCE, INTELLECTUAL MATURITY AND BODY IMAGE

A few studies have examined the intellectual abilities of people with albinism. Beckham (1946) studied 42 affected African American boys and girls in Chicago and concluded that they showed no intellectual deficit compared with their unaffected siblings. Later Stewart and Keeler (1965) studied a group of San Blas Cuna Indians and compared people with albinism and those without the condition. They used an intelligence test battery, which included a Draw-a-Person test; the IPAT culture-free test of "g," Scale 2; the Wechsler Adult Intelligence Scale (three subtests for arithmetic, similarities, and block design); and the Arthur Adaptation of International Performance Scale, Tray 2. Their results showed that there was no statistically significant difference between the two groups of subjects in the 10 scores recorded.

In South Africa, a study on intellectual maturity and body image differentiation was carried out by Manganyi et al. (1974). There were 28 young people with albinism (mean age 11.8 years, range 7–16) and 28 unaffected matched controls (mean age 11.9 years, range 7–16), from the black population in Soweto, Johannesburg, in the sample. For the Draw-a-Person test, three drawings were required, a male and female figure and a self-portrait. These figures were scored using a scale for intellectual maturity and a body-concept scale to study the sense of separate identity. Generally body concept is progressively experienced, in individual development, as having definite boundaries, which help a person to experience himself/herself as more or less separate from his/her environment. The results of this study showed differences (significant at the 2% level) between the affected and unaffected groups, the former showing a slightly higher level of intellectual maturity and, therefore, intelligence well within the normal range. These findings are supported by Keeler (1964) who stated that the IQ of the moon-child (a Cuna Indian with albinism) was not in the intellectually disabled range, but they suffered psychiatrically. He added that these children led an abnormal life,

> "restrained in many activities by their physical condition. They are largely rejected by the opposite sex. The inability to compete physically often leads them to sedentary and intellectual pursuits. Because of their weakness and rejection they develop anxieties in which they react by over-compensation or by regression" (Keeler, 1964, p. 1).

The findings on body image (Manganyi et al., 1974) showed that the subjects in the experimental group were slightly more differentiated in terms of their body image boundary characteristics than the controls, suggesting that the body image results may be an artifact of the intellectual maturity dimensions. However, the self-portrait results showed that those in the albinism group had more problems drawing a self-portrait than they did with the other two figures, while the control group did better creating the self-portrait than with the other drawings. These data indicate that the affected subjects found themselves facing a negative self-evaluation and identity and were unable to significantly affect their drawings. The authors conclude that albinism, in itself, must constitute a very complicated mode of being-in-society and, in conditions where skin color is a basic societal concern, their state must be one of real marginality. Manganyi (1972) has suggested that cultural factors have possible significance in the socialization and development of body image. The body is a way of "being-in-the-world," of experiencing both the subjective and objective aspects of being and of saying who one is.

So although people with albinism show intelligence within the normal range and appropriate intellectual maturity, they may experience some negative self-evaluation and identity and may find that they have to adjust to living life on the margins of society.

VII. EDUCATION ISSUES

There has been an ongoing debate in South Africa and other sub-Saharan countries about whether it is best for children with albinism to be educated in schools for the partially sighted, where they can receive appropriate attention and be protected, or in normal community schools. The author is aware of many affected children in Johannesburg who are sent away to special schools for the partially sighted and in such a school in the Kwazulu/Natal province of South Africa there are many affected learners (Elizabeth Louw, personal communication, 2016). Also, children with low vision are often educated in special schools for the blind in Malawi (Lynch et al., 2014). Principals in such special schools take in many scholars with albinism because, there, they can be instructed by special education teachers, receive the understanding and the visual aids that they need, and usually be housed in the school boarding hostel. However, because these schools are often situated in rural areas far from the child's home, this practice leads to isolating affected children from their community and makes for more difficult integration when they return home having completed their schooling. Proponents for inclusion and the community school placement of these children state that then the child becomes an accepted member of his peer group, is better known and more easily accepted in his community, and can establish friendships, networks, and contacts for job opportunities later on in life.

Lynch et al. (2014) have investigated strategies to improve the educational inclusion of visually impaired children with albinism in Malawi. They describe the barriers to full educational access, which affected children face. They also list low-cost strategies that could be used in mainstream schools, in developing countries, to reduce the effect of the limited vision associated with albinism. The study aimed to collect data about children's experiences living with their condition in various educational settings in Malawi. They included children with albinism, their parents and teachers in mainstream schools, special schools for the blind, and resource centers in Malawi, in the study. In 2010 the Malawi Integrated Education Program supported 90 children with low vision, in mainstream schools, and of these 66 (73%) had albinism. Findings from this study showed that children had heard about the myths surrounding their condition and its causes and this had created anxiety. Some had even heard about the killings of affected people and the use of body parts, by traditional healers, as magic potions to make people rich. They had also experienced name-calling and bullying. However, they were curious about the causes and wanted more information about the condition (those in resource or special education centers knew more than those in other schooling situations, probably because of the better specialized training their teachers had received). Very few had had a proper assessment of their vision and the uptake of visual aids was poor. Nevertheless, children learned and used various techniques to manage effectively the little residual vision they had. Some teachers mistakenly believed that as children aged their vision would become worse, and this myth had to be corrected.

Findings showed that the parents who had sent their children to mainstream schools were often not satisfied with the level of care and protection provided for their children (Lynch et al., 2014). Some removed their children and sent them to resource schools, where they thought better education was available. However, the system of supplying itinerant remedial teachers to mainstream schools, where there were children with special needs, was found to be worthwhile. These teachers believed in providing better access to the curriculum by offering large-print text books, class notes, and examination papers, moving children closer to the chalkboard, and encouraging them to participate in all activities inside and outside the classroom. They also played a role in supporting, informing, and advising the class teacher and emphasized the fact that, although the eyesight of affected children varies, they should all be allocated extra time to complete projects and exams. From a practical point of view, the teachers were also advised to encourage affected children to wear protective clothing to prevent sunburn, to keep their hats on in the classroom to protect their eyes, and to sit far from the windows in any room to reduce exposure to the bright light.

The provision of low-vision aids was also considered in this study (Lynch et al., 2014). It was noted that many of the children had been provided with various aids but most of them did not use them or had lost them. Although these aids would be most useful for certain types of work, such as map reading and drawing of graphs, and potentially could enhance the education experience, schoolchildren appeared to be managing without them. However, the authors suggest that if simple hand-held and low-cost magnifiers are to be introduced to those with low vision, their distribution should be negotiated with the target group of children and their wishes should be considered.

Another school-based study was carried out by Lund and Gaigher (2002) in South Africa. One of the aims of this project was to ask children, attending a special school, to describe their eye and skin problems and how they managed and adapted to their condition. An intervention program, focusing on sun protection habits, with the active participation of the children themselves, was then initiated. The school was well set up to cope with affected children, having shutters on the classroom windows, mobile chalkboards, and informed teachers. However, the children were found to have the usual skin and eye (most of which had been assessed) problems but were generally caring for themselves (together with some help from a nurse when necessary) in this protected environment. Although most wore hats, they still had some facial sunburn and the authors suggested that the immediate benefits of avoiding the sun and protecting oneself need to be stressed to effect behavioral change in these children. Also, peer group discussions on ways of managing the condition would likely reinforce positive behavior patterns and lead to sustainable coping strategies.

The findings of a multicenter study (Tumwesigye et al., 2009) showed that many children including those with albinism, in countries such as Kenya, Malawi, Tanzania, and Uganda, were placed inappropriately in schools for the visually impaired. Altogether 1062 children in these schools were examined and about 12% were found to have normal visual acuity. Many could learn to read print with the right training and simple visual aids (they were often only taught Braille in the special school), attend normal schools, and be integrated into society with all the associated benefits. Placement in a special school was unnecessary for some of these children and even detrimental to their education. The authors conclude that the reasons for these placements are multiple, and cooperation between the ministries of health and education is required to improve this situation.

In general, it appears that the mainstreaming of children with albinism is being promoted in several sub-Saharan African countries, such as Malawi and Zimbabwe (Lund, 2001). In South Africa, schools are generally understaffed and crowded with poor infrastructure so that, although mainstreaming has been promoted, integration appeared to be difficult to implement (Lund and Gaigher, 2002). The advantages of mainstream schooling seem to outweigh the disadvantages. However, the benefit of having remedial teachers, such as the itinerant teachers of Malawi, available to mainstream schools as supports for the children and their teachers is recognized and such a program should be strongly and widely recommended.

VIII. QUALITY OF LIFE

QOL has been described by the World Health Organization (WHO) as:

> individuals' perceptions of their position in life in the context of the culture and value systems in which they live and in relation to their goals, expectations, standards, and concerns *WHO Quality of Life (QOL) group (1996)*

The outcome of the psychosocial and health issues faced by persons with albinism in Africa is that their QOL may be reduced. QOL is a global and multifaceted concept, which includes the following functional domains (after Flanagan, 1982): intrapersonal (e.g., health, perception of life satisfaction, feelings of well-being); interpersonal (family life, social activities); and extra personal (work activities, housing). In terms of adjustment to albinism, the family and affected person need to attain adapted person–environment (reality) congruence. A good QOL is linked to positive self-concepts and body image in the affected person, whereas poorer QOL is associated with perceived stress and feelings of loss and grief (Liveneh and Antonak, 2005).

The QOL experienced by people with albinism has not been studied in much detail in Africa. However, in one study QOL in affected people and their exposure to stigma and discrimination, in Malawi, were investigated (Braathen and Ingstad, 2006). The study focused on a disability that was visible rather than on one that was physically or mentally limiting, and the authors hoped it would increase awareness of what visual appearance means cross-culturally. Altogether 25 people were interviewed and the findings showed that stigmatization was an issue. Stigma often occurred as first appearances took effect. However, the reactions of strangers dissipated as they got to know the affected person and found that he or she was not so different from themselves. Away from the familiar environment stigmatization was again experienced. Several factors were identified as playing a role in the QOL of people with albinism including their poor eyesight, skin problems, and relationships with other people. The researchers found that because Malawi is a very poor country, many people have a poor QOL and people with albinism were in a similar situation. Because many people in Malawi have never seen an affected person, there is a general lack of awareness. The words, such as "mzungu" or stranger, often used for affected people suggest that they are unfamiliar and do not belong in the local society. Regarding the physical problems, these are difficult to manage because protective clothing, sunscreen, and visual aids are often not available (and/or unaffordable). Also, the awareness that these things are necessary to improve the QOL is often poorly understood by the individuals themselves, their families, and health-care workers. Awareness programs are required to create an understanding of what albinism is and what those affected need to function well and contribute in their community. The authors conclude that people with albinism in Malawi have the right to be included in society and the right (as they quote Martin Luther King, speaking in 1963 in a different context) "not to be judged by the color of their skin, but by the content of their character" (Braathen and Ingstad, 2006, p. 610).

QOL refers to an individual's sense of overall well-being, including the physical, psychological, emotional, social, and spiritual aspects of life (Cohen and Biesecker, 2010). However, what actually contributes to QOL from the patient's perspective needs to be clarified because living with a genetic condition can profoundly affect QOL on many levels. The future aim of health care should be to promote adaptation and facilitate an improvement in the QOL for these patients; as patients adapt to their condition their QOL should improve. Empowerment is considered an important aspect of this concept and this will be discussed in a future chapter (see Chapter 13).

The term health-related QOL may be used in health-care settings. There it includes "one or more of the following: general health (physical and/or mental), functioning (physical, emotional, cognitive, sexual), social well-being, coping adjustment, and existential issues" (McAllister, 2016, p. 50). This aspect of QOL should probably be a focus when considering the lives of people with albinism, who have health-related needs, which may not be met in African settings and which impact detrimentally on their well-being.

Recently, a research group from Brazil (Maia et al., 2015) has focused on the topic of QOL, suggesting that the social reality of affected individuals needs more investigation because the myths and social segregation they face are linked with medical and psychosocial implications. They added that these individuals face prejudice and discrimination and as a result tend to show emotional instability and a less assertive personality than unaffected people, so that QOL is compromised.

In their project Maia et al. (2015) studied 40 people with albinism (who were treated at the authors' medical institution in Sao Paulo; 89% had a visual deficit and 60% had skin cancer) and 40 unaffected controls, matched for age and sex, using a WHO QOL questionnaire. Results indicated that there was a statistically significant difference between the groups in the physical domain. This domain included sections such as pain and discomfort, energy and fatigue, sleep, activities of daily living, dependence on medicines, and work capacity. This finding suggested that the affected persons' QOL was altered from a physical point of view because of their skin problems and reduced visual acuity. The researchers stated that affected people, at their clinic appointments, also made frequent complaints about personal and family problems and work relationships and mentioned feeling like the victims of prejudice. However, these complaints did not influence their responses to the QOL questionnaire. Possibly the instrument used in this study was not sufficiently sensitive to detect these issues, or the sample was too small. They concluded that it is necessary to provide better health care for people with albinism. Considering the biopsychosocial aspects of the condition, such care should focus not only on skin and eye care but also on family, genetic, and employment guidance, as well as disseminating information and demystifying the disorder.

Another project, undertaken in the United States, on QOL and albinism focused on vision-specific factors. The authors (Kutzbach et al., 2009) acknowledged that impaired vision affects many aspects of daily life. Measurements of visual acuity in people with albinism do not reflect their visual functioning, health status, and ability to cope with the activities of daily living. The National Eye Institute Visual Functioning Questionnaire was used with 44 affected participants to assess the effects of their visual deficits on their QOL. The findings showed that 48% of subjects were able to drive, but they reported problems with distance acuity, vision-specific mental health, and vision-specific roles.

Ezzedine et al. (2015) investigated patients with vitiligo, a rare acquired idiopathic skin disease associated with a progressive loss of skin color, in France. Their results indicated that although all their patients showed impaired QOL, increased self-perceived stress levels, and reduced self-image, those with fairer skin were more concerned with skin cancer risks and those with darker skin were more worried about their physical appearance. Subsequently, the same team (Salzes et al., 2016) studied QOL in their patients. They found that vitiligo has a major impact on health-related QOL. They developed a scale (The Vitiligo Impact Patient Scale) to assess the burden of having the condition. During this development they showed that three factors were problematic (some in fair-skinned patients, some in dark-skinned patients, and some in both): psychological effects on daily life; relationship and sexuality and economic constraints; and care and management of the condition.

The few published reports available show that QOL is probably compromised, in people with albinism, not only in Africa but also elsewhere. However, individuals affected by many different chronic illnesses and genetic disorders show comparable levels of adjustment, and these are more positive than those of patients who are depressed (reviewed by Biesecker and Erby, 2008). In this minireview, the researchers observed that most individuals adapt well to their illness over time and this is possibly because of the general tendency that humans have toward positive self-evaluation. Also chronic illness may be accepted as having an impact that is circumscribed, rather than global, for most people.

IX. COMMUNITY ATTITUDES

An attitude has been defined as a "mental and neural state of readiness organized through experience exerting a directive or dynamic influence upon the individual's response to all objects and situations with which it is related" (Allport, 1935, p. 798). Triandis (1971, p. 2) states that "an attitude is an idea charged with emotion which predisposes a class of actions to a particular class of social situations." The three interactive components involved include cognitive, affective, and behavioral components. Furthermore, attitudes are evaluations that can affect almost any aspect of life (Baron and Branscombe, 2012). They are often acquired from other people by means of social learning, social networking, and conditioning. Attitudes can influence and predict behavior, and shape perceptions and interpretations of situations. They can be changed by persuasive messages focused on the source, the message, and the targeted audience. However, although researchers are beginning to understand the cognitive processes that occur in persuasion and the way in which the message is processed, the audience may be resistant to change. People, also, may maintain their attitudes by "selective avoidance" and disregarding information that does not support their current view (Baron and Branscombe, 2012).

Historically people with albinism have been treated with a great variety of reactions and attitudes (some of which have been covered in Chapter 1), which have been reported as ranging from extremely negative to very positive. Attitudes such as fear and rejection, being thought unlucky and/or killed at birth, in countries such as Gabon, Cameroon, Botswana, Malawi, and South Africa have been described. Livingstone (1857) presented one of the few detailed reports of a black mother's dilemma in Botswana. This woman cared for her affected child for many years, but she was rejected by the father of the child and excommunicated from her village (where negative attitudes prevailed), while doing so. Eventually she tired of living apart from her family, killed the child, and returned to the father. However, in other parts of Africa, according to old reports (Pearson et al., 1913), attitudes were more positive and individuals with albinism were considered as sacred and inviolate and were under divine protection; whether this is still the case is unknown.

Beckham (1946) also mentioned the negative community reactions and discrimination against individuals with albinism in the urban communities where he worked in Chicago.

While Barnicot (1952) admitted that, although his sociological information was superficial, he had observed, during his Nigerian research work, that the mothers of affected children considered the condition as a misfortune. He added that the attitudes toward albinism seemed to have varied considerably in different parts of West Africa and that it was impossible to tell for how long a particular attitude had existed.

Among the Hopi Indians in Arizona affected people were said to be good luck charms and have some religious significance (Woolf and Grant, 1962; Woolf and Dukepoo, 1969). However, after further investigation into local culture these researchers found that individuals with albinism had no supernatural influence and were not considered to bring good luck; they were generally viewed as any other individual with a congenital defect and were well integrated into Hopi society (where the incidence is high and the condition is familiar to most members of the community). Nevertheless, they noted changing attitudes: the younger Hopi Indians who learned the condition was a genetic defect might show discrimination, whereas the older traditional people showed only positive attitudes. They commented that affected people were generally regarded affectionately in Hopi villages and there was admiration for their whiteness.

Attitudes, both of other people toward people with albinism and the affected themselves toward each other, were investigated in one African study (Kromberg and Jenkins, 1984). Attitudes were assessed by means of a schedule of questions, used in an interview with 35 people with albinism and 35 unaffected matched controls. The results showed that for the items measuring social distance there was very little difference between the two groups, although there may have been some element of social desirability bias in the subjects' responses. The majority in both groups agreed that they would travel with, work and eat with, and make friends with affected people. They also agreed, in principle, that unaffected people should marry people with albinism and they themselves would allow their children to do so. One of the few significant differences was regarding finding employment and the attitudes among the subjects with albinism were more negative than those of the controls. Although the subjects were young, those with albinism appeared to be anticipating that obtaining work would be a serious problem for them. Another significant difference was to do with the item on death. And, in support of the local popular belief, more controls than affected subjects said that individuals with albinism would die in mysterious circumstances or just disappear.

Also, in this study, a few items explored the extent of the knowledge on albinism of the affected and unaffected subjects. The findings showed that there was much ignorance about the causes of albinism, the hereditary process, and the associated physical problems. The ideas on causes included a gift or punishment from God and maternal impression, but most did not know the cause. Some had observed that the child of an affected person with a black partner would usually be black, but most did not understand the inheritance pattern at all. Most subjects in both groups thought the problems were chiefly physical, but a few (one subject with albinism and six controls) mentioned psychosocial problems too. One subject with albinism mentioned the attitudes of parents and said:

> Lack of love and tender care from parents and relatives leads to skin and eye problems *Kromberg and Jenkins (1984, p. 104)*

Examples of responses from four controls are worthwhile reporting here as they illustrate observations, which could affect the attitude of the observer and the community or alternatively could have been affected by community attitudes:

> 'Being black by birth but not by colour may be one of their major problems; this deprives them of the companionship of other black children'
> 'They cannot easily make friends because of skin and eye problems'
> 'They keep to themselves'
> 'They are never relaxed in company' *Kromberg and Jenkins (1984, p. 104)*

In general, findings from this study showed that community attitudes toward people with albinism appear to be reasonably positive and they seem to be quite well accepted in this South African urban community. Again the problem of social desirability bias might be occurring in the subjects' responses to the questions they were asked, or alternatively, urbanization and familiarity with the disorder might be associated with increased tolerance of albinism. Many members of the urban black community know the condition and are accustomed to affected people appearing among them occasionally. However, observation suggests that these positive attitudes do not extend to marriage, and reported South African data show that out of 20 women and 28 men of marriageable age (>20 years or more), only 5 females and 5 males were married (Kromberg and Jenkins, 1984). It is accepted that attitudes inferred from a person's statements about a theoretical situation or object and how the person behaves when confronted with that situation are not always consistent. Behavior is determined by what people would like to do, by what they think they should do (i.e., social norms), by what they have usually done (habits), and by the expected consequences of that behavior (Triandis, 1971). Other studies, undertaken in Cameroon (Vallois, 1950) and on the Hopi Indians in the United States (Woolf and Dukepoo, 1969), have also found that attitudes regarding marriage with a person with albinism appear to be negative and affected people seldom marry. In the latter study, the authors commented that in a dark-skinned population albinism seemed to have such a repressive effect on the affected person that suitors were rejected. However, the real or subconscious bias concerning marrying a person with albinism, in that community, did not manifest in choice of a sexual partner, and in one local legend an old affected unmarried man was said to have had 15 children.

The attitudes of a sample of 37 mothers of children with albinism and 37 matched control mothers with normally pigmented infants were studied in South Africa (Kromberg et al., 1987). Maternal attitudes in the former group were found to change over time. Initially, 2 weeks after the delivery, the observed behavior of the mothers with affected infants was significantly different from that of the controls, suggesting some rejection of the infant, as well as negative attitudes. However, by 9 months after the birth, mothers showed more positive attitudes and their behavior regarding the infant did not differ significantly from that of the control mothers. At this time also, the mothers of affected children were interacting with the infants significantly more than they did at the first interview, showing some acceptance. Kennedy (1970) also recognized that recovery from the effects of having an infant with a birth defect, or one that is not like the one that was expected, is gradual and takes time.

A further study (Kromberg, 1992), on attitudes of unaffected people to people with albinism as well as attitudes of affected people toward other affected people, showed that attitudes fell into the neutral to positive range and affected people were quite well accepted in their urban South African community (where the prevalence is high, 1 in 4000, Kromberg and Jenkins, 1982). However, there was a significant difference in the male group and the controls were less accepting of albinism than the affected males themselves. On the social distance scale the results showed that the closer the social distance became the more negative the attitudes became. The relationship between attitudes and behavior appears to be a function of the level of social involvement with the attitude object and the amount of prior experience with it (Linn, 1965).

X. CULTURAL BELIEFS, MYTHS AND SUPERSTITIONS

The term myth as used by Lestrade (1937, p. 292) covers the types of prose narrative, which "deal with the creation of the world, the gods, and spirits of Bantu belief and the origins of things and natural phenomena." Such myths may be an integral part of the local worldview, which is "essentially an (cognitive) attempt to make sense of the world and to impose meaning on it" (Hammond-Tooke, 1989, p. 33).

Many culturally related beliefs, myths, and superstitions surround albinism in Africa. Pearson et al. (1913) mentioned several of these myths in their 1911–13 monograph on albinism and some of these have been discussed in Chapter 1. For example, Vossius described people with albinism living along the Nile in 1666 and said they were believed to be exceptionally strong, although nearly blind (Pearson et al., 1913). In 1688 de la Croix reported that although affected people were kept in the king's court in Loango, they were considered to be "monsters"; while in the 1700s, Equini, an African himself, reported that they were regarded as deformed in Africa (Pearson et al., 1913). On the other hand people with albinism were sometimes favored as doctors by certain chiefs (Livingstone, 1857). Recently, in the author's experience, one highly successful traditional healer living in Johannesburg, South Africa, believed her success was due to her having a child with albinism. She attended a genetic counseling clinic because she wanted her second child to have the condition too.

Further reports included in the monograph by Pearson et al. (1913) stated that: in the Sudan, people were expected to care for affected people, never refuse them anything or expect them to work; in Ghana (previously Ashanti) affected people were believed to be special and under divine protection; while in the Congo they were given to the king to be educated and trained as sorcerers. In other communities fear predominated, particularly after the birth of an affected child, and because they were considered unlucky infanticide often occurred. Also, fear was reported when marriage was discussed and young men would neither marry an affected girl nor would a father sell his daughter to a man with albinism.

Because these myths varied from country to country, the author (Kromberg, 1992) set up a study to investigate myths and superstitions among the local black community in South Africa. Open-ended questions were used, by a trained interviewer (who communicated in the vernacular with the participants), to obtain information on their beliefs regarding the causes of albinism and its associated problems. Because the researcher was aware, from her many years of working with affected people, that there was a belief that people with albinism did not die naturally, an open-ended question (What happens to albinos at the end of their lives?) was added to the interview schedule. The responses to this question showed that 15 controls (43%) believed that affected individuals vanished at the end of their lives; although no affected person agreed, they showed much uncertainty, 11 (31%) stating that they were not sure what happened and 7 (20%) that they die but they had never heard of the death of a person with albinism; only 13 (37%) responded confidently that they died as other people did. Two controls stated that:

> "I always hear people say albinos are never buried, they walk away and disappear to die far from home"
> "They fall ill, recover quickly and after this disappear to die away from their people. Whether death follows or not is still a mystery" Kromberg (1992, p. 163)

The widely held belief in this death myth could be limiting the full acceptance of people with albinism in this community. As this study showed, some of the affected people themselves are aware of this myth and uncertain about their humanity and mortality. This could distort their interactions with other people leading to a reduction in normal communication and interaction. Positive attitudes develop when the majority of people have familiarity with the affected person, and if these individuals choose to limit such exposure, their behavior could affect community attitudes and lead to an increased development of negative attitudes.

Why the death myth should have become so widespread is poorly understood. However, from local anthropological studies (Hoernle, 1937), the color white has been shown to have special significance in local cultures. White beads, white skins, and white cow tails are used by traditional healers when attempting communication with deceased ancestors and their spirits. Also, at the same time, only white goats are sacrificed. Therefore, because people with albinism are white (and, by inference, "real" people are black, in Africa) perhaps affected individuals are seen as spirits (or the reincarnation of spirits as one diviner suggested). Because spirits cannot die, a belief that neither can affected people die could follow. To strengthen this argument, it has been reported that when the black people of Africa first saw white people, they said they were not human and attributed godlike powers to them (Mutwa, 1966). The origin of the myth, however, may be based in antiquity. If one considers a purely practical explanation, this might involve the possibility that when people with albinism develop unsightly skin cancers (often on the face), they no longer wish to show themselves in public, so they withdraw from society and their death is then unseen, unknown, and unreported. It is difficult to unravel the mystery associated with this widely held and curious belief. Albinism is the only genetic condition known to be linked with such a belief, and this belief appears to be prevalent only in Africa.

A more recent study (based in Zimbabwe and South Africa) of the myths surrounding people with albinism was conducted by Baker et al. (2010). They studied the ways in which the myths and superstitions impact on the lives of affected people, interfere with their normal functioning, and their access to education, employment, and marriage. Firsthand accounts were collected from people with albinism. Then suggestions were made on how the negative connotations associated with the myths can be challenged by scientific explanations.

The authors (Baker et al., 2010) of this article comment that, generally, myths are maintained to explain and account for phenomena, which are unusual and cannot be understood in familiar terms. Myths are required to meet the common human need for making sense of unexpected and unanticipated events. In many parts of Africa, the spiritual world is integrated into the present world, and spirits are believed to be capable of influencing happenings in the present world. It is well documented that people in South Africa and elsewhere often consult the spirits of their ancestors to explain the cause of disease, to find out why the spirits are angry, and/or to ask for blessing (Hammond-Tooke, 1989). The widely held belief in these myths is one of the greatest hindrances to affected people taking their full part in society, finding their true identity, and gaining social acceptance (Baker et al., 2010). Myths reported to these authors by study participants included: the cause being a curse on the family for marital misdemeanors or infidelity; albinism being infectious; the mother slept with a tikolosh (a small mythical evil creature, Hammond-Tooke, 1989), or a white man; and punishment for doing bad deeds.

Baker et al. (2010) also discuss the occurrence of infanticide, suggesting that, even in recent times, fear of and rejection of infants with albinism could be so severe that they could lead to midwives in Zimbabwe and Venda (in Northern South Africa), for example, leaving the child to die or strangling the affected infant and returning to the mother saying the "roof had fallen on its head" (Baker et al., 2010, p. 173). Also, children born in urban areas might be taken back to the rural area village where they die and are buried.

Another myth, or common misconception, discussed by Baker et al. (2010), is that associated with the perception that albinism is a contagious condition. This misconception can be so widespread that nurses might believe it and refuse to touch patients with albinism and passersby might refuse to buy vegetables and fruit from street vendors who have albinism. The fear of contagion can take many forms, e.g., refusing to shake hands, or sit next to, or share a taxi or utensils, or keeping a distance from affected people. Such beliefs can result in increased reclusive behavior in the affected person, withdrawal, and the development of inferiority complexes. Also, the consequence of such beliefs is social isolation of the person with albinism and the reduction of health, employment, and social opportunities for him/her. Modern myths are also mentioned and one suggests that sex with an affected person is a cure for AIDS. This myth makes life for women with albinism, who are already vulnerable in many ways, even more difficult.

However, the most threatening modern myth is that medicine (known to many Africans as "muti") made from the body parts or skin of individuals with albinism is very powerful. As a result, affected people are often the victims of crime and ritual mutilations or killings in several countries in Southern, Central, and Eastern Africa (Diale, 2016). The myth surrounding the power of medicines made of human body parts has its origins in the past. Body parts, particularly the skin, of chiefs were often used to make strong medicines; according to one report the chief's body, in several ethnic groups (Venda, Northern Sotho, and Lobedu of South

Africa), was not buried but kept and washed regularly until the skin peeled off and then the skin was used to make medicine, which was provided to the new chief to use in rainmaking rituals (Eiselen and Schapera, 1937). This tradition provided direct means of passing on the power from one leader to another and establishing close contact between the new rulers of the ethnic group and past generations of chiefs.

Another modern myth is being held by people who are beginning to understand the genetics of albinism and accept that it is an inherited condition. This myth suggests that if people with albinism do not reproduce, then the prevalence of albinism will decrease. This myth is based on a misunderstanding of the genetics of albinism (see Chapter 10) and the fact that most affected infants are born to parents with skin color in the normal range but who each carry one mutated gene for albinism. When that one gene mutation is passed to the child by both father and mother, that child will have a double dose of the mutated gene and will have the condition. Therefore preventing people with albinism from reproducing will not reduce the incidence of the condition.

The death myth is common in several African countries and has been reported in South Africa, Zimbabwe, and Cameroon. It appears that in Southern Africa the death myth might be associated with the intermediate nature of people with albinism who may be considered to be positioned between two worlds, to live among the living and the dead (Baker and Djatou, 2007), or to be neither black nor white. However, the status of the person with albinism could be compared, in some ways, with that of other marginal people, such as the colored (or mixed ancestry) people of South Africa (as described by Mann, 1957, in his doctoral research on the psychology of this group, who are neither black nor white.). People with albinism have also been called twilight people (and shadow people) in the Transkei (now the Eastern Cape Province) area of South Africa.

Because these widely believed myths impact on so many aspects of the lives of people with albinism, Baker et al. (2010) considered ways in which to challenge both the belief in the common myths and the stereotyping of albinism, as well as to change the prevailing attitudes. One way is to increase the education of people with albinism and their families and health workers (Lund, 2001), as well as their traditional healers. Counseling is another useful tactic, and in Johannesburg in the major hospitals all mothers who deliver an affected infant should be referred for counseling to the local genetic counseling clinic (Kromberg, personal communication, 2016). A third way is for families with an affected child to establish relationships with friends, relatives, and the community, to encourage acceptance of the condition, and to link up with the local support group or association for people with albinism. Such support groups, community work, and awareness programs will be covered in later chapters (see Chapters 12 and 13).

XI. SUMMARY AND CONCLUSION

The psychosocial issues covered in this chapter have focused on the person with albinism, his/her family, and community. The birth of an affected child affects not only the parents but also the siblings and grandmother. How the child adjusts depends on family support, on factors in his/her own personality, and on the way in which the self develops in response to the experiences encountered in daily living. Stigmatization is almost inevitable in African societies and it is one of the issues the affected person has to recognize and manage, with the help of the family and, if available, local health professionals and school teachers. The development of social identity and self-image is difficult in a rejecting society, but how this affects behavior is not yet apparent. Educational issues are central to adolescent development, especially since intelligence of affected people is within the normal range. Their education should be managed, preferably in the community, according to the needs of the child and family and with the aid of an enlightened teacher, so that the child can reach his/her potential.

QOL has not been carefully examined in Africa as yet. It appears that stigmatization, poor vision, and sensitive skin, with the high risk of skin cancer, as well as the lack of resources, community, and sometimes even family support, all detrimentally affect QOL. These factors also influence the individual development of feelings of security, personal identity, self-concept, and self-image in people with albinism. However, most people are resilient and most of those with genetic disorders can work through their challenges and adapt adequately to their situation.

Community attitudes, cultural beliefs, myths and superstitions have been discussed briefly here. Attitudes appear to be more positive in urban than in the rural areas, where myths abound, superstition is still powerful and traditional healers still hold sway. In some situations these culturally based beliefs put the lives of affected people in danger. African governments and nongovernmental organizations in several countries are recognizing the vulnerability and needs of their numerous disabled and disadvantaged subgroups (including those with albinism) and are starting to provide targeted health-care services and community awareness programs, which will be of benefit to such groups, and these will be covered in a later chapter (see Chapter 12).

References

Allport, G., 1935. Attitudes. In: Murchison, C. (Ed.), Handbook of Social Psychology. Clark University Press, Worcester.

Baker, C., Djatou, M., 2007. Literary and anthropological perspectives on albinism. In: Baker, C., Norridge, Z. (Eds.), Crossing Places: New Research in African Studies. Cambridge Scholars Publishing, Newcastle, pp. 63–75.

Baker, C., Lund, P., Nyathi, R., Taylor, J., 2010. The myths surrounding people with albinism in South Africa and Zimbabwe. J. Afr. Cult. Stud. 22 (2), 169–181.

Barnicot, N.A., 1952. Albinism in South West Nigeria. Ann. Eugen. 17 (1), 38–73.

Baron, R.A., Branscombe, N.R., 2012. Social Psychology, thirteenth ed. Pearson Education, New Jersey.

Beckham, A.S., 1946. Albinism in negro children. J. Genet. Psychol. 69, 199–215.

Biesecker, B.B., Erby, L.H., 2008. Adaptation to living with a genetic condition or risk: a mini- review. Clin. Genet. 74 (5), 401–407.

Bowlby, J., 1977. The making and breaking of affectional bonds. I. Aetiology and psychopathology in the light of attachment theory. Br. J. Psychiatry 130, 201–210.

Braathen, S.H., Ingstad, I., 2006. Albinism in Malawi: knowledge and beliefs from an African setting. Disabil. Soc. 21 (6), 599–611.

Childs, B., 1978. Priorities for research. In: Littlefield, J.W., De Grouchy, J. (Eds.), Birth Defects. Fifth International Conference Proceedings, Montreal. International Congress Series, vol. 432. Excerpta Med., Amsterdam.

Cohen, J.S., Biesecker, B.B., 2010. Quality of life in rare genetic conditions. A systematic review of the literature. Am. J. Med. Genet. Part A 152A, 1136–1156.

Cowen, L., Mok, J., Corey, M., MacMillan, H., Simmons, R., Levison, H., 1986. Psychologic adjustment of the family with a member who has cystic fibrosis. Pediatrics 77 (5), 745–753.

Diale, L., 2016. Albinism Still Has Stigma. Tuesday 31 May, 2. The New Age Newspaper.

Drotar, D., Baskiewicz, A., Irvin, N., Kennel, J., Klaus, M., 1975. The adaptation of parents to the birth of an infant with a congenital malformation: a hypothetical model. Pediatrics 56, 710–717.

Eiselen, W.M., Schapera, I., 1937. Religious Beliefs and Practices. In: Schapera, I. (Ed.), The Bantu Speaking Tribes of South Africa. Routledge, London, pp. 247–270.

English, H.B., English, A.C., 1970. A Comprehensive Dictionary of Psychological and Psychoanalytical Terms: A Guide to Usage. Longmans Group, London.

Estrada-Hernandez, N., Harper, D.C., 2007. Research on psychological and personal aspects of albinism: a critical review. Rehabil. Psychol. 52 (1), 263–271.

Ezeilo, B.N., 1989. Psychological aspects of albinism: an exploratory study with Nigerian (Igbo) albino subjects. Soc. Sci. Med. 29 (9), 1129–1131.

Ezzedine, K., Grimes, P.E., Meurant, J.-M., Seneschal, J., Leute-Labreze, C., Ballanger, F., Jouray, T., Taieb, C., Taieb, A., 2015. Living with vitiligo: results from a national survey indicate differences between skin phototypes. Br. J. Dermatol. 173, 607–609.

Flanagan, J.C., 1982. Measurement of quality of life: current state of the art. Arch. Phys. Med. Rehabil. 63, 56–59.

Fraiberg, S., 1977. Insights from the Blind. The New American Library, New York.

Fuller, R.L., Geiss, S., 1985. The significance of skin color of a newborn infant. Am. J. Dis. Child. 139, 672–673.

Gallo, A.M., Angst, D., Knafl, K.A., Hadley, E., Smith, C., 2005. Parents sharing information with their children about genetic conditions. September/October J. Pediatr. Health Care 267–275.

Goffman, E., 2006. Selections from stigma. In: Davis, L.J. (Ed.), Disability Studies Reader, second ed. Routledge, New York, pp. 131–140.

Hammond-Tooke, D., 1989. Rituals and Medicine. A.D. Donker, Johannesburg.

Heijnders, M., van der Meij, S., 2006. The fight against stigma: an overview of stigma-reducing strategies and interventions. Psychol. Health Med. 11 (3), 353–363.

Hoernle, A.W., 1937. Magic and medicine. In: Schapera, I. (Ed.), The Bantu-Speaking Tribes of South Africa. Routledge, London, pp. 221–245.

Houtzager, B.A., Grootenhuis, M.A., Caron, H.N., Last, B.F., 2004. Sibling self-report, parental proxies and quality of life: the importance of multiple informants for siblings of a critically ill child. J. Paediatr. Haematol. Oncol. 22 (1), 25–40.

Hutson, S.P., Alter, B.P., 2007. Experiences of siblings with Fanconi anemia. Pediatr. Blood Cancer 48 (1), 72–79.

Johns, N., 1971. Family reactions to the birth of a child with a congenital abnormality. Med. J. Aust. 1, 277–282.

Keeler, C.E., 1964. The Cuna moon-child syndrome. Dermatol. Trop. 3, 1.

Kennedy, J.F., 1970. Maternal reactions to the birth of a defective child. Soc. Casework 51, 410–416.

Kromberg, J.G.R., 1985. A Genetic and Psychosocial Study of Albinism in Southern Africa (Ph.D. thesis). University of the Witwatersrand, Johannesburg, South Africa.

Kromberg, J., 1992. Albinism in the South African Negro: IV. Attitudes and the death myth. Birth Defects Orig. Artic. Ser. 28 (1), 159–166.

Kromberg, J.G.R., Jenkins, T., 1982. Prevalence of albinism in the South African Negro. S. Afr. Med. J. 61, 383–386.

Kromberg, J.G.R., Jenkins, T., 1984. Albinism in the South African Negro. III genetic counselling issues. J. Biosoc. Sci. 16, 99–108.

Kromberg, J.G.R., Jenkins, T., 1997. Cultural influences on the perception of genetic disorders in the black population of Southern Africa. In: Clarke, A., Parsons, E. (Eds.), Culture, Kinship and Genes. MacMillan, London, pp. 147–157.

Kromberg, J.G.R., Zwane, E.M., Jenkins, T., 1987. The response of black mothers to the birth of an albino infant. Am. J. Dis. Child. 141, 911–916.

Kutzbach, B.R., Merrill, K.S., Hogue, K.M., Downes, S.J., Holleschau, A.M., MacDonald, J.T., Summers, C.G., 2009. Evaluation of vision-specific quality of life in albinism. J. AAPOS 13, 191–195.

Lestrade, G.P., 1937. Traditional literature (Chapter XIII). In: Schapera, I. (Ed.), The Bantu Speaking Tribes of South Africa. Routledge, London, pp. 291–308.

Linn, L.S., 1965. Verbal attitudes and overt behaviour, a study of racial discrimination. Soc. Forces XLIII, 353–364.

Liveneh, H., Antonak, R.F., 2005. Psychosocial adaptation to chronic illness and disability: a primer for counselors. J. Couns. Dev. 83, 12–20.

Livingstone, D., 1857. Missionary Travels. John Murray, London.

Lund, P.M., 2001. Health and education of children with albinism in Zimbabwe. Health Educ. Res. 16 (1), 1–7.

Lund, P., Gaigher, R., 2002. A health intervention programme for children with albinism at a special school in South Africa. Health Educ. Res. 17 (3), 365–372.

Lynch, P., Lund, P., Massah, B., 2014. Identifying strategies to enhance inclusion of visually impaired children with albinism in Malawi. Int. J. Educ. Dev. 39, 226–234.

Maia, M., dos Santos, G.A., Volpini, M.B.M.F., Rujula, M.J.P., 2015. Quality of Life in patients with oculocutaneous albinism. Ann. Bras. Dermatol. 90 (4), 513–517.

Manganyi, N.C., 1972. Body image boundary differentiation and self-steering behaviour in African paraplegics. J. Personal. Assess. 36, 45.

Manganyi, N.C., Kromberg, J.G.R., Jenkins, T., 1974. Studies on albinism in the South African Negro. 1. Intellectual maturity and body image differentiation. J. Biosoc. Sci. 6, 107–112.

Mann, J.W., 1957. The Problem of the Marginal Personality. A Psychological Study of a Coloured Group (Ph.D. thesis). University of Natal, Durban, South Africa.

Markus, H.R., Kitayama, S., 2010. Cultures and selves: a cycle of mutual constitution. Perspect. Psychol. Sci. 5 (4), 420–430.

McAllister, M., 2016. Genomics and patient empowerment. In: Kumar, D., Chadwick, R. (Eds.), Genomics and Society. Elsevier, Amsterdam, pp. 39–68.

Metcalfe, A., Plumridge, G., Coad, J., Shanks, A., Gill, P., 2011. Parents and children's communication about genetic risk: qualitative study learning from families' experiences. Eur. J. Hum. Genet. 19, 640–646.

Mintzer, D., Als, H., Trotnick, E.Z., Brazelton, T.B., 1992. Parenting an Infant with a Birth Defect: The Regulation of Self-Esteem. Zero to Three Classics: 7 articles on Infant/Toddler Development. National Center for Clinical Infant Programs, Arlington, VA, pp. 7–16.

Munyere, A., 2004. Living with a disability that others do not understand. Br. J. Spec. Educ. 31 (1), 31–32.

Murphy, R., 1995. Encounter: the body silent in America. In: Ingstad, B., Whyte, S.R. (Eds.), Disability and Culture. Henry Holt, New York, pp. 140–158.

Mutwa, V.C., 1966. Indaba My Children. Kahn and Averill, London.

Parker, R., Aggleton, P., 2003. HIV and AIDS-related stigma and discrimination: a conceptual framework and implications for action. Soc. Sci. Med. 57, 13–24.

Pearson, K., Nettleship, E., Usher, C.H., 1913. A monograph on albinism in man. Drapers Co, London. In: Research Memoirs Biometric Series, vol. VIII. Dulau, London.

Penn, C., Watermeyer, J., MacDonald, C., Moabelo, C., 2010. Grandmothers as gems of genetic wisdom: exploring South African traditional beliefs about causes of childhood genetic disorders. J. Genet. Couns. 19, 9–21.

Plumridge, G., Metcalfe, A., Coad, J., Gill, P., 2011. Parents' communication with siblings of children affected by a genetic condition. J. Genet. Couns. 20, 374–383.

Salzes, C., Abadie, S., Seneschal, J., Whitton, M., Meurant, J.-M., Joury, T., Ballanger, F., Boralevi, F., Taieb, A., Taieb, C., Ezzedine, K., 2016. The vitiligo impact patient scale (VIPs): development and validation of a vitiligo burden assessment tool. J. Investig. Dermatol. 136, 52–58.

Slutzky, H., 1969. Maternal reaction and adjustment to the birth and care of a cleft palate child. Cleft Palate J. 6, 425–429.

Stewart, H.F., Keeler, C.E., 1965. A comparison of the intelligence and personality of moon child albino and control Cuna Indians. J. Genet. Psychol. 106, 319–324.

Tajfel, H., Turner, J., 1997. An Integrative Theory of Intergroup Conflict. The Social Psychology of Intergroup Relations. Brooks-Cole, Monterey, CA, pp. 33–47.

Triandis, H.C., 1971. Attitude and Attitude Change. John Wiley and Sons, New York.

Triandis, H.C., 1989. The self and social behaviour in differing cultural contexts. Psychol. Rev. 96 (3), 506–520.

Tumwesigye, C., Musukwa, G., Njuguna, M., Shilio, B., Courtwright, P., Lewellan, S., 2009. Inappropriate enrolment of children in schools for the visually impaired in east Africa. Ann. Trop. Paediatr. 29 (2), 135–139.

Vallois, H.V., 1950. Sur quelques points de L'anthropologie des Noirs. L'anthropologie 54, 272–286.

Vernon, P.E., 1964. Personality Assessment, a Critical Survey (Methuen, London).

Wan, N., 2003. Orange in a world of apples: the voice of albinism. Disabil. Soc. 18 (3), 277–296.

Watkins, C.E., 1995. Pathological attachment styles in psychotherapy supervision. Psychotherapy 32 (2), 333–340.

Weil, J., 2000. Psychosocial Genetic Counseling. Oxford University Press, Oxford.

Wiener, L.S., Battles, H.B., Heilman, N., Sigelman, C.K., Pizzo, P.A., 1996. Factors associated with disclosure of diagnosis to children with HIV/AIDS. Pediatr. AIDS Infect. Fetus Adolesc. 7, 310–324.

Wonkam, A., Mba, C.Z., Mbanya, D., Ngogang, J., Ramesar, R., Angwafo, F.F., 2014. Psychosocial burden of sickle cell disease on parents with an affected child in Cameroon. J. Genet. Couns. 23, 192–201.

Woolf, C.M., Dukepoo, F.C., 1969. Hopi Indians, inbreeding and albinism. Science 164, 30–37.

Woolf, C.M., Grant, R.B., 1962. Albinism among the Hopi Indians of Arizona. Am. J. Med. Genet. 14, 391–399.

World Health Organization, Quality of Life group, 1996. What Quality of Life? World Health Forum. WHO, Geneva, pp. 354–356.

Wright, B.A., 1960. Physical Disability – A Psychological Approach. Harper and Row, New York.

Young, V.S., 2012. People with Albinism and Their Experience of Felt Stigma: Results of a Nationwide Survey (MA dissertation). New York University, New York.

10

Genetic Counseling and Albinism

Jennifer G.R. Kromberg

University of the Witwatersrand and National Health Laboratory Service,
Johannesburg, South Africa

I. INTRODUCTION

People with albinism, as well as every family with a child or member with albinism, should be offered genetic counseling. The fact that many people affected with a genetic disorder, as well as their families, do not understand or have a very limited understanding of the genetics of their condition (Morris et al., 2015) points to the need for such counseling. Furthermore, the benefits of genetic counseling have been recognized and documented (Aalfs et al., 2007). The main benefit is the provision of genetic and other relevant information, which can be empowering for those who need it (McAllister, 2016). In the case of albinism, the informed genetic counselor can also discuss the psychosocial and cultural issues (see Chapter 9), as well as debunking the myths around the condition, how it occurred, and what caused it. The evidence shows that the advantages of receiving such information are associated with improved feelings of personal control and empowerment, as well as with emotional gains, such as hope for the future. These benefits are best achieved through providing appropriate information in the context of a genetic counseling session. Such genetic counseling may be followed by positive behavior changes leading to important long-term health benefits (like the use of sun barrier creams to prevent skin cancer in people with albinism). As one mother of a child with albinism, who had had genetic counseling in Johannesburg, South Africa, recently stated:

> I think it [the genetic counseling service] should be advertised. People should be aware of it and have knowledge, because it is relevant for children with albinism… They also need to have knowledge about why their child is a certain way. What should they do. *Morris et al. (2015, p. 164).*

People from differing backgrounds, at different stages in their lives, and facing different genetic disorders will experience genetic counseling differently. Reactions and interactions in a counseling session vary according to personal, family and community beliefs, local healthcare settings, and cultural backgrounds. Motivations for attending for counseling also vary according to the short-term and long-term needs of the individuals and their families. For example, some will need information because they have just had a newborn child with a genetic condition (e.g., mothers with an infant with albinism); others will know of the family history and will want to find out what the risks are for their children; still others will seek support and understanding and want to discuss their adjustment and adaptation problems (e.g., adults with albinism), particularly when their partner, family, and/or community appear to be rejecting their affected child or themselves.

Ideally, genetic counseling should be provided by genetic counselors (trained at master's degree level) or by medical geneticists (medical doctors who have specialized in the field of medical genetics) (Kromberg et al., 2013a). However, when such health professionals are not available, nurses (the main group of health professionals working in primary health care in Africa), preferably with some training in genetics, can provide a basic service. The uptake of genetic counseling services appears to be best when the service is provided in a hospital setting, either in a genetic counseling clinic (where these are available) or attached to an appropriate treatment clinic. In the case of families with a member with albinism, genetic counseling provided at a dermatology clinic, where the skin is checked and sun barrier creams are available, or at an ophthalmology or low-vision clinic, where their visual problems are assessed and treated, would probably suit affected people best.

This chapter will cover a broad view of the field of genetic counseling because it is little known in Africa, its benefits are not yet recognized, and literature on the topic is very scarce. It will discuss what such counseling offers to people affected by genetic disorders in general, how it relates specifically to albinism, how affected people and their families can benefit from using a genetic counseling service, and what needs to be covered in a counseling session. It will include a discussion on what genetic counseling involves, how it is defined, the nature of the genetic counseling relationship, the role of the genetic counselor, communication, cross-cultural counseling, and decision-making issues. It will also outline the modes of inheritance of genetic conditions (putting albinism and its inheritance into context), briefly, and focus, more comprehensively, on recessive inheritance because this is the category into which most types of albinism fall. Furthermore, the risks that may have to be presented to patients and their relatives, as well as their perceptions and misconceptions about those risks, will be covered. Then some of the psychosocial aspects of genetic counseling will be discussed. Finally, the uptake of counseling services and the ethical issues they raise will be considered. Various terms for the person being counseled by a counselor will be used interchangeably in this chapter. These terms will include patient, client, and counselee.

Because there has been no specific research on genetic counseling for albinism and there is no available relevant published literature on the topic (apart from Kromberg and Jenkins, 1984), the general literature on genetic counseling has been consulted. In addition, where appropriate, the unpublished personal experiences and observations of some of the South African genetic counselors (consulted by the author) who have counseled individuals with albinism and their families have been included.

II. GENETIC COUNSELING

Definitions of Genetic Counseling

There is a wide variety of ways of thinking about genetic counseling even among the professionals working in the field. One definition states that genetic counseling is "the process of helping people understand and adapt to the medical, psychological, and familial implications of the genetic contributions to disease. This process integrates the following:

- Interpretation of family and medical histories to assess the chance of disease occurrence or recurrence.
- Education about inheritance, testing, management, prevention, resources, and research.
- Counseling to promote informed choices and adaptation to the risk or condition" (Resta et al., 2006, p. 79).

A more detailed but older definition states "genetic counseling is a communication process which deals with the human problems associated with the occurrence of a genetic disorder in a family" (Fraser, 1974, p. 663). According to Fraser (1974), the process should involve attempts by one or more trained persons to help the individual and/or family to:

1. understand the medical facts, diagnosis, prognosis, and management of the condition;
2. appreciate the genetics of the disorder and the risk of recurrence in relatives;
3. understand the options available for dealing with the recurrence risk;
4. select the appropriate course of action in view of the risks and family goals, and act according to this decision; and
5. make the best possible adjustment to the disorder in the family.

More recently, Harper (2010, p. 3) writes that genetic counseling is a "composite activity made up of a series of key elements that individually are very different, but which together constitute a process that is highly distinctive in its character and its ethos". These elements include: "Diagnostic and clinical aspects; documentation of family and pedigree information; recognition of inheritance pattern and risk estimation; communication and empathy with those seen; information on available options and further measures; support in decision-making and for decisions made" (Harper, 2010, p. 4).

Together these definitions give a broad insight into the process of genetic counseling, its role and scope, and what it involves.

Genetic Counseling and the Counseling Relationship

Genetic counselors are aware that genetic disorders, such as albinism, are often encompassed by strong, sometimes expressed and sometimes repressed, emotions. Also, the type of counseling required is determined by the mode of inheritance of the particular condition present in the family, the emotional well-being of the family, their educational background and ability to comprehend the information being given, and often by the severity of the condition itself.

Kessler (1999) writes that personal counseling is associated with three related tasks: understanding the client, sharing that understanding, and helping the client feel better about their situation and themselves. In this article, Kessler discusses how genetic counselors might empower their clients to cope more effectively with their issues. Genetic counselors have the opportunity to use both educational and counseling techniques to assist them to meet their clients' needs (see Fig. 10.1).

The working relationship between counselor and counselee is initially developed, in the opening phase of the session, through a discussion of and agreement on mutual expectations and goals to be achieved (Djurdjinovic, 2009; Wessels et al., 2015). When mothers with

FIGURE 10.1   Genetic counseling session with a man with albinism. *Photograph courtesy of Dr. J. Kromberg, University of the Witwatersrand, South Africa.*

newborn infants with albinism are counseled, the counselor needs to observe the maternal-infant interactions and be aware of the possibility of problematic bonding (Kromberg et al., 1987). Mutual respect, as well as empathy and understanding, should be apparent in the interaction. The counselor's and counselee's sense of satisfaction with the session is often dependent on these factors being evident and present throughout (Morris et al., 2015). Within this working relationship the counselor needs to facilitate a balanced discussion of all the items on the participants' agendas and maintain the integrity of the relationship. However, the counselor may need to raise some issues, such as recurrence risks and likely health problems (in children with albinism), if the counselee does not mention these aspects. There needs to be commitment to working empathically within boundaries and respecting confidentiality and patient autonomy. Also, counselors need to have self-awareness and an understanding of their own needs, values, beliefs, and possible emotional responses that could interfere in their providing nondirective and objective counseling and entering into empathic communication.

Throughout the process of genetic counseling, practitioners aim to bolster the client's self-esteem through showing decency, thoughtfulness, serious consideration of the clients' presenting needs, and through giving attention to their story. Such attention to self-esteem may be particularly pertinent in sessions with adults with albinism in Africa because in a small study on affected individuals living in Nigeria, some demonstrated poor self-esteem (Ezeilo, 1989). Expressing kindness, raising the spirits, giving rewards, teaching, and encouraging the application of coping skills also contribute to boosting the client's self-image. Positive reinforcement can be used when listening for information provided by the client, which can then be reflected back to him as showing strengths and accomplishments. In addition, consolation and comfort are required in many situations, including that of giving bad news, which has to be faced in many different genetic counseling sessions.

Unconditional positive regard, empathy, and genuineness are essential in any genetic counseling interaction (Weil, 2000). Unconditional positive regard "involves respecting and accepting the counselee as a complete individual, including his strengths and weaknesses and full range of feelings and behavior." Empathy "involves an understanding, in so far as possible, of the counselees' lived reality. This includes his or her past and present experiences, emotions and perceptions of the world and the roles these play in shaping behavior." Genuineness "involves the counselor's openness to her own emotional experiences in the interaction with the counselee, and a modulated but honest expression of this in her interaction with the counselee" (Weil, 2000, p. 54). These techniques of psychosocial counseling are also valuable in regular daily life and they underlie caring and supportive relationships within families, in general.

The approach taken by most genetic counselors in counseling, particularly regarding decision-making, is usually a nondirective one. Rogers (1942) presented this counseling approach as central. He stated that:

> The non-directive viewpoint places a high value on the right of every individual to be psychologically independent and to maintain his integrity. *Rogers (1942, p.127).*

There has been some debate about whether such nondirective counseling is always possible, whether it is a useful technique for genetic counseling, and/or whether directive (when a counselor attempts to influence and/or direct the client's attitudes or decisions) counseling is required in some situations. However, in the 1980s (Wertz and Fletcher, 1988), and again in the 1990s (Baumiller et al., 1996), this approach was endorsed by genetic communities around the world. Kessler in his thought-provoking essay on nondirectiveness has defined the approach as "procedures aimed at promoting the autonomy and self-directedness of the client" (Kessler, 1997, p. 167). In the case of providing genetic counseling for pregnant women at risk for having a child with albinism, it is especially important for the counselor to maintain an objective view and use the nondirective approach. In particular, the counselor should not make decisions for the client regarding major life events such as marriage, having children, and other important personal issues (Lynch, 1969). Such decisions are the responsibility of the client who should be the only one to exercise this right. Increased feelings of autonomy and the ability to cope should be the outcomes of the patient's counseling experience.

A workshop on nondirectiveness was held at the 2003 Annual Education Conference of the National Society of Genetic Counselors (Weil et al., 2006). At this workshop there was support for a flexible approach to genetic counseling and to nondirectiveness, based on the clinical situation and the needs and values of the clients. The need for the profession to more accurately determine the theoretical basis of genetic counseling was also recognized. However, Weil (2003) suggested that the central ethos should be to bring the psychosocial component into every aspect of counseling. This approach appears to be particularly important in cases of albinism, where the psychosocial issues might outweigh the genetic issues.

Many genetic counseling sessions are one-time visits, but nevertheless, whatever the cognitive aspects of the case, the counselee can still benefit from a therapeutic experience. Communications between counselor and counselee, both verbal and nonverbal, are complex and the context is often complicated (Djurdjinovic, 2009). The patient and the counselor each have their own unique backgrounds of beliefs, values, needs, and expectations, some of which overlap. Each brings to the session their own family and socioethnic community issues, and the counselor also represents an institutional medical system. The counselor's challenge is to offer the patient the fullest possible appropriate and relevant experience, which provides opportunities to discuss, comprehensively, all concerns and questions. This process requires counselors to be familiar with the psychological dynamics of the situation (in the case of albinism, particularly those psychosocial issues presented in the previous chapter of this book) as well as the required genetic information.

Addressing the psychosocial issues in a genetic counseling session depends on the goals set for the session, the experience, willingness and competence of the counselor, and the time available. In a medical setting where many health professionals have other roles, it becomes more essential that genetic counselors remain committed to and able to show the full scope of genetic counseling. The psychological aspects should be intertwined with the genetic and medical discussions in every genetic counseling session to make it a satisfying experience for all participants.

Role of the Genetic Counselor

The role of the genetic counselor has been studied in South Africa (Kromberg et al., 2013b), Europe (Skirton et al., 2015), the United States (National Society of Genetic Counselors, 2010), Australia (Kromberg et al., 2006), and elsewhere. The basis of the practice is good counseling skills, an in-depth knowledge of human genetics and genetic testing, and the ability to communicate complex genetic information to patients. Genetic counselors are trained health professionals with backgrounds in various disciplines, including nursing, social work, psychology, the basic sciences, and genetics. In Africa, they need a thorough knowledge of the genetic disorders that occur at an unusually high prevalence in black African people (such as albinism and sickle cell anemia, although this latter condition is more common in Central than in Southern Africa). They generally work in medical genetics teams, some of which are based in academic settings, some in public health settings and hospitals, and others in private hospitals or private practice. The key members of the team are the medical geneticist and the genetic counselor, who complement each other with their different skill sets and work together to provide a comprehensive service for the patient, together with the laboratory medical scientist and sometimes a genetic nurse.

In Europe, genetic counselors are employed in specialist genetics centers and in specialist units. They are expected to fulfill several roles including the "provision of information and the facilitation of the client to their genetic status and situation" (Skirton et al., 2015, p. 452). The primary role of medical genetic services, within which most counselors work, is to offer care to patients at risk for, or affected by, genetic disorders (Harper, 2007). However, a study of counselors in South Africa shows that, although their major role was patient counseling, they also had teaching, research, marketing, administrative, and community genetics roles (Kromberg et al., 2013b). These additional roles were partly due to the university-based setting in which they worked. These roles mean that these genetic counselors also teach students in the medical and other health professions at university level, conduct research projects on genetic disorders, are involved in community programs, such as workshops on medical genetics for practicing health professionals, and support groups, such as the South African Inherited Disorders Association (now the Genetic Alliance–South Africa), the Huntington disease group, the Albinism group, and others. The few counselors who work either in a laboratory setting or in private practice generally focus on their counseling role, although some involvement in laboratory testing (in the case of those with a scientific background) or marketing (in the case of those in private practice) may be required.

The experienced genetic counselors in South Africa are also involved in training student counselors and both the University of the Witwatersrand and the University of Cape Town run training programs (Greenberg et al., 2012). These are open to suitable applicants who have an appropriate basic degree (preferably a 4-year honors degree in biological science, genetics, social science, with psychology, or nursing). South African trained counselors are now working in many different countries such as the United Kingdom, United States, Australia, and Oman. Also, the first international student is being trained, in Cape Town, at present, with registration at both the University of Cape Town and the University of Oman.

Communication

In common practice in the United States, communication has been broadly defined as "the sharing of messages with the least possible distortions" (Oosterwal, 2009, p. 351). Some distortions are generally unavoidable because the sender and receiver of the message often differ in many ways, for example, in age, gender, education, socioeconomic and ethnic background, religion, personality, and personal experience. Nevertheless, good communicators recognize the needs and interests of others and can identify with them.

As in any interview, communication in genetic counseling sessions needs to be open and honest. Because emotions around genetic disorders may be intense and strong feelings may limit, distort, or prevent effective communication, genetic counselors need to be aware of strategies to improve communication. Communication, of course, can be affected by the cultural background of both the patient and the counselor, especially if they are of different backgrounds, as so often happens in South Africa.

Individual counselors have their own style of communicating, as do their patients. However, the counselor needs to create a caring and concerned atmosphere and a supportive relationship in the initial phase of the session to promote good communication (Davis, 1984). Active listening and appropriate responding encourage patients to communicate and express themselves, as well as their concerns and wishes. The use of open-ended questions further stimulates communication. Also, notably, communication occurs during silences and via nonverbal means, facial expressions, and body movement. Good counseling requires energy, empathy, and time. Counselors should work constantly to promote effective communication, which can enhance the health and well-being of the counselees and lead to more effective genetic counseling services.

In a recent study (Morris et al., 2015) one patient (a Black African mother with a child with Down syndrome) recognized the client/counselor relationship and the unique communication style of her genetic counselor (an English-speaking White female) by stating (in her broken English):

> Firstly, the people they know how to communicate with us and it was the first time seeing healthcare professionals who are like this…even if you don't understand English, they try to speak to you so that you can understand and tell you and explain it to you, read to you, show you pictures… *Morris et al. (2015, p. 162)*.

Effective communication about genetic risk is also essential in many counseling situations. Such communication may help the counselee to participate effectively in problem-solving, decision-making, and future planning. Various communication models have been assessed and, generally, where the genetic risk was communicated well, improvements in outcome for the users, such as the acquiring of knowledge, understanding, and risk information, were shown (Edwards et al., 2008).

Communication within the family is essential if the inherited nature of the condition is to be accepted and coping and adaptation among all family members is to proceed effectively. During a counseling session, genetic counselors will often become aware of the communication pattern occurring in the family. Open communication, generally, is associated with the sharing of information, better coping and planning skills, and improved emotional support within the family (Weil, 2000). Communication patterns vary, however, and some families have open and effective communication, while others will resist such communication, particularly where denial and suppression are used as defense mechanisms. In general, the way family members interact, their family rules, roles and beliefs, and the emotional and geographical distance between members will determine their communication. In Africa, the collectivist, as opposed to the individualistic, nature of many of the local cultures needs to be recognized because patients may not be willing to communicate openly, especially concerning making decisions (e.g., for genetic testing), until they have had discussions with key family members (Kromberg and Jenkins, 1997a).

Cross-Cultural Counseling

Cross-cultural or multicultural counseling "refers to a process whereby a trained professional from one ethno-cultural background interacts with a client from another for the purpose of promoting the client's cognitive, social, emotional and spiritual health and development. Generally, counselors who are competent in multicultural interactions are comfortable with differences in values, assumptions, behaviors and beliefs, and possess the skills to communicate effectively across existing or perceived cultural boundaries" (Oosterwal, 2009, p. 331).

Culture has been defined as integral to the total way of life of a group of people: "the way they act and think, organize themselves, relate and communicate, make or build things, express feelings and emotions, and respond to the world" (Oosterwal, 2009, p. 333). Culture is the most significant factor that determines values, beliefs, attitudes, and customs; it stems from the family and community and is passed on through the generations (Greb, 1998). For effective cross-cultural counseling, genetic counselors need to be aware of how their own cultural background shapes their values, beliefs, and attitudes, how it may influence their clients, and how the culture of the western medical health system within which they work may also affect the situation. They often need to learn to accept the clients' belief systems and tolerate ambiguity. Conflict can arise where there is failure to recognize the cultural differences between the counselor providing the service and the client, and the values of both need to be considered.

Many genetic counseling sessions in South Africa occur in a cross-cultural context because most of the medical geneticists and genetic counselors are White and of European extraction while most of the patients are from the South African Black population, and, in addition, patients may be from other African countries (Kromberg et al., 2013a). This dynamic has also been observed in the United States where, in 1998, ~93% of genetic counselors were White Americans (Weil, 2000) and where the need for greater ethnic diversity in the genetic counseling profession is recognized (Schoonveld et al., 2007), as it is in South Africa.

Genetic counseling services have been offered formally, in South Africa, since 1973 (and informally, on a small scale, for about 15 years before that) (Greenberg et al., 2012) and genetic research has covered many different genetic disorders, including albinism. While offering genetic counseling to Black patients, various cultural influences on the perception of these disorders have been observed (Kromberg and Jenkins, 1997a). The first of these is the traditional worldview and system of thought that favors collective rather than individual thinking (this core value of collectivism is also apparent in many immigrant groups now living and working in the United States, Oosterwal, 2009) and may commonly include an external locus of control. This system may lead to slower decision-making because consultation with various role players may be required and, in the genetic counseling context, may cause the rejection of prenatal diagnosis, even in high-risk situations (Soodyall and Kromberg, 2016). Another common philosophy is that of fatalism. If the belief is that absolute control over life's events is in the hands of powerful supernatural beings, the individual cannot alter what happens to him or her. This way of thinking may again affect decision-making. Also, integral to these systems of thought is a rather indistinct line between life and death, which influences (and could perpetuate), for example, the myth that people with albinism do not die but disappear. It also leads to the belief that the spirits of the ancestors can interfere with daily events for better or for worse; this affects both perception of genetic disorders and beliefs about their causes.

Ancestor worship is surrounded by ritual and tradition. If these rituals are ignored, problems may result including abnormalities in the offspring of the offender. In support of this belief, one of the infants with albinism living in Johannesburg was named "spirit of the ancestors"; her mother was a traditional healer and found that her practice prospered after the birth of this child. However, another common belief is that maternal impression (such as laughing at people with albinism or stepping over their feet, while pregnant) causes albinism in a child. Various other myths, superstitions associated with albinism, and cultural issues have been covered in a previous chapter of this book (see Chapter 9).

Another cultural tradition practiced among the South African Black population (as well as in other African populations), which may impact on how they are counseled in the genetics clinic, is that which influences mate selection. Some groups (such as the Tswana, Schapera, 1937) believe that the best partner is a close relative. In these unions there is a higher rate of albinism because both partners are more likely to carry the same recessive faulty gene than unrelated partners and the gene mutations for albinism are common in Southern African Black ethnic groups (Kromberg and Jenkins, 1982). Also, plural marriages, generally in the form of polygamy (one husband and two or more wives) often occur in Africa (Helman, 1994) and may have to be taken into account when family history is being collected for a counseling session (Greenberg et al., 2012).

Several attitudes toward reproduction, which are associated with cultural beliefs, can affect genetic counseling. The status of a woman may be partly determined by the number of children she has, especially in the rural areas. How this affects the family with a child with albinism is unclear. However, after having one affected child any couple might want to prove they can have an unaffected child and might go ahead with a pregnancy, regardless of the risk of recurrence. Alternatively, they might want a replacement (unaffected) child for the so-called unaffected one they "lost" (when the child with albinism was born). The desire for another child, in this context, might also mean that recurrence risks are not taken into account, when another pregnancy is considered. Cultural adherence and attitudes toward reproduction, however, are changing and women themselves now want to limit their families especially because many are breadwinners and the head of the household. In Soweto, an urban city of about 3,000,000 people, adjacent to Johannesburg, the number of births per woman fell in the 1990s to an average of 2.8 per women compared with 5.7 in deep rural areas, while more recently the total fertility rate for all Black females in the country was 2.38 children per woman (Statistics South Africa, 2009).

Attitudes toward health care are also tied to cultural beliefs in many ways. Many people will still consult the services of both the Western medical health professionals and the traditional healer (Greenberg et al., 2012). Altogether 48% of the mothers of children with Down syndrome, in a research study, claimed to use both services for their affected child (Kromberg and Zwane, 1993). According to these mothers, the traditional healers offered strengthening treatments for the infant, in the form of strings and pouches, to tie around the infant's neck, abdomen, wrist, or ankle and prescribed special herbs and potions, which the mothers accepted and used.

Cultural beliefs and practices, as well as various historical experiences, can create layers of complexity in the multicultural genetic counseling process. The specific cultural factors that can interfere with good genetic counseling include "language; health care beliefs and practices; attitudes toward time, fate, and human instrumentality; and obligations toward and expectations of children, family and society" (Weil, 2000, p. 214). Although the experience of this latter author was mainly in the United States, all these factors have been observed in the present author's experience while providing a genetic counseling service over a 30 year period in South Africa, and some have been mentioned in the recent article by Morris et al. (2015). Empowerment of the patient is particularly important in South Africa, where there are ongoing challenges due to many patients having been historically disempowered and affected by a patriarchal model of medicine and the divides of class, ethnic group, and culture.

Recently, a research project on genetic counseling for indigenous Australians has been completed (Kowal et al., 2015). The findings showed that flexibility, taking time to build rapport, acknowledging different family structures, decision-making processes, and socioeconomic disadvantages were important factors in multicultural interactions. Genetic counselors, therefore, have to be aware of the particular cultural issues and challenges that might impinge on their counseling sessions with patients from ethnocultural groups other than their own.

It appears that there are no available published reports from African countries on the topic of cross-cultural genetic counseling. However, research has shown that when communication takes place between people from the same cultural background it is only successful in 55%–60% of cases, but when counselor and client come from different backgrounds only 20%–25% of communications are effective (Oosterwal, 2009). Nevertheless, it is possible to learn to communicate across cultural barriers and to develop the necessary skills and insight, as well as an understanding of the ways people communicate. Also, obstacles can be recognized and methods of overcoming them learned, so that communication can become more effective. Competency in cross-cultural counseling only occurs when counselors know and understand the client's culture, use their own cultural values as a resource, and approach the task with humility (Oosterwal, 2009).

Decision-Making

Decision-making is a very complex process and in the genetic counseling field it is associated mostly with issues around genetic testing and reproduction. The process may take place over some time, and patients may attend a genetic counseling session with their key decisions already considered and made. Counselees use their values, beliefs, previous experiences, and decisions, as well as both correct and incorrect information, in the process of decision-making. These factors also have personal, family, and social meaning and are influenced by the counselees' community and cultural history (Weil, 2000), particularly in the case of albinism (Kromberg and Jenkins, 1984). The information offered in the genetic counseling session is only useful to counselees if it is integrated into their broader life experience in a meaningful way (Kessler, 1989). One of the major issues in decision-making is reproductive drive or the desire for a child; this may be for a first child when both parents are unaffected but at high risk, for a first child when one parent is affected, for a second child after the first one is affected (or died), for another child after a couple have some affected and some unaffected children, or for a child with a new partner. All these scenarios have been encountered by the writer regarding counseling for albinism. Many counselees who had intended to have a pregnancy before receiving counseling will go ahead and plan one after counseling, regardless of the risk information provided.

The philosophy of genetic counseling supports autonomy of the patient and promotes respect for their right to make autonomous decisions. The genetic counselor is there to facilitate the process and to stimulate dialog about the possible outcomes of various decisions and other associated issues. Counselors have a vital role in making sure that appropriate reflection and deliberation take place in a session (Hodgson and Spriggs, 2005). Autonomous decision-making involves (1) critical reflection: dialog is essential to allow the patients to explore and reason through their decision-making process; (2) a fundamental idea of how they want to live: dialog helps the patient to clarify the suitability of a decision in terms of their values and lifestyle and consolidate those values; (3) awareness of influences on deliberation: dialog helps the patient recognize influences on their thinking and clarifies misunderstandings; and (4) rationality: this is an essential element of autonomous decision-making because it excludes absurd decisions and those based on irrational thought; during dialog the patients' thoughts and reasons for choosing a specific option can be explored (Hodgson and Spriggs, 2005).

Dialog is a two-way interactive process, which is an essential part of genetic counseling as it promotes and enhances the patients' autonomy, helps them reflect on their reasons for making particular decisions, and corrects their misunderstandings.

Another factor that influences decision-making is the personal experience of the disorder itself (whether it is albinism or some other condition) in one's self, one's child, or a close relative. This experience provides insight into the impact the condition has on the affected person and the family. Research shows that the more serious the impact of the disorder the less likely the counselees will decide to go ahead with further reproduction and the more likely they are to request prenatal diagnosis and selective abortion (Wertz, 1992). However, the attitude toward the affected person can also influence the decisions. Some families' attitudes are based not only on their rich experience of living with an affected family member, with its complex emotional issues of love and recognition of what that individual contributed to the family, but also the pain and difficulties associated with the disorder. Some of these families feel that using genetic testing to avoid the birth of a subsequent affected child is devaluing the life of the affected person in the family (Weil, 2000). However, others feel that these two events must be kept separate and different decisions could apply in each individual case, bearing in mind the complex issues attached to such situations.

Some families come for genetic counseling without having made any reproductive decisions and for these patients genetic counseling is really useful and decision-making becomes an integral part of the counseling process (Weil, 2000). However, for others who have made decisions, genetic counseling can provide support for the decision, professional acceptance of it, and information on how to implement it. Alternatively, the new information provided in the counseling situation can lead to reviewing, reassessment, and revising of the decision. During this decision-making process, questions may have to be answered, such as what decision needs to be made, who will make it, and how will it be made? (Shiloh, 1996). Decisions often have to be made even though the risks are not clear cut, the situation may not be well structured, and the process is seldom orderly or sequential. Also the participants may simplify the information given to them and use heuristic reasoning, finding an acceptable rather than the optimal solution.

Furthermore, women may feel a need to assess the effect of the decision and its impact on their lives, as they are likely to be the major care provider after the birth of another child, whether or not that child is affected, but may be ambivalent because they also need to consider their partner's wishes (Rapp, 2000) (see Chapter 11 for further information on this topic).

Decision-making is a dynamic process, which often involves conscious and unconscious conflicts between contradictory wishes, expectations, and values (Kessler, 1980). It may also involve conflict between partners who have different agendas and these may be difficult to manage, depending on the quality of the relationship the couple have and their ways of coping, compromising, and making decisions. Other people may also influence decision-making, particularly in Africa, where many believe in shared decision-making among key family and community members, who may differ in their beliefs from the patient (Kromberg and Jenkins, 1997a). Alternatively, sometimes, decisions may be finalized only after consulting the local traditional healer, whose diagnosis and prognosis may differ from that of the Western medical professional, complicating and confusing the patient's decision-making process (Soodyall and Kromberg, 2016).

Issues to Be Covered in Genetic Counseling Sessions on Albinism

After conducting research on the adjustment of a sample of young people with albinism in Johannesburg, the researchers came to the conclusion that relevant genetic counseling should be offered to people with albinism to assist them in understanding their condition, its causes, implications, and recurrence risks, as well as the myths, cultural, and psychosocial issues that surround it (Kromberg and Jenkins, 1984). Such counseling sessions should cover:

1. A clear description of albinism and its basic defect. Recognition that the birth of an affected child can cause delays in maternal- and paternal-infant bonding.
2. An explanation of the genetic cause of albinism, its mode of inheritance, occurrence and recurrence risks, likely health complications, such as skin lesions and poor vision, and discussion of the associated educational issues, the fact that the intelligence is in the normal range, the prognosis, management, and the treatment necessary to minimize the complications (e.g., referral to ophthalmology and dermatology clinics is usually necessary).
3. Discussion of the belief in the death myth, so that the patient accepts that this belief is mistaken, death is a natural course of events and inevitable, and so that the myth should neither impact on his or her quality of life nor interfere with the understanding of the genetic factors involved in the condition. Discussion of the psychosocial complications, such as stigmatization and discrimination, and other cultural aspects.
4. Discussion of marriage because there is some disparity between what people say and what they do in relation to marriage with a person with albinism.

5. Discussion on, and if necessary referral for, vocational guidance because people with albinism sometimes feel that they are unemployable and they need to find employment, which will not involve too much exposure to sunshine.
6. Discussion on adjusting to the condition, how to cope with and manage it, and on many of the psychosocial and cultural issues described in detail in other chapters of this book (see Chapters 9 and 13).

These issues to be covered in genetic counseling sessions are offered here as a basic guide; however, it is the empowerment of these patients, which should take priority. Counselors need to provide relevant information, but also be willing to listen empathically to the patients' experiences and stories, both factual and emotional, to respond appropriately and to promote coping strategies. In this way the patient can be empowered to take control of the situation, cope better, behave effectively, and have hope for the future (McAllister, 2016). As Edwards et al. (2008) state the discussion of the psychosocial issues in genetic counseling sessions is just as important as the provision of genetic information, and this especially appears to apply to the counseling of people with albinism.

III. MODES OF INHERITANCE, RECURRENCE AND OCCURRENCE RISKS

Modes of Inheritance

Genetic disorders are inherited in several different ways and the mode of inheritance affects the recurrence risk. It is therefore important to establish exactly how the condition is inherited. Inheritance may be monogenic, i.e., caused by one gene mutation in either one or both parents, or polygenic, i.e., caused by many genes, which in combination cause the disorder, and/or multifactorial, i.e., caused by both genetic and environmental factors. The modes of inheritance are described here to place albinism and its recessive mode of inheritance into context.

The four most common types of inheritance of genetic disorders are:

1. Dominant inheritance in which, if one parent carries a gene fault for a dominantly inherited condition, his or her children will have a 50% chance of inheriting that faulty gene and also having the condition (e.g., achondroplasia, Huntington's disease). Dominant conditions may demonstrate partial penetrance, so that a few members of an at-risk family may have the faulty gene for the condition but show no obvious signs of the condition. They are also subject to variable expressivity of the faulty gene, so that some members of the family may have milder signs of the condition and others may have a severe form of the disorder. Where there is no family history of the condition, the disorder in a family may be caused by a new mutation in the affected individual. However, once it appears in a family member, it will then be inherited by his or her offspring with a 50% chance.
2. Recessive inheritance in which two people carry the same recessive gene fault, which, generally, causes no problem in the single dose in them. However, if they reproduce, their children have a 1 in 4 or 25% chance of inheriting a double dose of the faulty gene (i.e.,

one from each parent) and having the disorder (such as oculocutaneous albinism or sickle cell anemia). Also, their other unaffected children, have a 2 in 3 chance of being carriers like their parents. Because this is the mode of inheritance associated with most cases of albinism, including the common type, OCA2, found in Africa, it will be discussed further below.

3. X-linked inheritance in which a woman carries a faulty gene for a disorder (usually without showing any signs of having it) on one of her two X chromosomes, she then has a 50% chance of passing on this X chromosome with the faulty gene and 50% chance of passing on the normal X to her children. If a boy inherits the faulty X he will have the disorder, but if a girl inherits it she will only be a carrier (like her mother). The affected sons when they produce children will have sons who neither have the gene nor the disorder, but all their daughters will be obligate carriers. These daughters, in turn, will have affected sons with a 50% chance and daughters with a 50% chance of being a carrier like their mother. Some forms of ocular albinism are inherited in this way, as is the common form of hemophilia.

4. Polygenic or multifactorial inheritance in which the condition in the family is caused by a combination of many genes, each with a small effect, and environmental factors (which are not often well understood). In these disorders there might be one or some predisposing genetic factors, which do not produce the disorder in an individual unless he or she is exposed to the environmental factors. The disorders inherited in this way are common and they include most forms of spina bifida, club feet, and cleft lip and palate.

Recessive Inheritance and the Occurrence of Albinism

Couples who have a child with a recessive disorder have a high risk of recurrence (25%). Generally, risks have been considered high if they are 1 in 2 (50%) to 1 in 4 (25%) and low if they are less than 1 in 25 (4%) of having an affected child (Carter et al., 1971). Alternatively, according to another study, they have been considered high or low, if above or below 10%, respectively (Lippman-Hand and Fraser, 1979). However, these descriptive terms may not be helpful and perceptions as to what is a high risk and what is a low risk may differ according to the counselor's and the client's subjective views of the situation. Furthermore, perception of risk depends on the type of disorder, its severity, prognosis, and perceived burden, as well as what the needs of individuals are and how they define risk and determine risk-taking behavior.

Scientists believe that every single person carries about three faulty recessive genes that could cause a genetic disorder in their children, if the child inherits the same gene fault from both mother and father (Jones and Bodmer, 1974). Fortunately, the recessive gene pool is very large and therefore people with the same recessive gene rarely mate and reproduce with each other. However, in some populations specific faulty genes (gene mutations) are more common than they are in other populations. This is the case with the gene for albinism, which is more common, generally, in Black African populations than in other populations, although it occurs in every population. On the other hand, the cystic fibrosis gene is more common in White populations, such as the British, than in other populations, whereas the thalassemia genes are more common in populations of Mediterranean and Asian descent. In some populations culturally determined marriage patterns favor consanguineous unions (marriage between relatives). Where this is the case, couples who are consanguineous are more likely to carry the same recessive faulty gene, having inherited it from their common ancestor. Several populations in Southern Africa, such as the Tswana and Bakgatla of Botswana (among whom cross-cousin marriages, i.e., with the daughter of a maternal uncle or paternal aunt were preferred or if such a girl was not available, at least she should be from the same village or ward, Schapera, 1940), fall into this category, and in these groups the rates of recessively inherited conditions, such as albinism, are increased.

In Africa, the prevalence rate of albinism averages at about 1 in 4000–5000 and about 1 in 30 people carry the gene for albinism (Stevens et al., 1995). Therefore, ~1 in 900 (1 in 30 × 1 in 30) unions are between two people who both carry one albinism gene. Then, if that couple reproduces, 1 in 4 of their children will be affected. So that, doing the calculations (approximate estimates only) for any Southern African population group, 1 in 900 × 1 in 4 = 1 in 3600, showing that ~1 child in 3600 children in the community will be affected. Therefore the risk that any random Black African couple, without a history of albinism, would have a child with albinism would be about 1 in 3600 (or 0.028%); with a relevant history the risk would be much higher.

Risks of Recurrence for Siblings and Other Relatives of People With Albinism

Various questions arise among persons with albinism and their family members, as well as others with a history of albinism, in genetic counseling sessions. These include the following:

1. Will the siblings of the person with albinism be affected? The chance that the full siblings (with the same father and mother as the child with albinism) will be affected is 1 in 4 or 25%. The risk is 1 in 4 for every pregnancy, and this statistic does not mean that if a couple have four children only one will be affected. A couple may have 10 children and 5 may be affected (as occurred in one family known to the writer) or 10 children and only 2 are affected. Also, there is no order associated with this risk, i.e., if the first born child has albinism, the next three will not necessarily be unaffected; every pregnancy carries the same risk of 1 in 4, regardless of what has happened previously (see Fig. 10.2). Brothers and sisters have equal chances of being affected.

2. Will a person with albinism have affected children? The risk to an affected person's offspring is determined by the carrier status of the partner. Unless he or she has children with a close relative (who is more likely to carry the faulty gene for albinism than someone in the general population) the children will probably be unaffected. If the affected person has children with an unrelated person, of Black African origin, the chance that the partner is a carrier for the OCA2 faulty gene is about 1 in 30. The carrier rate in the non-African Black population is likely to be much lower, probably in the region of 1 in 60. The person with albinism has two faulty genes for the condition and will always pass on one to his/her child. So the calculation, where the partner is unrelated and from a Black ethnic group, is 1 × 1 in 30 that they are both carriers and 1 × 1 in 30 × 1 in 2 = 1 in 60 (or 1.8%) that they will have an affected child. Then, if the partner is tested and found to be a carrier the chance of passing on the gene is 1 in 2, so the risk of having affected children would be 1 × 1 in 2 = 1 in 2 (50%) . Alternatively, without knowing the carrier status of the partner, the risk is less than 1.8% that a person with albinism will have an affected child with a Black partner with no known family history of albinism.

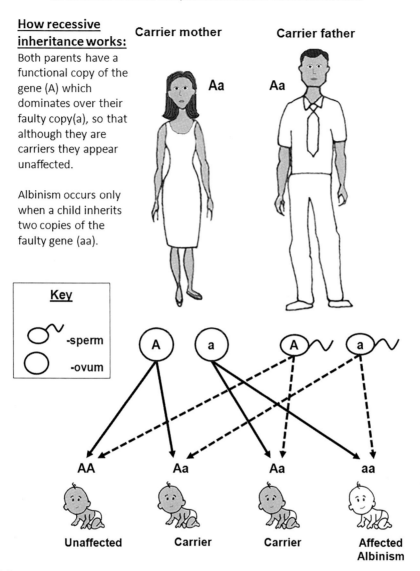

How recessive inheritance works:

Both parents have a functional copy of the gene (A) which dominates over their faulty copy(a), so that although they are carriers they appear unaffected.

Albinism occurs only when a child inherits two copies of the faulty gene (aa).

FIGURE 10.2   Recessive inheritance. *Image courtesy of Dr. J. Kromberg, University of the Witwatersrand, South Africa.*

3. What if a person with albinism has children with a close relative? If the affected person has children with a first cousin, for example, the risk would be higher than if the partner is unrelated. The first cousin of an affected person has a carrier risk of 1 in 4. So the risk that an affected person and his/her first cousin would have an affected child would be 1×1 in 4×1 in $2 = 1$ in 8. So their risk of having an affected child would be about 12%. Once this couple has had an affected child, this would confirm the partner's carrier status and that he/she carries the faulty gene and the risk would go up to 1 in 2 (50%).

4. Will the unaffected children of a person with albinism have affected children? The unaffected children of an affected person are obligatory carriers of the faulty gene for albinism because they always inherit one gene for albinism from their affected parent. When these children have their own children, with an unrelated and unaffected partner, their risk of having affected children would be low (1×1 in $30 = 1$ in 30, that they are both carriers, then 1 in 30×1 in $4 = 1$ in 120) that their child would have albinism. If their partner was a first cousin then, again, the risk would be higher.

5. Will the unaffected siblings of the person with albinism have affected children? These siblings, because they are not affected, have a 2 in 3 chance of being carriers of the faulty gene. Their risk of having an affected child with an unrelated unaffected partner, however, would be small: 2 in 3×1 in $30 = 1$ in 45 and then 1 in 45×1 in $4 = 1$ in 180. The risk with a related partner would again be higher.

6. Will the siblings of the parents of a child with albinism have affected children? The siblings of parents, who have an affected child (i.e., the aunts and uncles of the affected child), could also have inherited the gene from their common parents, so they would have a risk of 1 in 2 (50%) chance of carrying the gene. If they married an unrelated unaffected person their combined risk of both being carriers would be 1 in 2×1 in $30 = 1$ in 60. Then the risk would be 1 in 60×1 in $4 = 1$ in 240 of having children with albinism.

A pedigree has been drawn up illustrating a family with a member with albinism and the risks for various relatives (see Fig. 10.3).

It is important to note that with molecular testing now available for many families, the affected person can be tested and if the faulty gene causing his/her albinism can be identified, then the partner can be tested for the same faulty gene and informed regarding whether or not he/she is a carrier. The risks to their offspring can then be refined (see Chapter 11 for a discussion on such testing).

Families with a child with any recessive condition, including albinism, have many at-risk relatives. A South African study on cystic fibrosis showed that each couple with an affected child had on average 14 at-risk relatives (Macaulay et al., 2012). According to the World Health Organization, an individual who is aware of a genetic condition in him/herself or in his/her family has the ethical duty to inform blood relatives that they may be at risk (WHO, 1998). It should not be necessary for the genetic health professionals to make contact with these at-risk relatives, but they should encourage their patients to inform their own relatives. An informative letter or pamphlet on the condition helps in this situation and if disseminated should increase awareness in relatives of their risks.

Perception of Risk

How Do at-Risk Patients Perceive the Risk?

Various "sociological and anthropological studies have shown that perception and acceptance of risk have their roots in social and cultural factors" (Slovic, 1987, p. 281). Responses to risks are mediated by social influences and the opinions of family members, friends, work colleagues, and health professionals. Research shows that disagreements about the level of the risk might not change when evidence is presented and that strong initial views may

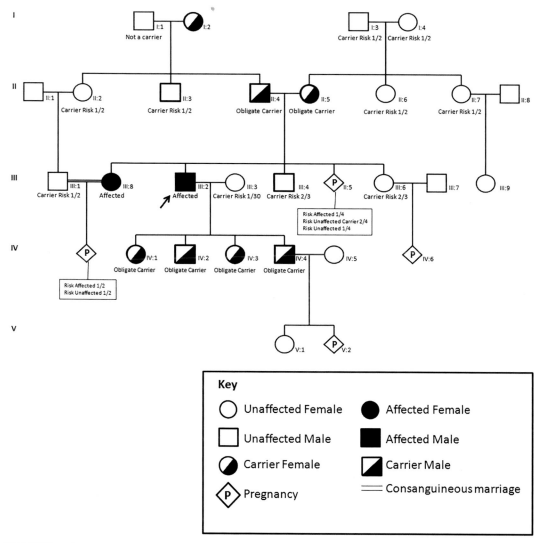

FIGURE 10.3 Pedigree showing recessive inheritance and risks for relatives. *Photograph courtesy of Dr. J. Kromberg, University of the Witwatersrand, South Africa.*

resist change because they influence the way in which subsequent information is understood (Slovic, 1987).

It is well recognized that perception of risk is very personal and the numeric figure given in the counseling session contributes to only part of the patients' understanding of risk. The writer has personally counseled patients who felt that a 1% risk for a particular condition was extremely high and other patients who perceived the same risk for the same disorder as very low. Another couple, known to the writer, who had a child with albinism was very pleased to hear that their recurrence risk was only 1 in 4 (25%) when, after understanding that the condition was inherited and genetic, they had thought all their children would be affected. The perception is influenced by how the family manage with the affected member they have, how much past experience they have had, how severe the condition is in the individual, how accepting the parents, the extended family, and community are, and many other factors, such as how numerate the family is. McAllister (2016, p. 41) states that inaccurate risk recall may be affected by "heuristic thinking," whereby people use shortcuts to help them simplify the nature of the risk, make decisions, and solve problems. These shortcuts may include educated guesses, intuitive judgments, and common sense decision-making, and they are all prone to errors because they reduce the time spent in thinking through the problem.

Perception of risk is also affected by various extraneous factors. Studies have shown how perception of risk might be influenced by the social and geographical proximity of the affected family member and that interpretation of risk may change over time and, therefore, is not static (McAllister, 2002, 2003).

The findings from another research study showed that risk tended to be interpreted in a binary form: either it will happen or it will not happen (Lippman-Hand and Fraser, 1979). With, for example, risks of 1 in 4 or 1 in 100, counselees might focus on the numerator "one" and disregard the denominator. Then, while concentrating on the "one" (the outcome that was undesirable) they might convert their risk into binary form. These researchers observed that the interaction of risk and consequence was emphasized by both counselor and counselee. They found the risk was inextricably intertwined with the possible consequences of taking the risk, and, often, the consideration of the consequences was used to put the problem of how to interpret the risk into context. Counselors and counselees did not think the two could be separated, as one geneticist (when discussing the risk of a dominantly inherited condition with a 50% recurrence risk) stated:

> So that would mean that each of your children would have a 50-50 chance, and you know better than I do, I guess, what being affected means in terms of altering your way of life. *Lippman-Hand and Fraser (1979, p. 121).*

In other words the important issue in that session became what having an affected child would mean. The same situation has been observed in counseling parents with a 1 in 4 risk of having a child with albinism (see Chapter 11). Some families will go on to ask for advice about reproductive behavior and what choices they might have.

> This is a high risk, one in four. Are there others like us? What do people do with this risk? Do they take the chance? What do doctors suggest they do? We've never had to face this kind of thing before, so we want to know what others say. *Lippman-Hand and Fraser (1979, p. 121).*

However, this couple did not want advice, as such, they wanted a background against which to assess their own options. Although counselors usually try to provide patients with precise recurrence or occurrence risks and used to assume these were useful as a basis for action, it has to be acknowledged that these figures are not the only factor on which decisions are made and actions taken.

Do people remember the risks given to them in the genetic counseling situation? Research shows that the recall of genetic risk information is not straightforward (McAllister, 2016) and that patients may return to the beliefs about risks that they held before receiving counseling.

Prior to receiving counseling many families would have developed their own ideas and predictions about who in the family was at risk and who would have an affected child. The writer has counseled a family, with a history of albinism, who had allocated albinism carrier status to several members of the family, depending on their skin color and position in the family (J. Kromberg, personal communication, 2016). Such ideas may be based on the pattern of occurrence in the family and, although often numerically distorted, they are sometimes helpful in coping and coming to terms with the risk. Another factor is the parents' need or desire for another child. If this need is very strong, the risk may become distorted and minimized and, in the case of the recessive conditions (such as albinism), the parents might focus on the 3 in 4 (75%) risk for an unaffected child, while ignoring the 1 in 4 (25%) risk for an affected child.

Some Common Misconceptions About Risks

Misconceptions about risk and the interpretation of risk are common. People may believe that if a child looks like the parent who is the carrier of a genetic condition that child will also be a carrier. Furthermore, people may feel that birth order is important and that if the first (or the last) child is the affected one, the first (or the last) child of other members of the family will also be affected. Also, families may think that a recessive condition is sex-specific, for example, if their three affected children are all of the same sex, and that only girls (or only boys) are affected. The fact that "chance has no memory" needs to be understood because some couples believe that once they have had their one in four chance of having a child with albinism, the next three children will be unaffected. Another misconception is that because the mother carried the infant throughout pregnancy, it is her fault that something has gone wrong (this is not relevant to recessive conditions where both parents are carriers and pass on the faulty gene mutation). A further aspect of this erroneous belief is that if the father has the family history of albinism, he is the sole cause of the child being affected, which again is untrue.

A misconception that is closer to the truth is that the lighter-skinned members of a family with a member with albinism will be carriers. This is partly true, and skin color in siblings and other relatives of affected people has been measured and compared with people with albinism and a random sample of controls from the black population (Roberts et al., 1986). However, the findings showed that, although skin color was significantly paler in the unaffected siblings than in the controls (at several wavelengths), the measurements could not be used to identify carriers reliably since there is such a wide range of skin color in the Black African population.

IV. PSYCHOSOCIAL ASPECTS OF GENETIC COUNSELING

Families who present for genetic counseling are often not sure why they are there, especially if they have been referred through their local hospital system (Morris et al., 2015). Genetic counseling sessions in this context have to start with why the patient was referred and what they can expect from the session. The counselor also has to bear in mind that the patients come with their own, often unspoken, agendas, their own psychosocial and cultural issues, values, norms, ideas, and family pressures, and their own experience with their genetic disorder and with the local health-care system, its disappointments, helpfulness, or otherwise. Some mothers with infants with albinism, even in the newborn period, have already come across stigmatization of themselves and their infant because of the albinism. As one mother with an affected newborn, during her first genetic counseling session, stated (in her broken English):

> It's just that … some people, you know, discourage people and speak foul words. *Morris et al. (2015, p. 164).*

However, as found in the same study, some of the psychosocial factors that contributed to creating a positive experience in the genetic counseling clinic included the showing of respect and empathy by the counselor for the patient, evidence of compassion and support, the fact that questions were answered, and the provision of adequate information and appropriate referrals. Negative experiences were associated with language barriers, lengthy consultations, delayed referral for genetic counseling (some mothers described their rising anxiety and the fact that they received information only long after the child's birth) and, after the counseling, difficulties in explaining the information about the condition to their relatives. The mothers in this study also emphasized the community need for education both on albinism and on genetic counseling because they had found a lack of knowledge not only in the community but also among health professionals (Morris et al., 2015). Another earlier South African study also showed that many local patients benefited from having genetic counseling and felt they had had their expectations met (Levy, 1992).

In a genetic session, counselors have to establish rapport, collect family histories and pedigrees, give information on the disorder and the recurrence risks, discuss options open to deal with the risk, and help the family make a good adjustment in their circumstances. However, one of the most important psychosocial aspects has been shown to be the emotional support, empathic communication, and nonjudgmental acceptance of the patient, provided in the session; some research findings suggest that patients gain more benefit from this side of genetic counseling than they do from the educational side that provides information (Edwards et al., 2008).

For these reasons, genetic counselors need to be aware of the psychosocial issues that may arise, or may need to be aired, in any counseling session and particularly those being raised by a mother with a newborn child with albinism. These issues include: possible feelings of disappointment and sadness about the infant's condition; feelings of anxiety about how to cope; concern about what the husband, family, and community will think and say; concern about the cultural issues and myths that surround the condition; and anxiety about what the future holds for the child. There are psychological issues relating to the mother's own individual reaction to the event, her bonding with her child (often delayed, Kromberg et al., 1987), her attitudes, beliefs, values, parenting skills, childcare patterns and coping strategies, and social issues relating to her social interactions and how she is going to manage within her home, family, and community.

Recent studies undertaken in the United Kingdom suggest that the value of genetic counseling can be assessed by examining patient-reported outcomes, which cover decision-making, perceived personal control, coping, and quality of life (McAllister, 2016). Findings showed that the patients' benefits from using genetic counseling services could be summarized by using the term "empowerment" (McAllister et al., 2011). In this context this concept comprises five dimensions:

1. Decisional control (knowing the options and feeling enabled to make informed decisions).
2. Cognitive control (knowing the scientific explanation).
3. Behavioral control (being able to make effective use of health systems to manage the disorder).
4. Hope (having hope for a rewarding family life).
5. Emotional regulation (feeling able to cope with the condition and to manage the distress, guilt, and anxiety surrounding the situation).

Genetic counselors providing counseling need to understand the way in which families adapt to chronic illness and disability because many genetic conditions fall into this category. In addition, families faced by inherited disorders have to learn to cope with many psychosocial and physical challenges. These include dealing with stress, crises, loss, grief, damaged body image and self-concept, stigma, uncertainty, unpredictability, and an altered quality of life (Liveneh and Antonak, 2005). Many aspects of this psychosocial adaptation process are observed when families with a member with albinism present for counseling, as well as when people with albinism attend a counseling session themselves. Liveneh and Antonak (2005) discuss three broad domains associated with psychosocial adaptation to chronic illness and disability, and although they focused on the adaptation to acquired disabling conditions, there are many parallels with the experiences of families with genetic disorders and those with a member with albinism. The domains include the dynamics of the adaptation process (basic concepts, psychosocial responses, and coping strategies), assessment of the process, and intervention strategies to facilitate coping and adaptation.

The basic concepts associated with adaptation to having a disability and/or genetic condition, such as albinism, or to having a child affected with the condition include: stress (which has been observed in mothers after the birth of an infant with albinism, Kromberg et al., 1987); crisis (also observed after the birth, especially if this is an unexpected occurrence and a first affected child in the family); loss and grief (when the mother consciously or unconsciously recognizes that she has lost the unaffected child she was hoping for); chronic sorrow (when the parents realize what their child's future holds and the visible differences are a constant reminder of the permanency of the condition); and stigma (when the parents recognize the impact of community attitudes and the nature of the prejudice and stigma toward people with albinism). If it is the affected person him or herself who is adapting to having the condition, further basic concepts need to be considered. These include the impact of body image or the unconscious mental representation of one's own body (Manganyi et al., 1974), which develops gradually, influenced by sensory (e.g.,visual, auditory, and kinesthetic), interpersonal (e.g., attitudinal), environmental (e.g., physical conditions), and temporal factors (Liveneh and Antonak, 2005). Successful adaptation to having albinism may "reflect the integration of physical and sensory changes into a transformed body image and self-perception" (Liveneh and Antonak, 2005, p. 13), whereas unsuccessful adaptation may be evidenced by physical and psychiatric symptoms, such as strong constant feelings of anxiety and depression, chronic fatigue, social withdrawal, and possibly cognitive distortions (Liveneh and Antonak, 2005).

It is helpful here to examine, briefly, some of the literature on coping because the family with an affected individual and the individuals with albinism themselves need coping strategies. Coping has been defined as a psychological strategy used to decrease the impact of stress generating life events (Billings and Moos, 1981). Two types of strategies have been described:

(1) disengagement coping strategies (when passive, indirect, avoidance-oriented activities, such as denial, wish-fulfilling fantasy, self and other blame, and even substance abuse, may be used); often higher levels of distress are observed when these strategies are used and (2) engagement coping strategies (when strategies to defuse the stressful situation are used, for example, goal-oriented activities, such as seeking information and social support, problem-solving, and planning) (Tobin et al., 1989). These coping strategies are often associated with higher levels of well-being, acceptance of the condition, and better adaptation.

Methods of assessing interventions to assist people to adapt successfully to their situation have also been described (Liveneh and Antonak, 2005). Some of the clinical interventions discussed would already be included in most genetic counseling sessions. Most counselors would provide clients with relevant medical information and supportive family and group experiences (by referring them to an albinism support group) and assist clients to explore the personal meaning of their disorder, while some would also discuss with clients various adaptive coping skills for successful community functioning. This group of interventions offers the counselor opportunities to apply a variety of approaches that can be individualized to meet the psychosocial needs of the specific client with albinism.

V. UPTAKE OF GENETIC COUNSELING SERVICES

As far as is known, there are genetic services in Cameroon, Egypt, Sudan, and Tunisia, but none, outside South Africa, in Southern Africa. Some patients who need the service come to South Africa or go to Europe (if they can afford to do so). Uptake of genetic counseling, in general, at the Johannesburg hospitals where regular weekly clinics are held, is good; for example, 3365 patients, with many different genetic disorders, were counseled in 2008, and 65% were from the Black population (Kromberg et al., 2013b). About one-third of these patients were counseled by genetic counselors and two-thirds by clinical geneticists. However, uptake by families with a member with albinism is limited. Initially (at the introduction of genetics clinics in the 1970s), such families were seen soon after the birth of the affected child, after referral by the hospital midwife or medical professional or pediatrician. Observation suggests that the number of referrals was higher while research programs focusing on albinism were being conducted (in the 1970s, 1980s, and 1990s) and awareness of the condition and of the availability of genetic counseling clinics was perhaps greater. More recently, the number of referrals appears to have fallen. It is possible that many new families are not being referred by busy nurses and doctors, or possibly they are being referred to the St. John's Eye Hospital, their general medical practitioners, or to the Albinism Society of South Africa, where they may gain some answers to their questions. General practitioners, being more aware of albinism and more knowledgeable about genetics (due to the increased training in genetics at medical schools in South Africa), may be better able to respond to their patients' queries and therefore may not refer the patients on to a genetic clinic.

Because there appeared to be no data available from any genetic counseling clinics in Africa regarding the uptake of genetic counseling services by families with a member with albinism, an audit, over a 4-year period, of affected families presenting at the Johannesburg genetic counseling clinics, was completed by the author and her colleagues (Kromberg et al., 2015). This audit showed that 38 families (i.e., averaging 9 cases per annum) were counseled between the years 2011–14. Most index patients (73%) were between 20 and 39 years of age and about one-third (34%) had a family history. Most families had had a previous affected child and this was the main reason for using the genetic clinic services, but five families had two or more affected children and one person was affected herself.

The majority of the couples who presented for counseling, in this series, had a 1 in 4 risk of recurrence for albinism in their next child. The exceptions were those mothers (usually, but occasionally fathers) who had an affected child but were divorced or separated from the father of the child, had a new partner, and wanted a revised risk assessment or those who were affected themselves (see Chapter 11 for discussion of this issue).

The data from this study showed that on average only 9 families presented for genetic counseling per annum, whereas ~12 infants with albinism are born at one large Johannesburg hospital (Chris Hani Baragwanath) alone every year. These findings suggested that genetic counseling services in Johannesburg are underutilized by people with albinism and their families. The reasons, given by the study participants, for attending the clinics were to discuss causes, recurrence risks, and management of albinism (in couples with newborn affected infants), and to have carrier testing, to seek prenatal diagnosis and counseling (in the other participants). The poor uptake of counseling services may be due to the lack of referrals of affected newborn infants and their parents from the hospital wards, as well as from ophthalmology and dermatology clinics, and also to the lack of awareness of genetic counseling services and their benefits for patients with albinism, among the general public and health professionals. The poor health literacy of the general public, local inequalities and challenges relating to recent urbanization, and high rates of unemployment probably also have an effect on the utilization and uptake of health services such as these.

The low uptake of genetic counseling services by families with a child with cystic fibrosis has been studied in South Africa (Macaulay et al., 2012). The findings showed that uptake was initially, from 1990 to 2005, about six families a year. However, once a specific service was initiated by a genetic counselor, at the cystic fibrosis clinic at a large local hospital, uptake improved nearly tenfold (to 58 cases in 2006). Similarly, it might be worthwhile initiating a genetic service for families with a member with albinism, at either or both the state hospital clinics dealing with ophthalmology and dermatology. Access to genetic counseling will then become easier for affected families and can be combined with a visit to a treatment clinic.

Another way to improve utilization of genetic services is to educate health professionals and the lay public about the availability of services and the benefits for patients. A study on the knowledge of general practitioners regarding hereditary cancers, in Johannesburg, showed that many were not fully aware of the availability of genetic counseling and testing services (Van Wyk et al., 2016). Also many were not confident enough to offer a genetic counseling service themselves and 86% were interested in learning more about these cancers, so that they could refer patients appropriately for genetic services.

VI. ETHICS AND GENETIC COUNSELING

Those who provide genetic information to patients should be aware of the ethical issues involved (Kromberg and Jenkins, 1997b). The fundamental ethical principles that underlie genetic counseling include voluntariness, informed consent, and confidentiality; these originate from respect for the individual, his autonomy, freedom of choice, equity, and privacy.

A code of ethical principles has been drafted for genetics professionals (Baumiller et al., 1996). This code covers responsibilities to patients and their families, to society, and to the profession. Also the National Society of Genetic Counselors has drawn up a code of ethics for genetic counselors in the United States (Schmerler, 2009), which attempts to guide their conduct so that the goals of the profession are achieved. This code covers the relationships, established by counselors in the course of their professional duties, with themselves, clients, colleagues, and society.

Various ethical issues are raised by the delivery of genetic services. These include the requirement that the services should be accessible, voluntary, of high quality and of impeccable ethical standards (Kromberg and Jenkins, 1997b).

If the nature of a genetic counseling service for people with albinism is considered, then this is associated with various ethical issues at the different levels at which the service is offered:

1. Prenatal diagnosis for albinism is surrounded by ethical issues and these will be covered in detail in the following chapter (see Chapter 11).
2. The testing of unaffected children for the genes for albinism is ethically contentious. Generally, when the testing of children is proposed the best interests of the child and the preservation of the possibility of future choice, when an adult, should be considered. In the case of albinism, there would be no justification for testing children because being a carrier does no harm and there would be no benefit for the child in knowing he/she was a carrier. Carrier testing in children is not performed for the curiosity of the parents or simply at the request of the parent or their doctor or other health professional. Counselors should assist parents in how and when to inform their children that they are at risk of being carriers and can be tested. Once the child can understand the issues (usually at around 18 years of age or older) he/she can request testing and give informed consent. The psychological effects of being a carrier for the gene for albinism have not yet been investigated. Furthermore, the effects on family dynamics of testing a child and giving results to the parents are still poorly understood. Unfortunately, because of the development of private laboratories, which may provide testing to anyone on request, and because of medical practitioners, who may not fully recognize the ethical issues, testing is sometimes carried out on children. When clinicians become aware of the issues, they might acknowledge that they have been undertaking testing without realizing the implications (Harper and Clarke, 1990).

   In 1995 the American Society of Human Genetics (ASHG) with the American College of Medical Genetics and Genomics published a statement on genetic testing in children and adolescents recommending that genetic "testing should be deferred until adulthood, particularly for adult-onset conditions or for carrier status, for reproductive decision making" (Botkin et al., 2015, p. 7). However, since then several new tests have been introduced and clinicians have gained much relevant experience. Recently, in 2015, the ASHG decided to publish a new statement entitled "Points to consider: Ethical, Legal, and Psychosocial Implications of Genetic Testing in Children and Adolescents" (Botkin et al., 2015). In this document the ASHG offered the following similar recommendation: "Unless there is a clinical intervention, appropriate in childhood, parents should be encouraged to defer predictive or pre-dispositional testing for adult-onset conditions until adulthood or at least until the child is an older adolescent who can participate in decision making in a relatively mature manner." However, the 1995 statement is more relevant to issues associated with albinism as it states that testing for carrier status should also be deferred until adulthood; this applies to the testing of unaffected at-risk children for albinism mutations. Occasionally parents of a child with albinism might wish to know whether their unaffected children carry the gene for albinism, but they should be encouraged to wait until the child him or herself can understand the implications of such testing and of receiving results and can, then, give informed consent to being tested.

3. Population screening in groups at high risk for albinism has been discussed briefly previously (Kromberg and Jenkins, 1997b). However, such screening raises ethical issues and it might lead to more anxiety (perhaps even some stigmatization and discrimination) than benefit to those found to be carriers. Also, what effect will such screening have on community attitudes and the acceptance and integration of people with albinism, either in the community and/or as marriage partners? Furthermore, carriers themselves could find difficulty in accepting and coping with the psychological impact of carrier status and might feel confused about how to select a marriage partner. Their carrier status, however, should be put in context using the information that scientists believe that every person, on average, carries about three faulty recessive genes (in single dose), which could be the cause of a disorder in a child who inherits the same faulty gene from both parents (Jones and Bodmer, 1974).

Because testing is available for the common OCA2 mutations (in the National Health Laboratory Service in Johannesburg, South Africa), screening could become an issue in at-risk target population groups, in future. Such screening has been carried out for other recessive conditions such as Tay-Sachs disease (a disease that is untreatable and lethal in childhood) in the Ashkenazi Jewish population. That population has a high rate of the condition and about 1 in 25 people in South Africa is a carrier. (Jenkins et al., 1977). At present, the only members of the general population who are screened for OCA2 are the new partners of people who carry a mutation (and generally already have an affected child). Population-wide screening is usually only promoted when the disorder is seen as a serious one, treatment is costly, there is a high rate of carriers in a recognized population group, the carrier test is available, reliable and inexpensive, screening is cost-effective, genetic counseling services are accessible and available to carriers, and research on the psychosocial issues has been completed.

It is unethical to offer a test to a high-risk community just because the test can be done and is available. In the case of screening for albinism, if there is no low-cost test for the OCA2 mutations and if only 80% of mutations can be detected, it may not be worthwhile to offer screening to the public, especially if there are competing claims on the scarce resources available for health care, in a middle-income country such as South Africa. Although some sections of the target population may want population screening for albinism, much research has to be carried out before such screening is offered. "Genetic testing should be considered in the same way as a new drug. It can have efficacy and it can have toxicity" (Frances Collins, as quoted in Nowak, 1994, p. 464).

An alternative to general population screening would be cascade screening, which might be preferable in the case of OCA2. Cascade screening is undertaken by testing close high-risk relatives in families with a member with OCA. This has been quite successful in the case of cystic fibrosis and some future cases in at-risk families have been prevented in this way.

VII. CONCLUSION

In conclusion, the provision of genetic counseling services to families with a member with albinism should be promoted. The issues that should be covered in such counseling sessions have been discussed in this chapter and they represent a unique set of circumstances, which is specific only to genetic counseling for albinism in Southern Africa. However, further research may show that other countries elsewhere in Africa can benefit from the insights presented here.

Presently, uptake of the existing services in South Africa is limited, although there are many benefits to receiving such counseling from properly trained health professionals and genetic counselors. These benefits include the gaining of genetic information and knowledge and the feelings of empowerment achieved through the cathartic experience of discussing personal concerns with a trained counselor. Such counselors are skilled communicators as well as empathic listeners and knowledgeable responders.

During genetic counseling sessions a family history is collected, the mode of inheritance and the occurrence and recurrence risks of the genetic disorder are explained, and the psychosocial and cultural elements of the patient's situation are discussed. In addition, management of the disorder, available genetic tests, and options for dealing with the recurrence risk are outlined, in a culturally sensitive way. Counselors are aware of the psychosocial and ethical issues associated with having a genetic disorder, such as albinism, and the difficulties that surround decision-making in a genetic context.

Genetic counselors can calculate recurrence risks for various members of the family with a member with albinism, as well as for the affected person him or herself. Further they can assess the risks of members of the general population and recommend genetic testing in certain situations. In cases where one partner has a child with albinism but wishes to reproduce with a new unaffected and unrelated partner, such partner may wish to be tested for carrier status, to determine his or her risk. The understanding of such risks is influenced by the personal perceptions of risk of the individuals involved and by the common misconceptions associated with various types of risk held by people in general.

Although the availability of genetic counseling services in Africa is limited, patients from several sub-Saharan African countries are traveling to South Africa (and elsewhere) to access a service. However, training of medical geneticists from a few African countries, e.g., Cameroon, is occurring in the United Kingdom, United States, and South Africa, and it is hoped that in the near future training of student genetic counselors from other African countries will commence. Furthermore, capacity building in human genetics for developing countries is being discussed by the African Society of Human Genetics, Wellcome Trust, National Institute of Health, United States, and other key stakeholders, and some of their initiatives and perspectives have been reviewed by Wonkam et al. (2010). As a result, the first medical genetic service in Cameroon has been initiated, and it is hoped that developments in other African countries will follow. In future, therefore, people with albinism and their families, living in Africa, should be able to benefit from accessible, local, and informed genetic services.

References

Aalfs, C.M., Oort, F.J., de Haes, J.C.J.M., Leschot, N.J., Smets, E.M.A., 2007. A comparison of counselee and counselor satisfaction in reproductive genetic counseling. Clin. Genet. 72, 74–82.

American Society of Human Genetics Board of Directors; American College of Medical Genetics, Board of Directors, 1995. Points to consider: ethical, legal and psychosocial implications of genetic testing in children and adolescents. Am. J. Hum. Genet. 57, 1233–1241.

Billings, A.G., Moos, R.H., 1981. The role of coping responses and social resources in attenuating the stress of life events. J. Behav. Med. 4, 139–157.

Baumiller, R.C., Cunningham, G., Fisher, N., Fox, L., Henderson, M., Lebel, R., McGrath, G., Pelias, M.Z., Porter, I., Seydel, F., Wilson, R., 1996. Code of ethical principles for genetic professionals: an explication. Am. J. Med. Genet. 65, 179–183.

Botkin, J.R., Belmont, J.W., Berg, J.S., Berkman, B.E., Bombard, Y., Holm, I.A., Levy, H.P., Ormond, K.E., Saal, H.M., Spinner, N.B., Wilfond, B.S., McInerney, J.D., 2015. Points to consider: ethical, legal, and psychosocial implications of genetic testing in children and adolescents. Am. J. Hum. Genet. 97, 6–21.

Carter, C.O., Fraser Roberts, J.S., Evans, K.E., Buck, A.R., 1971. Genetic clinic: a follow up. Lancet 1, 281–285.

Davis, J.G., 1984. Communication skills for MD and PhD geneticists. In: Weiss, J.O., Bernhardt, B.A., Paul, N.W. (Eds.), Genetic Disorders and Birth Defects in Families and Society: Toward Interdisciplinary Understanding, vol. 20(4). March of Dimes Birth Defects, Original Article Series, New York, pp. 41–44.

Djurdjinovic, L., 2009. Psychosocial counselling. In: Uhlmann, W.R., Schuette, J.L., Yashar, B.M. (Eds.), A Guide to Genetic Counselling. Wiley-Blackwell, New York, pp. 133–175.

Edwards, A., Gray, J., Clarke, A., Dundon, J., Elwyn, G., Gaff, C., Hood, K., Iredale, R., Sivel, S., Shaw, C., Thornton, H., 2008. Interventions to improve risk communication in clinical genetics: a systematic review. Patient Educ. Couns. 71, 4–25.

Ezeilo, B.N., 1989. Psychological aspects of albinism: an exploratory study with Nigerian (Igbo) albino subjects. Soc. Sci. Med. 29 (9), 1129–1131.

Fraser, F.C., 1974. Genetic counseling. Am. J. Hum. Genet. 26 (5), 636–659.

Greb, A., 1998. Multiculturalism and the practice of genetic counseling. In: Baker, D.L., Schuette, J.L., Uhlmann, W.R. (Eds.), A Guide to Genetic Counseling. Wiley, New York, pp. 171–198.

Greenberg, J., Kromberg, J., Loggenberg, K., Wessels, T.-M., 2012. Genetic counseling in South Africa. In: Kumar, D. (Ed.), Genomics and Health in the Developing World. Oxford University Press, New York, pp. 531–546.

Harper, P., Clarke, A., 1990. Should we test children for adult genetic diseases. Lancet 335, 1205–1206.

Harper, P.S., 2007. A Short History of Medical Genetics. Oxford University Press, Oxford.

Harper, P.S., 2010. Practical Genetic Counseling, seventh ed. Hodder Arnold, London.

Helman, C.G., 1994. Culture, Health and Illness, third ed. Butterworth Heineman, Oxford.

Hodgson, J., Spriggs, M., 2005. A practical account of autonomy: why genetic counseling is especially well suited to the facilitation of informed autonomous decision making. J. Genet. Couns. 14, 89–97.

Jenkins, T., Lane, A.B., Kromberg, J.G.R., 1977. Tay Sachs disease screening and prevention in South Africa. S. Afr. Med. J. 51, 95–98.

Jones, A., Bodmer, W.F., 1974. Our Future Inheritance: Choice or Chance? a Study by a British Association Working Party. Oxford University Press, London.

Kessler, S., 1980. The psychological paradigm shift in genetic counseling. Soc. Biol. 27 (3), 167–185.

Kessler, S., 1989. Psychological aspects of genetic counseling. VI. A critical review of the literature dealing with education and reproduction. Am. J. Med. Genet. 34, 340–353.

Kessler, S., 1997. Psychological aspects of genetic counseling. XI. Nondirectiveness Revisited. Am. J. Med. Genet. 72, 164–171.

Kessler, S., 1999. Psychological aspects of genetic counseling. XIII. Empathy and decency. J. Genet. Couns. 8, 333–344.

Kowal, E., Gallacher, L., Macciocca, I., Sahhar, M., 2015. Genetic counseling for indigenous Australians: an exploratory study from the perspective of genetic health professionals. J. Genet. Couns. 24, 597–607.

Kromberg, J.G.R., Jenkins, T., 1982. Prevalence of albinism in the South African Negro. S. Afr. Med. J. 61, 383–386.

Kromberg, J.G.R., Jenkins, T., 1984. Albinism in the South African Negro. III. Genetic counselling issues. J. Biosoc. Sci. 16, 99–108.

Kromberg, J.G.R., Zwane, E., Jenkins, T., 1987. The response of black mothers to the birth of an albino child. Am. J. Dis. Child 141, 911–916.

Kromberg, J.G.R., Zwane, E., 1993. Down syndrome in the black population in the Southern Transvaal. S. Afr. J. Child. Adolesc. Psychiatry 5 (1), 30–33.

Kromberg, J.G.R., Jenkins, T., 1997a. Cultural influences on the perception of genetic disorders in the black population of Southern Africa. In: Clarke, A., Parsons, E. (Eds.), Culture, Kinship and Genes. Towards Cross-cultural Genetics. MacMillan, London, pp. 147–157.

Kromberg, J.G.R., Jenkins, T., May 1997. Genetic services in South Africa and common ethical issues related to genetic information. Spec. Med. 54–60.

Kromberg, J.G.R., Parkes, J., Taylor, S., 2006. Genetic counselling as a developing health care profession: a case study in the Queensland context. Aust. J. Prim. Health 12, 5–12.

Kromberg, J.G.R., Sizer, E., Christianson, A.L., 2013a. Genetic services and testing in South Africa. J. Community Genet. 4, 413–423.

Kromberg, J.G.R., Wessels, T.-M., Krause, A., 2013b. Roles of genetic counselors in South Africa. J. Genet. Couns. 22 (6), 753–761.

Kromberg, J.G.R., Essop, F.E., Rosendorff, J., 2015. Prenatal diagnosis for oculocutaneous albinism. In: Southern African Society for Human Genetics 16th Congress, 16–19 August 2015 (Abstract book).

Levy, B., 1992. Expectations of Genetic Counselling MSc dissertation. University of the Witwatersrand, Johannesburg, South Africa.

Lippman-Hand, A., Fraser, C., 1979. Genetic counseling provision and reception of information. Am. J. Med. Genet. 3, 113–127.

Liveneh, H., Antonak, R.F., 2005. Psychosocial adaptation to chronic illness and disability: a primer for counselors. J. Couns. Dev. 83, 12–20.

Lynch, H.T., 1969. Dynamic Genetic Counseling for Clinicians. CC Thomas, Springfield Illinois.

Macaulay, S., Gregersen, N., Krause, A., 2012. Uptake of genetic counseling services by patients with cystic fibrosis and their families. S. Afr. Fam. Pract. 54 (3), 250–255.

Manganyi, N.C., Kromberg, J.G.R., Jenkins, T., 1974. Studies on albinism in the South African Negro. 1. Intellectual maturity and body image differentiation. J. Biosoc. Sci. 6, 107–112.

McAllister, M., 2002. Predictive genetic testing and beyond: a theory of engagement. J. Health Psychol. 7, 491–508.

McAllister, M., 2003. Personal theories of inheritance, coping strategies, risk perception and engagement in hereditary non-polyposis colon cancer families offered genetic testing. Clin. Genet. 64, 179–189.

McAllister, M., Dunn, G., Todd, C., 2011. Empowerment: qualitative underpinning of a new patient reported outcome for clinical genetic services. Eur. J. Hum. Genet. 19, 125–130.

McAllister, M., 2016. Genomics and patient empowerment. In: Kumar, D., Chadwick, R. (Eds.), Genomics and Society. Ethical, Legal, Cultural and Socioeconomic Implications. Elsevier Academic Press, London, pp. 39–68.

Morris, M., Glass, M., Wessels, T.-M., Kromberg, J.G.R., 2015. Mothers' experience of genetic counseling in Johannesburg, South Africa. J. Genet. Couns. 24, 158–168.

National Society of Genetic Counselors, 2010. Professional Status Survey: Executive Summary. Available at: www.nsgc.org.

Nowak, R., 1994. Genetic testing set to take off. Science 265, 464–467.

Oosterwal, G., 2009. Multicultural counseling. In: Uhlmann, W.R., Schuette, J., Yashar, B.M. (Eds.), A Guide to Genetic Counseling, second ed. Wiley-Blackwell, New Jersey, pp. 331–361.

Rapp, R., 2000. Testing Women. Testing the Fetus: The Social Impact of Amniocentesis in America. Routledge, New York.

Resta, R., Biesecker, B.B., Bennett, R.L., 2006. A new definition of genetic counseling. National society of genetic counselors task force report. J. Genet. Couns. 15 (2), 77–83.

Roberts, D.F., Kromberg, J.G.R., Jenkins, T., 1986. Differentiation of heterozygotes in recessive albinism. J. Med. Genet. 23, 323–327.

Rogers, C.R., 1942. Counseling and Psychotherapy. Houghton Mifflin, Cambridge Mass.

Schapera, I., 1937. The Bantu Speaking Tribes of South Africa. An Ethnographic Survey. Routledge, London.

Schapera, I., 1940. Married Life in an African Tribe. Faber and Faber, London.

Schmerler, S., 2009. Ethical and legal issues. In: Uhlmann, W.R., Schuette, J.L., Yashar, B.M. (Eds.), A Guide to Genetic Counseling. Wiley-Blackwell, New Jersey, pp. 363–399.

Schoonveld, K.C., Veach, P.M., LeRoy, B.S., 2007. What is it like to be in the minority? Ethnic and gender diversity in the genetic counseling profession. J. Genet. Couns. 16 (1), 53–69.

Shiloh, S., 1996. Decision-making in the context of genetic risk. In: Marteau, T., Richards, M. (Eds.), The Troubled Helix. Social and Psychological Implications of the New Human Genetics. Cambridge University Press, Cambridge.

Skirton, H., Cordier, C., Ingvoldstad, C., Taris, N., Benjamin, C., 2015. The role of the genetic counselor: a systematic review of research evidence. Eur. J. Hum. Genet. 23, 452–458.

Slovic, P., 1987. Perception of risk. Science 236, 280–285.

Soodyall, H., Kromberg, J.G.R., 2016. Human genetics and genomics and sociocultural beliefs and practices in South Africa. In: Kumar, D., Chadwick, R. (Eds.), Genomics and Society. Ethical, Legal, Cultural and Socioeconomic Implications. Elsevier, London, pp. 309–319.

Statistics South Africa, 2009. Mid-year Population Estimates. Mortality, Expectation of Life at Birth, and Fertility, pp. 6–7. Available at: http://www.statssa.gov.za/publications/P0302/PO3022009.pdf.

Stevens, G., Van Beukering, J., Jenkins, T., Ramsay, M., 1995. An intragenic deletion of the P gene is the common mutation causing tyrosinase-positive oculocutaneous albinism in Southern African negroids. Am. J. Hum. Genet. 56, 586–591.

Tobin, D.L., Holroyd, K.A., Reynolds, R.V., Wigal, J.K., 1989. The hierarchical factor structure of the coping strategies inventory. Cogn. Ther. Res. 13 (4), 343–361.

Van Wyk, C., Wessels, T.-M., Kromberg, J.G.R., Krause, A., 2016. Knowledge regarding basic concepts of hereditary cancers, and the available genetic counseling and testing services: a survey of general practitioners in Johannesburg, South Africa. S. Afr. Med. J. 106 (3), 268–271.

Weil, J., 2000. Psychosocial Genetic Counseling. Oxford University Press, Oxford.

Weil, J., 2003. Psychosocial genetic counseling in the post-nondirective era: a point of view. J. Genet. Couns. 12 (3), 199–211.

Weil, J., Ormond, K., Peters, J., Peters, K., Biesecker, B.B., LeRoy, B., 2006. The relationship of nondirectiveness to genetic counseling: Report of a Workshop at the 2003 NSGC Annual Education Conference. J. Genet. Couns. 15 (2), 85–93.

Wertz, D.C., Fletcher, J.C., 1988. Attitudes of genetic counselors: a multinational survey. Am. J. Hum. Genet. 42, 592–600.

Wertz, D.C., 1992. How parents of affected children view selective abortion. In: Holmes, H.B. (Ed.), Issues in Reproductive Technology. I. An Anthology. Garland, New York, pp. 161–189.

Wessels, T.-M., Koole, T., Penn, C., 2015. 'And then you can decide' – antenatal foetal diagnosis decision making in South Africa. Health Expect. 18, 3313–3324.

Wonkam, A., Muna, W., Ramesar, R., Rotimi, C.N., Newport, M.J., 2010. Capacity-building in human genetics for developing countries: initiatives and perspectives in sub-Saharan Africa. Public Health Genom. 13, 492–494.

World Health Organisation (WHO), 1998. Proposed Information Guidelines on Ethical Issues in Medical Genetics and Genetic Services (Report of a WHO Meeting on ethical issues in medical genetics).

Genetic Testing, Postnatal, and Prenatal Diagnosis for Albinism

Robyn Kerr, Jennifer G.R. Kromberg
University of the Witwatersrand and National Health Laboratory Service,
Johannesburg, South Africa

I. INTRODUCTION

Genetic testing is now available to families with a member with OCA, in many different situations. Such testing can be used to confirm the diagnosis in a person suspected of being affected and to determine the type of albinism in an affected person. Testing whether an at-risk individual is a carrier for the family disorder and, prenatally, whether a fetus is affected can be done. These issues will be covered in the first part of the present chapter.

The second part of the chapter will further explore the issue of prenatal diagnosis for OCA, whether and when it is acceptable, whether at-risk couples would want it, how it is performed, what supportive counseling should be supplied, what the outcomes might be, whether there is a demand, and what the associated ethical issues are.

II. DIAGNOSTIC TESTING FOR OCA

Although the phenotype for most individuals with albinism would be obvious, especially in a Black African population, and genetic testing for diagnostic reasons would not be indicated, there are certain circumstances in which determining the genotype would be useful. These include delineating the type of OCA allowing for clarification of a type with possible treatment options and determining the mutation profile allowing for carrier and prenatal testing. Furthermore, in populations of European extraction some fair-haired children have nystagmus and they may require diagnostic testing to differentiate between isolated congenital nystagmus and oculocutaneous albinism.

Delineating the Type of OCA

As discussed previously (see Chapter 5) albinism exhibits locus heterogeneity—each different type of albinism is caused by recessive mutations at a different gene locus. Once the locus involved is known, clarification of the type of albinism is possible. Knowing the type segregating in the family may have implications for genetic counseling and possible treatment options.

Four main types of nonsyndromic OCA were initially described and the associated genes identified—OCA1 (*TYR*), OCA2 (*OCA2*), OCA3 (*TYRP1*), and OCA4 (*SLC45A2*). Ocular albinism (OA) is usually caused by a single gene, *GPR143*, which maps to the X chromosome. More recently, a further three types of OCA have been described: OCA5 (linked to 4q24), OCA6 (*SLC24A5*), and OCA7 (C10orf11). Types of albinism and associated genes have been reviewed by Montoliu et al. (2014).

Syndromic phenotypes have been described where albinism is only part of the clinical spectrum and other, usually more severe, tissue/organ involvement is seen—Hermansky–Pudlak (HPS), Chediak–Higashi syndrome (CHS) and Griscelli syndrome (GS). HPS presents with oculocutaneous albinism, whereas CHS and GS present with decreased pigmentation of the hair and eyes only, as opposed to the classic OCA phenotype. To date 10 different types of HPS are known and are caused by mutations in their corresponding genes (see Chapter 5). CHS is caused by mutations in the *LYST* gene, while three loci have been implicated in GS: *MYO5A* (GS1), *RAB27A* (GS2), and *MLPH* (GS3).

Because of the number of possible genes involved (21 to date, as delineated here), molecular diagnostic testing for albinism may be complex. If a phenotype is highly suggestive of type, testing can start with a focused single-gene approach. In most populations the mutation spectrum is wide (see Human Genome Mutation Database: www.hgmd.org for a list of all known pathogenic variants described for each gene) and common mutations have not been described. Consequently, single-gene mutation screening usually involves Sanger sequencing of the entire coding region (all exons) plus flanking regions of the target gene. Sanger sequencing can detect point mutations plus small (<20 bp) insertions or deletions. Larger deletion/duplication events involving the pigment genes can be detected by array comparative genome hybridization. This methodology will detect copy number changes and rearrangements of >400 bp (therefore insertions/deletions in the range 20–400 bp may be missed, irrespective of testing approach).

Technological advances over the last decade in the area of DNA sequencing have opened up new options for both research and diagnostic testing. Collectively referred to as next-generation sequencing (NGS) technologies or high-throughput technologies, these now offer patients with albinism an option for testing that was, in practical terms, not possible before. Several genes, a whole exome, or even a whole genome can be sequenced, looking for disease-causing mutations. If the albinism phenotype is clinically unclear, current testing approaches would then involve a NGS panel approach whereby targeted sequencing of the coding regions of several (or all) of the 21 pigment-related genes listed above are interrogated. Laboratories worldwide, which offer genetic testing for albinism, can be found at www.genetests.org.

Treatment Options and the Importance of Establishing Genotype

At present, management of individuals with OCA is essentially symptomatic. Visual issues must be monitored and spectacles prescribed where they could help. In Africa, management revolves largely around skin issues, educating parents and affected individuals with regard to skin cancer prevention. Besides confirming a diagnosis and delineating type of albinism present in a family, or establishing carrier status and being able to undertake prenatal diagnosis, genetic testing for albinism may have clinical application in the future: it is noteworthy that there is a pharmacogenetics trial taking place for a drug, nitisinone. As discussed previously (see Chapter 5), nitisinone is currently only being tested for use in individuals with OCA type 1B. It is not appropriate for treatment of classical OCA1 (nor OCA2, the most common type in Africa) but may be useful for OCA types where tyrosinase activity is low but not absent. While the drug is currently undergoing clinical trials, it may become a viable treatment option in the future for OCA1B. Clinicians wanting to use this drug would need to know a patient's mutation status to prescribe the drug only in appropriate cases.

Diagnostic Testing for OCA2 in Johannesburg, South Africa

The vast majority of individuals seen with OCA in South Africa are of Black ancestry. The reasons for this are twofold: firstly, it is now well established that the prevalence of OCA is far higher in black Africans than in any other population group worldwide; secondly, due to population demographics—the South African population is ~80% Black, 9% mixed ancestry, 8% White, and 3% Asian (Census, 2011)—there are also simply more Black individuals than members of any other population group in the country.

A common OCA2 mutation is found in Black African patients affected with OCA. This mutation is a 2.7 kb deletion, which completely removes exon 7 of the OCA2 gene. A relatively simple polymerase chain reaction assay was designed to test for this deletion (Durham-Pierre et al., 1994). It was found that the 2.7 kb deletion mutation accounted for 78% of OCA2 mutations in the South African Black population, and it was not found in any White or Asian individuals, confirming that the deletion is in fact common in, and specific to, Black African populations (Stevens et al., 1995). Consequently, the Molecular Diagnostic Service of the Division of Human Genetics at the National Health Laboratory Service (NHLS) in Johannesburg set up its testing service for OCA to serve the population at hand, and offers diagnostic testing for this 2.7 kb deletion mutation. Data have been collected for the 17-year period, 1999–2016, during which 101 test requests were received (from all over South Africa and a few from neighboring territories) by the Molecular Diagnostic Laboratory. A summary of type of test requested and ethnic background of patients/families is given in Table 11.1.

Only 50 diagnostic tests were processed for this period (17 years). This is a reflection of the fact that the phenotype is an obvious one and a clinical diagnosis is easily made. Genetic testing is usually only requested when a family requires either carrier or prenatal testing. Up to and including 2016, carrier test requests totaled 43, and 8 prenatal tests were performed. Results of testing are given in Table 11.2.

TABLE 11.1   OCA2 Test Requests by Patient Ethnic Background

	Black	Mixed Ancestry	Malagasy	White	Indian	Total
Carrier	26	4	0	11	2	43
Diagnostic	34[a]	5	2	9	0	50
Prenatal	7	1	0	0	0	8
Total	67	10	2	20	2	101

[a] *Two diagnostic test requests came from outside South Africa—one from Namibia and one from Uganda. Both patients were Black African and were thus included here.*

TABLE 11.2   Results of Testing for the OCA2 2.7 kb Deletion Mutation in 101 Individuals

	Black	Mixed Ancestry	Malagasy	White	Indian	Total
M/M[a,c]	15[b]	1	1	0	0	17
M/U[a,c]	10	3	0	0	0	13
M/N[a,c]	18	5	0	0	0	23
N/N[c]	13	1	1	14	2	31
M/0[a,c]	1	1	0	0	0	2
Not tested	7	0	0	6	0	13
Unsuccessful	2	0	0	0	0	2
Total	66	11	2	20	2	101

[a] *0, absence/deletion of allele; M, 2.7 kb del mutation; N, normal allele; U, mutation present but unidentified.*
[b] *This figure includes the two non-South African individuals, namely the Ugandan and the Namibian.*
[c] *M/M, homozygous for the 2.7 kb deletion; M/U, a compound heterozygote with one mutation being the 2.7 kb deletion and the second mutation remaining unidentified; M/N: heterozygous for the 2.7 kb deletion; N/N, zero copies of the 2.7 kb deletion mutation present; M/0, the individual is hemizygous for the 2.7 kb deletion.*

Individuals homozygous for the 2.7 kb deletion (M/M) were only found in Black individuals or individuals of mixed ancestry. The 2.7 kb deletion was not found in any White or Indian individuals. Excluding, then, the White and Indian individuals from further analysis: of the 30 individuals sent in for carrier testing, 23 (77%) were found to be carriers of the 2.7 kb deletion (M/N), in keeping with the literature (Stevens et al., 1995). The group of individuals where no 2.7 kb deletion mutation was identified (N/N) included the following individuals: all White and Indian individuals tested (N = 16); one individual from Madagascar; four affected Black individuals; nine obligate carrier parents, and one fetus (prenatal testing).

In the early years of diagnostic testing in Johannesburg, all ethnicities were offered 2.7 kb deletion testing, in some cases simply to exclude the possibility of the pathogenic mutation being this deletion. Again, in keeping with findings in the literature, the 2.7 kb deletion was identified only in Black African populations or in individuals of African ancestry (for example, in an individual from Madagascar) or where African admixture was present in the family. Therefore pedigree analysis and genetic counseling is important—testing for this mutation in a White individual has no utility. From 2001 onward, individuals with albinism with a White family background, who attended the Genetic Counseling Clinic in Johannesburg, were not offered 2.7 kb deletion testing anymore and their DNA was simply banked for possible future use. Currently, non-African patients seeking genetic testing for albinism are offered NGS/panel testing in overseas laboratories.

Thus certain individuals were "Not Tested" because the individual or family was not of Black African ancestry. In some instances, the wrong blood tube had been sent into the laboratory and/or DNA could not be extracted from the sample. In two cases, the test was "Unsuccessful" because of the poor quality of the DNA following extraction. Therefore, tests completed where a result was available totaled 86 (101–15).

Two individuals, in this series of patients, presented with Prader–Willi syndrome (PWS) as well as OCA; one was of Black African origin and the other of mixed ancestry. PWS is a genomic imprinting disorder resulting from lack of expression of paternally inherited genes lying on chromosome 15q11-13. Most cases of PWS are caused by a deletion of this region on the chromosome 15 inherited from the paternal parent. The deleted region can include the OCA2 gene. If the maternally derived chromosome in the individual carries a mutated OCA2 allele, the individual will have no functional copy of the OCA2 gene and will be affected with albinism. The individual will then present with both PWS and OCA. In the two cases tested here, the patients with PWS carried a deletion on one chromosome, which removed the entire OCA2 gene from that chromosome, creating a null allele, and it was found that the pathogenic mutation on the other allele was in fact the 2.7 kb deletion mutation. As these individuals have only one copy of the OCA2 gene (and this gene carries the 2.7 kb deletion mutation), they are said to be hemizygous for the 2.7 kb deletion mutation.

III. CARRIER TESTING

All forms of albinism (except OA which is generally X-linked) are autosomal recessive and Mendelian laws of inheritance apply: parents of an affected individual are obligate carriers and they have a 25% (1 in 4) recurrence risk of having another affected child in each subsequent pregnancy. Unaffected siblings of an affected individual have a 66% (2 in 3) risk of being carriers (see Chapter 10). Parents, siblings, or closely related individuals may seek carrier testing. Ideally, family mutations should be identified in an affected individual and then focused carrier testing can be offered to relatives. This would have to be the approach in all non-African families as the mutation spectrum is diverse. Black Africans (particularly in Southern and East Africa) are in a unique position as the 2.7 kb deletion mutation is common. This has two immediate implications: firstly, carrier testing can be offered, even in the absence of an affected individual (although in this circumstance, a negative result for the 2.7 kb deletion would not then necessarily mean the individual is not a carrier of another OCA2 mutation and should be carefully counseled) and secondly, population screening is a viable option, from a theoretical point of view.

Where the genetic cause of albinism is known in a family, carrier testing can be offered. In general, if there is an affected individual in a family and genetic testing is undertaken and either only one or neither of the disease-causing mutations is identified, then carrier testing for at-risk individuals in that family becomes complicated. For the side of the family where the mutation is unknown, carrier testing cannot be offered. However, it is often possible to undertake linked marker (linkage) analysis for a prenatal test, if an at-risk couple request prenatal testing, but this would require the availability of DNA from an affected member of the family and a genetic work-up of the family, prior to prenatal testing.

When counseling families with a member with albinism, particularly those of non-African descent, it is important to remember that OCA exhibits locus heterogeneity—different genes can be mutated in different individuals but can result in the same (or a very similar) phenotype. It is therefore possible that two individuals could both be carriers but of different types of albinism. In this scenario, their risk of having an affected child would be zero (as opposed to 25% if they were carriers of the same type). Similarly, if two affected individuals who have different types of albinism have offspring, their risk of having affected children would also be zero (as opposed to 100% if they both had the same type).

When offering/undertaking carrier testing for albinism, it is important to consider some issues that apply to genetic carrier testing in general (discussed in Ross et al., 2013). The individual's/family's belief and value system must be respected; the individual's right to privacy and confidentiality with respect to results must be adhered to; and steps to avoid stigmatization and discrimination should be taken. The clinician/counselor has an obligation to offer education and information to explain that limited (or no) treatment options are usually available for inherited conditions such as albinism and no cure is possible, at this time. For carrier parents, few reproductive choices are available. These couples could opt to refrain from childbearing; use donor sperm/eggs; undertake prenatal diagnosis and request termination for an affected fetus; or undertake preimplantation genetic diagnosis to select for unaffected embryos (a service only available in certain countries and at considerable cost). The issue of informed consent is also paramount and carrier testing for most disorders requires legal informed consent to be given by the individual requesting the test. In South Africa informed consent can only be given once the individual is 18 years or older.

IV. PRENATAL GENETIC DIAGNOSIS

Historical Background

Prenatal diagnosis for many different genetic conditions has been offered in Johannesburg (as well as in some of the other major cities in the country, e.g., Cape Town) for several decades. The first two conditions to be diagnosed prenatally in South Africa were Down syndrome (particularly in the older at-risk pregnant woman) and neural tube defects (mostly in those with a family history and/or a previously affected child). Then diagnostic testing for the recessively inherited lethal condition Tay-Sachs disease, in at-risk couples of Ashkenazi Jewish origin, who had a previously affected child or where both parents were carriers, was offered (Kromberg et al., 1989). Fetal samples for the testing of these conditions were collected, initially, by means of amniocentesis or, later, by chorionic villus sampling (CVS).

In a series of 4554 cases of amniocentesis for prenatal genetic diagnosis over a 10-year period (between 1976 and 1985), in Johannesburg, Kromberg et al. (1989) found that the demand for the service increased fivefold. A correct fetal diagnosis was made in 99.9% of cases (the four missed diagnoses were due, in three cases, to twins, one of whom was normal and the other had a neural tube defect in each case and mosaic Down syndrome in a fourth case), and abnormalities were detected in 3.2% of pregnancies. The rate of "spontaneous" abortion after the procedure was 0.7%, while the total fetal loss rate from 16 to 28 weeks gestation was not much increased over that in any other series of second trimester pregnancies. These figures were in keeping with internationally reported risks for amniocentesis (Leschot et al., 1985). The procedure, performed by specialist obstetricians, had therefore become a safe, reliable, and successful one in expert hands, in Johannesburg. The majority of the women included in this study were South African; however, samples of amniotic fluid were also sent in from Zimbabwe and Namibia and occasionally patients came from other African countries, such as Botswana, Malawi, and Kenya. The patients either had their amniocentesis carried out through the genetic counseling and antenatal clinics in Johannesburg hospitals or through their private obstetricians or radiologist. The samples were then sent in to the Human Genetics laboratories (at the South African Institute for Medical Research, now the National Health Laboratory Service) where the indicated genetic tests were performed.

Initially all the tests, for the collection of fetal cells, were undertaken by means of second trimester amniocentesis (Fig. 11.1) at 16 weeks gestation, but by the mid-1980s it became possible to perform first trimester CVS at 9–11 weeks gestation, for some conditions. This technique slowly became the technique of choice for prenatal diagnosis in most high-risk pregnancies. However, although a result could be obtained earlier in pregnancy (with the benefit of having an earlier and simpler termination of pregnancy, if that became necessary), there was a slightly higher risk of spontaneous fetal loss (about 1.5% after CVS, instead of 0.7% following amniocentesis). Neural tube defects could not be diagnosed by CVS because amniotic fluid was required for making the diagnosis. Furthermore, many women only presented for antenatal care later in pregnancy, when it was too late for CVS, so that amniocentesis was required for those late comers who were identified as at high risk.

Later, a method of collecting fetal tissue samples was developed. This method was known as fetoscopy, and the obstetrician visualized the fetus using a fetoscope, at 20-weeks gestation,

FIGURE 11.1 Amniocentesis procedure. *From https://en.wikipedia.org/ wiki/Amniocentesis (By Bruce Blaus).*

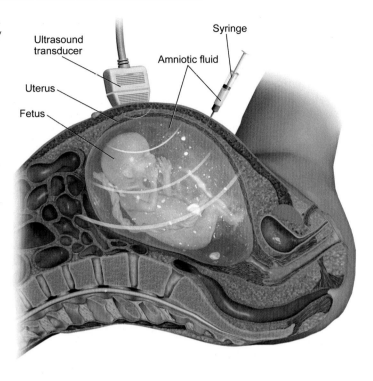

and collected skin samples, often from the fetal scalp, so that various genetic skin conditions could be diagnosed. Follow-up of these fetuses showed that the wound healed quickly and the scar at the site could hardly be detected when the fetus was examined at birth.

Another noninvasive method of prenatal assessment, ultrasound scanning of the fetus, has been refined. Scanning can now provide a high-definition image, so that abnormalities, such as many neural tube defects, some bony and limb abnormalities, and other congenital defects, can be strongly suspected or detected. Also, in the case of a fetus with Down syndrome the thickness of the folds of skin around the neck and other minor signs can be measured to assess the likelihood that the fetus is at high risk for Down syndrome, in which case an invasive test such as amniocentesis would be recommended to make a diagnosis.

Recently, a new technique has been developed by which cell-free fetal DNA can be collected from maternal serum samples taken during pregnancy and tested for various genetic conditions. These tests are presently not yet available in South Africa (nor elsewhere in Africa), but maternal serum samples can be collected locally and shipped to the United States or elsewhere for testing, depending on the genetic tests required and where they are performed.

Prenatal Genetic Testing for Albinism

Prenatal diagnosis for albinism has been possible since 1983 (Eady et al., 1983). Initially it was used, in London, United Kingdom, for prenatal diagnosis of OCA1. The technique used was electron microscopic examination of fetal skin samples. These samples were collected, by specialist obstetricians, during fetoscopy at 20 weeks gestation. The melanocytes of persons with OCA1 were known to possess only stage 1 and 2 melanosomes, showing that there was a lack of melanin produced (Witkop et al., 1978) and the condition could be diagnosed in this way.

Eady et al. (1983) described the case of a woman and her husband from a Middle Eastern country, who requested prenatal diagnosis for OCA1, in London. They had a previous child with OCA1 and felt that the condition, in their society, caused both social and economic hardship. The results of the testing showed an affected fetus, in utero, and the parents requested termination of the pregnancy.

In 1999 Rosenmann et al. described 34 cases of prenatal diagnosis for OCA1 carried out in Israel. In the majority of cases (31) prenatal diagnosis testing had been performed following fetal scalp biopsies. In 26 of these, normal melanogenesis was found, while a diagnosis of OCA1 was made in five cases. In a further three cases molecular testing was undertaken and results showed that two fetuses were unaffected while one was affected. No follow-up appears to have been reported on these cases and no further information is available. However, Fasihi et al. (2006) reported on 25 years of experience of prenatal diagnosis for various skin diseases (such as epidermolysis bullosa, icthyosis, and albinism). Altogether 191 fetal skin biopsies were examined and 12 were for albinism. They reported that in all cases (except two), where the fetus was found to be affected, termination of the pregnancy was requested.

In Europe both OCA1 and OCA2 have been reported in individuals with albinism. However, in sub-Saharan Africa, the large majority of affected families have OCA2 and the common 2.7 kb deletion. This mutation can be tested for in the diagnostic setting at the NHLS Human Genetics laboratories in Johannesburg. In those who do not have this mutation, prenatal diagnosis cannot be offered, at present, unless the family specific mutation(s) has previously been identified.

Demand for Prenatal Diagnosis for Albinism in South Africa

To assess demand for prenatal diagnosis of OCA in Africa, a small study was undertaken in South Africa (Kromberg et al., 2015). The aim of the study was to investigate the demand for and the outcomes of such testing at the Division of Human Genetics in Johannesburg. Records for 33 families, who had genetic counseling for OCA from 2011 to 2014, and five families who requested prenatal diagnosis, between 1999 (when it was first offered) and 2014, were reviewed retrospectively. The majority of the 33 women had recently had a child with OCA and only two were pregnant.

Of the 33 women who participated in the above study and who had attended a genetic counseling session, some (18) were informed and some (15) were not informed about prenatal diagnosis. Counselors of the 18 women believed that families should receive all the necessary information on which to base their future choices, such as those regarding reproduction. Of these 18 patients, 16 were not pregnant and two were pregnant (see Table 11.3). Among the 16 patients, 7 (39%) volunteered the fact that they would request prenatal testing in a future pregnancy. The reasons they gave included: to prepare themselves for another affected child (5) or to terminate a pregnancy if the fetus was found to be affected (2). One of these latter two participants stated that she had OCA herself, she had had a hard time as a child being teased and stigmatized, and she would not want any child of hers to undergo the same traumatic experiences. The other woman had three affected children and she felt it had been difficult

TABLE 11.3   Opinions on Prenatal Diagnosis (PND) in 18 at-Risk Women

Opinions	No.	%
Not pregnant	16/18	89
PND accepted (in theory)	7	39
Reasons		
To prepare themselves (would not request TOP)	5	28
Would request TOP[a]	2	11
Pregnant	2/18	11
PND declined	2	11
Reasons		
Not worried about having a child with OCA	2	11

[a] TOP, *termination of pregnancy.*

and costly bringing them up and attending to their skin and eye problems, so she would not want to care for another affected child. On the other hand, the 2/18 at-risk women who were pregnant stated that they did not want prenatal diagnosis, as they were not worried about having another affected child.

The fact that 39% (7/16) of these women thought that they would request prenatal testing in future pregnancies suggests that demand for such testing might increase in future, as the relevant information is disseminated. However, any offer of prenatal diagnosis will be met with a variety of responses from at-risk women, as shown in the above study.

Most of the women participating in this study who were offered prenatal diagnosis (61%, 9/16) stated that they would refuse it. However, most were not pregnant at that time, had a new born infant with the condition, had not yet been through the experience of raising a child with albinism, or faced the stigmatization and community rejection that often ensues. The decision therefore was a theoretical one, which, as is well known, can change when the situation is faced in reality.

Among the five couples (all had a one in four risk for an affected child) who requested prenatal diagnosis one couple had had two pregnancies tested, one by means of amniocentesis, and one by CVS. The outcomes of these six tests showed that in four cases the fetus was a carrier like the parents, in one case the fetus was not carrying an OCA mutation at all, and in one case the fetus was affected. The parents of the affected fetus decided to continue the pregnancy and a healthy affected infant was born at full term. This family had been treated in private practice and had not had genetic counseling, so that it is not known what they were told about their situation and the risks, prior to testing. If their decision regarding termination had been considered and/or made before the prenatal procedure (which carries a very small procedure-related risk) was carried out, they might have refused to have it. Alternatively, they might have agreed to the procedure for reassurance, hoping the results would show an unaffected fetus or to prepare themselves for another affected infant.

The facts and figures, from this small sample, suggest that, at present, demand for prenatal diagnosis for albinism appears to be low in Johannesburg. However, this could be due to the fact that many affected families are not aware that prenatal diagnosis for OCA is possible and do not question their health professionals about this situation. Furthermore, health professionals themselves might be unaware of the availability of the service and might not offer the information to their affected patients or their at-risk parents. Also, such professionals might have their own views on whether or not prenatal diagnosis for albinism should be offered and if they have prejudged this situation they might withhold the information from pregnant women at risk for albinism in a child. Alternatively, at-risk women may be fatalistic, may believe that the course of pregnancy cannot be altered, may not want to have prenatal diagnosis, and may fear the invasive nature of the procedure. Studies of prenatal diagnosis for genetic conditions in Johannesburg (Schreier et al., 1986; Kromberg et al., 1989) showed that demand was increasing, but even among older women (35 years or more) at risk for Down syndrome and other chromosome abnormalities it was only 35% in the 1980s. Reasons for the low-utilization rate included: ignorance of the availability of prenatal genetic testing, booking late for antenatal care, religious objections to termination of pregnancy, previous obstetric problems, infertility, needle phobia, and fears concerning the very small procedure-related risk (<1%) of miscarriage.

A study of school children with albinism in Zimbabwe (Lund, 2001) showed that many (99%) wanted more information on albinism while some questioned having children and queried "medical investigations" asking:

Is there a medical investigation to find out whether I will bear albinos? *Lund (2001, p. 4)*

This suggests that with increasing education of at-risk people on their condition, the demand for services such as prenatal genetic diagnosis might increase.

Genetic Counseling for Prenatal Genetic Diagnosis

Ideally, all women who request prenatal diagnosis for a genetic condition should have genetic counseling prior to having the procedure (see Chapter 10). In the case of couples at risk for albinism in their children, there are various aspects of their situation that need to be discussed in a counseling session, various points that should be raised for discussion, choices that have to be considered, and decisions that have to be made. Genetic counselors are trained to offer this service.

Basically, genetic counselors provide the necessary information in a nondirective and nonjudgmental manner and this type of counseling should result in autonomous decision-making. The main activities in the prenatal counseling encounter, with at-risk pregnant women, include making decisions concerning: the risk faced; the invasive procedure required to make a diagnosis; and the possible options in the event of an abnormal result.

In a recent qualitative study such decision-making was investigated in South Africa (Wessels et al., 2015). Genetic counseling consultations with 14 pregnant Black women of advanced maternal age (35 years or more) at risk for chromosome abnormalities in the fetus were recorded. Each woman was informed about her genetic risk, the amniocentesis procedure, and the very small procedure-related abortion risk, and further discussion covered the possibility of having an infant with abnormalities and having an abnormal test result. Then the woman was invited to make a decision regarding whether or not she wanted prenatal diagnosis. The results showed that genetic counseling sessions in this context had six discernible phases. These included an opening phase, information-gathering and information-giving phases, and decision-making, counseling, and closing phases. The counselors used several strategies to facilitate decision-making. Questioning was employed to encourage reflection, introspection, and the sharing of thoughts and feelings, as well as thinking about the hypothetical future.

This study (Wessels et al., 2015) illustrates some of the choices women have and the decisions they might make, when faced with the option of prenatal diagnosis for a genetic condition that can be detected in the fetus in utero. In this case the women were in the older age group, considering their risk of having a child with Down syndrome (~1 in 100 at age 40 years) with its associated developmental delay. They therefore had issues that were somewhat different to those faced by couples who have a one in four risk of having a child with OCA. Nevertheless, it is useful to consider the dilemmas of the women in this study because some will be similar to those faced by women who are considering prenatal diagnosis for OCA. As the authors (Wessels et al., 2015) mentioned in their discussion, the counselors in their study took great care to present the required information in an unbiased way and to assist the women to make well-informed and autonomous decisions. Counselors dealing with women considering prenatal diagnosis for OCA should provide counseling in the same way.

In another study (Aalfs et al., 2007) the aim was to assess counselee and counselor satisfaction in a reproductive genetic counseling setting in the Netherlands. Altogether 151 women and 11 counselors participated in the research run by the Department of Clinical Genetics in Amsterdam. Although OCA is not mentioned in this study, the counselees were at risk for many different genetic conditions, such as Down syndrome, cystic fibrosis, muscular dystrophy, and neural tube defects. The findings showed little difference between the counselees and counselors overall visit-specific satisfaction and both groups reported high satisfaction with the genetic consultation (suggesting that there is good quality reproductive counseling at this center). The counselees' satisfaction was positively associated with being pregnant and with the personal control they felt they had gained from the counseling. Genetic counselors generally aim to decrease uncertainty and increase these feelings of control in their clients, by providing relevant information about genetic risks and testing options and assisting them in reaching well-informed decisions. Both counselees and counselors, in this study, felt this aim was achieved. Providing emotional support is another aim of counseling and this activity scored lowest in this study, suggesting that the participants were not satisfied with this aspect of counseling (Aalfs et al., 2007). Further investigation is required to determine the nature of this discontent and how this situation can be better managed.

Decision-Making About Prenatal Diagnosis

Whether or not to have prenatal diagnosis is often a difficult decision to make, especially if the condition for which the request is made might have variable expressivity, be severe or mild in the fetus, and might be associated with physical or intellectual disabilities or both. In the case of OCA, the decision-making is particularly difficult because the visual defects associated with the condition might be severe in some cases and mild in others and the stigmatization might or might not occur and might or might not cause psychological trauma for the affected child and the family.

Several studies have examined women's decision-making strategies in connection with prenatal screening (noninvasive prenatal testing for genetic conditions to assess the risk status of the fetus) and these have been reviewed by Green et al. (2004). Although these studies examined screening (mostly for either Down syndrome in the fetus or for carrier status for other genetic conditions in the mother and/or her partner; in both cases the outcome of a positive test would be the offer of prenatal fetal diagnosis) during pregnancy, the results reflect, to a certain extent, on issues associated with prenatal diagnosis as well. Although these studies were of varying quality, research design, and biases, some broad patterns were identified. The following questions were covered (and these could also be put, for discussion, to a group of women at risk for OCA in the fetus):

1. Do women want prenatal screening?
   About 60%–75% (this figure would be lower, if prenatal diagnosis was being studied) of women stated that they valued prenatal genetic screening as it empowered them and helped them to make informed decisions. However, a smaller group of participants (about 10%) felt that prenatal testing might medicalize pregnancy, raise worries, create a false sense of control, and lead to increased stigmatization of the condition in the community.
2. Do we know why women choose to have or not have prenatal screening?
   Various factors were identified as having contributed to the decisions of many women to have testing: to gain information; to inform later decisions regarding termination or continuation of the pregnancy and preparation for the affected child; to know for certain whether or not the child had an abnormality; and to gain reassurance that everything was alright. Factors involved in the decisions of fewer women included having no reason not to have testing and following the recommendation of health professionals. In one study the viewpoint of those delivering the testing services and women using the service were shown to differ and the obstetrician sought to detect abnormalities, while the patient sought reassurance that there was no abnormality (Green and Statham, 1996).

   Women chose not to have testing for the following reasons, they: would not act on the result or did not want to worry about it; did not want to know; would not have an abortion; believed that screening tests results do not provide a definite answer and are unreliable; were convinced the pregnancy was at low risk; believed the abnormality would not be serious; had a poor previous experience with screening; and/or they lacked resources to access testing.

   Further studies (described by Green et al., 2004) found that women who had testing for a specific condition differed from those who did not by: having more negative attitudes toward the condition; perceiving the risk of subsequent interventions as lower; perceiving others as thinking they should have the test; and being more likely to intend to have an abortion for an affected fetus. It is possible that these verbalized reasons for having or not having testing do not reflect the actual thought processes and cognitions that underpin the decision.

It should also be noted that, although the studies mentioned above are about prenatal screening in pregnancy, similar patterns of thinking have been found in studies on prenatal diagnosis decision-making. However, the findings on screening should not be taken as predictive of diagnostic or termination choices because many other factors become apparent as the situation develops. Also, aspects of decision-making vary over time and in different circumstances.

3. Are women making informed decisions?

When the relevant information on the advantages and disadvantages of all the courses of action available is evaluated, in the context of the woman's personal beliefs and values, prior to making a decision and a choice, decision-making is said to be informed (Green et al., 2004). Women value making such decisions and, when asked, most will state that their choice was informed. However, it is evident that their knowledge of prenatal screening is poor, few thought carefully about the information before making a choice, some were influenced by third parties (health professionals or relatives or friends) and did not make their own decisions, some found it difficult to refuse the health professional's offer, and some found the decision-making difficult and sought advice (Green et al., 2004). Most women therefore are not making informed choices and they are not fully aware of what an informed choice is and do not have the skills or understanding to make one.

4. Are informed decisions good for decision-makers?

The limited literature on this topic suggests that more informed decision-making results in better postdecision outcomes (Green et al., 2004). Studies show that most (over 92%) women were satisfied that they made the right choice, but a few (3%–30%) of those screening positive regretted their decision. Making more informed decisions appears to be associated with greater anxiety levels, which is also shown regarding effective decision-making in the literature. However, such decision-making is related to less false reassurance and increased decisional satisfaction.

It is worth considering these studies (as reviewed by Green et al., 2004) in detail because they show that, in general, women approve of prenatal screening, although they have some reservations regarding its role and the information it provides to individuals and society. Such reservations might be extended to the screening of at-risk individuals for OCA genes and decision-making regarding prenatal diagnosis for OCA in high-risk pregnancies. However, research also shows that a significant proportion of women are not making informed choices about prenatal screening and/or carrier testing; many are relying on simple heuristics or recommendations from others, instead of considering the available detailed relevant information, to make their own choices. It is possible that such findings could apply, at present, to women who are considering prenatal testing for OCA, in Africa.

There are ways of facilitating the decision-making process in this situation and guidelines regarding the genetic counselor's role have been provided (Weil, 2000). There are several goals (after Weil, 2000, p. 145):

1. The decision must be based on adequate consideration of all the options and their potential consequences
2. The decision must be consistent with the counselee's beliefs and values, wishes, and life circumstances

3. The decision should be made through a process that brings a sense of resolution, and, if the outcome is undesirable, minimizes feelings of guilt, blame, and regret because the decision was based on all the relevant factors and was the best that could be made in the circumstances

4. The process facilitates implementation of the decision

The needs of the counselee seeking prenatal diagnosis must be assessed because many will have made their decision prior to the genetic counseling session, in which case the counselor should focus on support and implementation. However, new information provided in the session may reflect on the decision and reopen the decision-making process. This process may be impeded by a variety of factors; these include stress, ambivalence, concern about the opinions of others, disagreement with a partner, confusion concerning all the options, and lack of correct information (Weil, 2000). Such sources of difficulty need to be identified and addressed by the counselor, and reviewing and clarification of the issues are essential.

The discussion around decisions should also include covering the internal and external, as well as the subjective and objective, resources available to the counselees (Lippman-Hand and Fraser, 1979). Some couples might not have considered these factors and might be uncertain about how to proceed with reproductive decision-making. Another aspect that has to be recognized is the constant interaction of risk and consequence, which is integral in decision-making in prenatal counseling sessions and needs to be processed; in this context the question "Will I be able to handle the situation?" sometimes arises (Lippman-Hand and Fraser, 1979). An example of this issue was provided by a woman who had had three children with OCA and frankly admitted that she would decide on prenatal diagnosis in the next pregnancy and selective termination because she would not be able to cope with another affected child (Kromberg et al., 2015).

Reproductive decision-making is obviously a broader field than prenatal diagnosis decision-making, but some of the influential factors overlap. For example, women with a new partner sometimes want a child by that partner, regardless of the risk. This has been a factor in decision-making in couples at risk for OCA in the writers' experience. Also, reproductive decisions are made over long periods and the desire for more children is determined by psychosocial, socioeconomic, relationship stability, and many other factors. Genetic counseling, with genetic risk information, is just one of the factors involved in influencing the decisions of families with genetic disorders. Research shows that reproductive outcomes are often not strongly impacted by such counseling and in fact there appears to be a net increase in desire for more children after counseling (Kessler, 1989). Precounseling reproductive intentions were identified as a major factor in determining postcounseling outcome. However, it is often the "personal meaning" of the genetic information provided at a counseling session that is stored, by counselees, for future use in later reproduction decision-making (Kessler, 1989, p. 341).

The role of men in prenatal decision-making does not appear to have been much studied and women often attend prenatal genetic counseling sessions alone. However, a common reason for rejecting prenatal diagnosis is that "My husband would not let me" (Rapp, 2000). On the other hand, some women stated that their husbands wielded the authority and had told them to have testing. Conflicts between partners can arise and show the different concerns men and women might have regarding childbearing and childrearing. Rapp (2000) suggests that decision-making, in a couple considering amniocentesis and prenatal diagnosis, exposes the existing gender-related negotiations within which a pregnancy is undertaken. There is a complex interchange between the couple, which includes manipulations, negotiations, domination, and sometimes resistance, according to the stories women tell about their decisions to accept or reject this reproductive technology.

Continuing a Pregnancy After Prenatal Diagnosis of a Genetic Disorder

In the study by Kromberg et al. (2015), a few women at risk for OCA stated that they would like to have prenatal diagnosis, hypothetically (because most were not pregnant at the time), to prepare themselves for having an affected child. Furthermore, one pregnant woman after receiving the information that the fetus was affected, decided to continue the pregnancy.

Although this issue has not been studied in the context of OCA, it has been the subject of research in connection with Down syndrome (Hurford et al., 2013). These authors, working in San Francisco, studied the decision to continue a pregnancy when the fetus was diagnosed with Down syndrome, the timing of the decision and the satisfaction with receiving a prenatal diagnosis. A survey was completed by 56 mothers of children prenatally diagnosed with Down syndrome. The questions covered decision-making after receiving a prenatal diagnosis. About one-third of the participants knew before they became pregnant that they would not request termination for any reason. Furthermore, half the women stated that they did not decide to continue the pregnancy until after the diagnosis was made. The majority (82%) stated that receiving the diagnosis, during their pregnancy, made them more anxious. However, most (88%) of the women also reported that if they could do it over again they would still request prenatal diagnosis. The personal factors that appeared to have most impact on decision-making included religious and spiritual beliefs, as well as feelings of attachment to the infant. Although anxiety was increased, because of the diagnosis, most participants felt they benefited by having prenatal diagnosis because it gave them time to process the information and prepare for the birth of the affected child.

Generally, in the case of Down syndrome, as well as other chromosome abnormalities associated with intellectual disabilities, where the fetus is found to be affected, the termination of pregnancy rate is high. Rates vary from 61% to 93% in different studies (Hurford et al., 2013). Differences in the severity of the phenotype affect decisions to terminate, with higher continuation of pregnancy for milder conditions. One study found that younger women were more likely to continue the pregnancy than older women and that, in the United States, Hispanic women were more likely to continue than other women (Schaffer et al., 2006). Another study showed that most women who had an amniocentesis for prenatal fetal diagnosis were either unsure how they would react to the diagnosis or confident that they would not request termination, rather than definite about making such a request (Skotko, 2005). This may also apply to the situation experienced by pregnant women at risk for OCA.

Selective Abortion

Prenatal diagnosis aims to investigate whether or not a fetus has an abnormality and to give parents information on which they can make decisions. However, research shows that although women request prenatal diagnosis, about 30% say they will not consider termination on grounds of fetal abnormality (Green and Statham, 1996). For example, in one study on a sample of non-pregnant women, 66% said they would request prenatal diagnosis for Down syndrome, but only 40% thought they would request termination of an affected pregnancy (Evers Kiebooms et al., 1993). Several factors may contribute to the decision to terminate and these include:

> cultural and individual attitudes to abortion; the nature of the abnormality; when it is detected; the certainty of the diagnosis and prognosis; parental age and reproductive history; and who counsels the couple.
> *Green and Statham (1996, p. 149)*

If a woman at risk for OCA in a child decides to have prenatal genetic diagnosis and if she receives a diagnosis that the fetus in the uterus is affected, then a decision regarding either termination or continuation of the pregnancy has to be made, usually within a specified short time limit. If the woman has had genetic counseling, this decision would have been discussed and probably made prior to testing. However, requesting abortion of a desired pregnancy, after receiving a positive diagnosis about halfway through the pregnancy, makes this type of abortion different from others. Rapp (2000) interviewed many women in this situation and found that three themes emerged:

1. Many women had abortions for previous pregnancies but, nevertheless, were ambivalent about aborting the present one, to which they had committed themselves. Most women (95%), if they choose to abort, do so before 15 weeks gestation. However, the stage at which genetic testing can be done and the time taken for results mean the pregnancy is more advanced when the result is obtained and a more difficult decision has to be made. This decision means giving up the pregnancy and going through a version of a labor. Also by this stage the fetus is often considered as a baby.
2. Selective abortion is ethically different from abortion, in general. Ending a pregnancy because of a specific diagnosed disability forces a woman to act as a "moral philosopher" (Rapp, 2000, p. 131), assessing the quality of life of the affected child (as well as that of her own, her other children, and her family) and judging the standards by which life is worthwhile, as well as interrogating her own biases and stereotypes.
3. Questions around selective abortion provoked a direct discussion about the knowledge, attitudes, and beliefs that surround disabilities. Many women commented that the disabled child (mostly those with Down syndrome) might have a reasonable life, but other family members paid the price. They also noted that if they had an affected child they would manage somehow and everyone would rally round, but if one had the option of prenatal diagnosis and selective abortion one would take it.

How does the woman with a positive diagnosis of OCA in her fetus decide what to do? There are very few reports available on what couples actually do in this situation. One early report focused on one couple who said life was too hard for an affected child in their society, the sun was too hot, the cancer risk too high, and the visual problems too difficult to manage, and therefore, after receiving a prenatal genetic diagnosis of OCA1 in the fetus, they requested termination (Eady et al., 1983). However, in another more recent report, a woman went through prenatal diagnosis in South Africa but when given the positive (OCA2) result decided to continue the pregnancy (Kromberg et al., 2015). So in the small sample recorded here, there are two completely different opinions. Selective abortion for albinism in the fetus is, therefore, not an easy decision. Many factors have to be considered, even those that are external, such as geographical, because they affect the amount of sun exposure experienced and the associated development of skin cancer and/or those that are cultural which influence stigmatization, as well as the threat of being targeted for muti (traditional medicine) mutilations or murders, by traditional healers.

Women who use prenatal diagnosis services need to consider not only what their response would be to a positive result but also what the effects on them themselves would be if they had an abortion for an affected fetus. Often this decision reflects not only on the woman herself and her husband but also on the other members, as well as the children, in the family. From the woman's own experience she might have unexpected emotional reactions after the procedure and might take some time to come to terms with it. One of the most difficult situations occurs when a husband and wife disagree regarding the abortion decision. In the writer's (JGRK) experience, in one particular case, the husband insisted on abortion (the diagnosis was Down syndrome) but the wife adamantly resisted it. However, the husband acknowledged that he could not and would not live with such a child, in addition to their four other children, and, therefore, he would separate from the family if the pregnancy went ahead. The wife then very reluctantly agreed to request a termination of the pregnancy, but she regretted her decision later and was very unhappy for years after the event. According to another report, from Scotland, in some women, distress tended to persist even 2 years after the termination; the most vulnerable groups were those who were young and immature and those who had secondary infertility (White van Mourik et al., 1992).

The psychological response to abortion for genetic reasons has been investigated. Blumberg et al. (1975) found that depression was common in both parents, posttermination; they had guilt feelings about their decision-making and loss of self-esteem because of the conceiving of a handicapped child.

However, many other couples have abortions for genetic abnormalities in the fetus, are able to justify their decisions, cope with the emotional and physical after-effects of the experience, and consider their decision the right one for them and their family, in their circumstances at the time.

Ethical Issues and Prenatal Diagnosis

Ethical issues are of concern to all medical geneticists and genetic counselors especially since they started offering genetic services, which include prenatal genetic diagnosis, to the public.

Ethics is a form of philosophy, which refers to a systematic reflection on the moral aspects of life and the associated conflicts that may arise (McConkie-Rosell and Spiridigliozzi, 2004), while biomedical ethics involves the ethical issues that may occur in medicine, medical research, and society. Traditionally, the system of principle-based ethics is used as a guide by medical geneticists and genetic counselors. The primary principles include respect for autonomy, beneficence, nonmaleficence, and justice, but autonomy appears to be central. The health professional may need to balance the demands of competing principles when selecting a course of action and sometimes might be considered to be paternalistic (Kromberg and Wessels, 2013). These principles are based on the ethics of care, which involve insight into and understanding of the circumstances, needs and feelings of another person, and responsiveness on the part of the counselor to that person's situation (Schmerler, 1998).

Genetic services have been shown to be cost-effective and worthwhile, even in developing countries such as those in Africa. About 8 million (90% of all those affected) infants with severe malformations or genetic disorders were born in low- and middle-income countries such as South Africa (World Health Organisation, 2005). Many genetic disorders, such as the hemoglobinopathies (e.g., sickle cell anemia), can now be prevented or ameliorated with appropriate programs. Several decades ago the WHO (1985) recommended that such services should be promoted. It is, therefore, unethical not to provide them as an integral part of a comprehensive health care program.

An international survey (36 countries participated) was carried out on the common ethical issues arising from the practice of medical genetics (Wertz et al., 1995; Kromberg and Jenkins, 1997). After the survey several guidelines were highlighted: genetic information should only be supplied by those aware of the ethical issues involved; patients must be fully informed of the limitations of genetic tests (for example, prenatal genetic diagnosis testing) and receive supportive genetic counseling; and children should not be tested unless they will benefit from such testing. Other ethical issues raised included equal access to services, the provision of high-quality services, the offering of options and freedom of choice, and the use of non-directive counseling methods by well-trained counselors, respectful of human diversity and individual differences (Kromberg and Krause, 2013).

In the context of OCA and prenatal diagnosis, the ethical issues include the provision of prenatal genetic diagnosis and selective abortion services and recognizing the problem of access to these services and whether they are provided equitably and fairly, so that all at-risk families can make use of them.

Also, in the context of testing children (see Chapter 10), it might be seen as ethical if, on prenatal testing, the result on the fetus is given (to the parents) as affected or unaffected, but the unaffected result is not further specified as carrier or noncarrier. This approach would maintain the child's autonomy to find out later, at a suitable time (when the information becomes important to him/her and he/she can give informed consent), whether or not he/she is a carrier. The ethical issue that arises regarding identifying carriers is that those so labeled could become stigmatized, either as children in the community or later as marriage partners.

Some might suggest that offering prenatal diagnosis to at-risk couples is unethical because the outcome could be selective abortion. However, once a genetic test is available those who want to use it have a right to do so, their freedom of choice should be maintained, and they have the right to act on the results if they wish to do so. In South Africa, the Choice on Termination of Pregnancy Act (1996) states that termination can be carried out when two medical practitioners believe that the continued pregnancy would result in a severe malformation in the fetus; the decision to request the termination is made by the woman and her partner (Todd et al., 2010).

V. CONCLUSION

Both genetic testing and prenatal testing services for people with albinism, their parents, and at-risk relatives are available in Johannesburg. Due to the fact that the majority of those in the Black population who carry a mutation for albinism have the common OCA2 mutation caused by a 2.7 kb deletion, testing is possible in most cases. Patients (from all over South Africa and a few from other sub-Saharan countries) are generally managed, clinically, at genetic counseling and other health care clinics in the cities where they live, but their samples are sent to Johannesburg for testing. Although this service is underutilized at present, with the appropriate community education and awareness programs that are evolving this should change and demand should increase in future.

Diagnostic, carrier, and prenatal testing, for those who do not have the common mutation, is somewhat complex and for that reason should probably be centralized in an experienced laboratory where expertise can be built up, as is presently the case in South Africa. The clinical service for the patients, however, remains regional.

The demand for prenatal testing for OCA, although it has been available for nearly 17 years in Johannesburg, is low but likely to expand as affected families and their doctors become aware of the availability of the necessary techniques. Such services should ideally be accompanied by genetic counseling, especially considering that there are psychosocial issues associated with the procedures and the end result could be decision-making regarding termination of pregnancy for an affected fetus. However, some patients might wish to continue the pregnancy even though the fetus is affected and this should be understood by the health professionals involved. Decision-making in this context is difficult. It is affected by many current and past factors in the lives and experiences of the parents, by how they deal with their reproductive needs, cope with problems, find solutions, and perceive and manage their situation, at the time.

There are also ethical issues that need to be recognized, such as respect for patient autonomy, rights to full information, equitable accessible services, freedom of choice, and those issues that surround prenatal genetic testing and selective abortion. The genetic counselors and other health professionals who deal with these patients and their needs should be aware of these issues so that they can offer those who seek it an ethically based, quality service.

References

Aalfs, C.M., Oort, F.J., De Haes, J.C.J.M., Leschot, N.J., Smets, E.M.A., 2007. A comparison of counselee and counselor satisfaction in reproductive genetic counseling. Clin. Genet. 72, 74–82.

Blumberg, B.D., Golbus, M.S., Hanson, K.H., 1975. The psychological sequelae of abortion performed for a genetic indication. Am. J. Obstet. Gynecol. 122, 799–808.

Census, 2011. Census in Brief. Statistics South Africa 2012, Pretoria. ISBN: 9780621413885.

Durham-Pierre, D., Gardner, J.M., Nakatsu, Y., King, R.A., Francke, U., Ching, A., Aquaron, R., del Marmol, V., Brilliant, M.H., 1994. African origin of an intragenic deletion of the human P gene in tyrosinase positive oculocutaneous albinism. Nat. Genet. 7, 176–179.

Eady, R.A.J., Gunner, D.B., Garner, A., Rodeck, C.H., 1983. Prenatal diagnosis of oculocutaneous albinism by electronmicroscopy of fetal skin. J. Investig. Dermatol. 80, 210–212.

Evers Kiebooms, G., Denayer, L., Decruyenaere, M., van den Berghe, H., 1993. Community attitudes towards prenatal testing for congenital handicap. J. Reprod. Infant Psychol. 11, 21–30.

Fasihi, H., Eady, R.A.J., Mellerio, J.E., Ashton, G.H.S., Dopping-Hepenstal, P.J.C., Denyer, J.E., Nicolaides, K.H., Rodeck, C.H., McGrath, J.A., 2006. Prenatal diagnosis for severe skin disorders: 25 years' experience. Br. J. Dermatol. 154, 106–113.

Green, J., Statham, H., 1996. Psychosocial aspects of prenatal screening and diagnosis. In: Marteau, T.M., Richards (Eds.), The Troubled Helix. Social and Psychological Implications of the New Human Genetics. Cambridge University Press, Cambridge, pp. 140–163.

Green, J.M., Hewison, J., Bekker, H.L., Bryant, L.D., Cuckle, H.S., 2004. Understanding decision-making about prenatal genetic screening (Chapter 8) Psychosocial Aspects of Genetic Screening of Pregnant Women and New-Borns: A Systematic ReviewHealth Technol. Assess. 8 (33), 51–55.

Hurford, E., Hawkins, A., Hudgins, L., Taylor, J., 2013. The decision to continue a pregnancy affected by Down syndrome: timing of decision and satisfaction with receiving a prenatal diagnosis. J. Genet. Couns. 22, 587–593.

Kessler, S., 1989. Psychological aspects of genetic counseling: VI. A critical review of the literature dealing with education and reproduction. Am. J. Med. Genet. 34, 340–353.

Kromberg, J.G.R., Bernstein, R., Jacobson, M.J., Rosendorff, J., Jenkins, T., 1989. A decade of mid trimester amniocentisis in Johannesburg. S. Afr. Med. J. 76, 344–349.

Kromberg, J.G.R., Jenkins, T., May 1997. Genetic services in South Africa and common ethical issues related to genetic information. Spec. Med. 55–59.

Kromberg, J.G.R., Krause, A., 2013. Human genetics in Johannesburg, South Africa: past, present and future. S. Afr. Med. J. 103 (12 Suppl. 1), 957–961.

Kromberg, J.G.R., Wessels, T.-M., 2013. Ethical issues and Huntington disease. S. Afr. Med. J. 103 (12 Suppl. 1)), 1023–1026.

Kromberg, J.G.R., Rosendorff, J., Essop, F., August 2015. Prenatal Diagnosis for Oculocutaneous Albinism. South Africa Society for Human Genetics Congress, Pretoria, South Africa. Abstract book.

Leschot, N.F., Verjaal, M., Treffers, P.E., 1985. Risks of mid-trimester amniocentesis: assessment of 3000 pregnancies. Br. J. Obstet. Gynaecol. 92, 804–807.

Lippman-Hand, A., Fraser, F.C., 1979. Genetic counseling: provision and reception of information. Am. J. Med. Genet. 3, 113–127.

Lund, P.M., 2001. Health and education of children with albinism in Zimbabwe. Health Educ. Res. 16 (1), 1–7.

McConkie-Rosell, A., Spiridigliozzi, G.A., 2004. Family matters: a conceptual framework for genetic testing in children. J. Genet. Couns. 13 (1), 9–29.

Montoliu, L., Gronskov, K., Wei, A.H., Martinez-Garcia, M., Fernandes, A., Arveiler, B., Morice-Pikard, F., Riazuddin, S., Suzuki, T., Ahmed, Z.M., Rosenberg, T., Li, W., 2014. Increasing the complexity : new genes and new types of albinism. Pigment Cell Melanoma Res. 27 (1), 11–18.

Rapp, R., 2000. Testing women, testing the fetus. In: The Social Impact of Amniocentesis in America. Routledge, New York.

Rosenmann, E., Rosenmann, A., Ne'eman, Z., Lewin, A., Bejarano-Achache, I., Blumenfeld, A., 1999. Prenatal diagnosis of oculocutaneous albinism type 1: review and personal experience. Pediatr. Dev. Pathol. 2, 404–414.

Ross, L.F., Howard, M., David, K.L., Anderson, R.R., 2013. Technical report: ethical and policy issues in genetic testing and screening of children. Genet. Med. 15, 234–245.

Schaffer, B.L., Caughey, A.B., Norton, M.E., 2006. Variation in the decision to terminate pregnancy in the setting of fetal aneuploidy. Prenat. Diagn. 26, 667–671.

Schmerler, S., 1998. Ethical and legal issues. In: Baker, D.L., Schuette, J.L., Uhlmann, W.R. (Eds.), A Guide to Genetic Counseling. Wiley Liss, New York, pp. 249–275.

Schreier, A., Kromberg, J.G.R., Hofmeyer, J., Fishman, A., 1986. Prenatal risks in the older woman. In: Paper Presented at the 5th Conference on Priorities in Prenatal Care. 25–28 February. Rustenburg, South Africa.

Skotko, B., 2005. Prenatal diagnosed Down syndrome: mothers who continued their pregnancies evaluate their health care providers. Am. J. Obstet. Gynecol. 192, 670–677.

South African Choice on Termination of Pregnancy Act No 92, 1996. Available at: http://www.info.gov.za.

Stevens, G., van Beukering, J., Jenkins, T., Ramsay, M., 1995. An intragenic deletion of the P gene is the common mutation causing tyrosinase-positive oculocutaneous albinism in Southern African Negroids. Am. J. Hum. Genet. 56, 586–591.

Todd, C., Haw, T., Kromberg, J., Christianson, A., 2010. Genetic counseling for fetal abnormalities in a South African community. J. Genet. Couns. 19, 247–254.

Weil, J., 2000. Psychosocial Genetic Counseling. Oxford University Press, Oxford.

Wertz, D.C., Fletcher, J.C., Berg, K., Boulyjenkov, V., 1995. Guidelines on ethical issues in medical genetics and the provision of genetic services. In: World Health Organisation Hereditary Diseases Program. WHO, Geneva.

Wessels, T.-M., Koole, T., Penn, C., 2015. 'And then you can decide' – antenatal foetal diagnosis decision making in South Africa. Health Expect. 16 (6), 3313–3324.

White van Mourik, M.C.A., Connor, J.M., Ferguson-Smith, M.A., 1992. The psychosocial sequelae of a second trimester of termination of pregnancy for fetal abnormality. Prenat. Diagn. 12, 189–204.

Witkop, C.J., Quevedo, W.C., Fitzpatrick, T.B., 1978. Albinism. In: Stanbury, J.B., Wyngarden, J.B., Frederickson, D.S. (Eds.), The Metabolic Basis of Inherited Disease, fourth ed. McGraw-Hill, New York, pp. 283–316.

World Health Organisation, 1985. Community Approaches to the Control of Hereditary Disorders. WHO, Geneva. Report of WHO Advisory Group.

World Health Organisation, 2005. Control of Genetic Diseases. Report of the Secretariat.

Albinism and Social Marginalization

Sam Clarke, Jon Beale

Standing Voice, London, United Kingdom

I. INTRODUCTION

The human rights predicament of people with albinism varies globally. Almost universally stigma exists; in Europe and North America, children with albinism are routinely teased in school, while popular culture often depicts people with albinism as villainous and threatening (Barton, 2006), mysterious and alien-like (KatyPerryVEVO, 2011), and objects of suspicion, ridicule, and disgust (Webb, 2013). What has transpired in Africa in recent years has been the extreme instantiation of a discriminatory logic that recurs globally, different in degree, certainly, but perhaps not in essence.

The severity of human rights abuses enacted against Africans with albinism is nonetheless without geopolitical parallel and commands our urgent attention and resources. In what follows, we delineate some of the core challenges faced by persons with albinism in Africa, as well as tested strategies for responding to these. We begin, in Section II, by establishing terms and arguing for a qualified definition of albinism as a socially produced disability in the context of African societies. We proceed, in Section III, to map the landscape of interventions around this issue and the revolutionary turn toward stakeholder unification that has gripped and invigorated this field in recent years. Section IV communicates some of the key lessons to have been generated by this process and Section V accordingly imparts best practices.

This contribution—its insights, parameters, assumptions, and hopes—is informed by the work of Standing Voice (SV). SV is an international NGO operating in Tanzania (see Fig. 12.1). The SV team has over 10 years' experience defending the rights of persons with albinism in Africa and works holistically across health, education, advocacy, and community to empower this population at local, national, and international levels. Our analysis correspondingly ranges in geographic scope and scale, but the recurring axis of enquiry is Tanzania: a decision that reflects the gravity (and quantity) of abuses that have taken place against people with albinism there; the early embroilment of Tanzania in this increasingly continental issue, and its utility, therefore, as a case study of how and how not to intervene; and the operational fact of our expertise in that country.

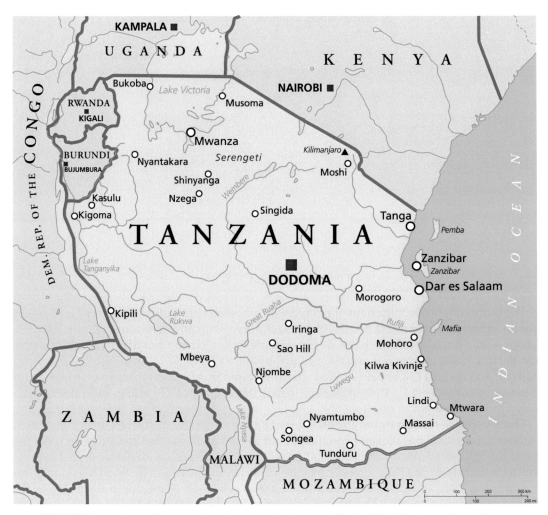

FIGURE 12.1   Map of Tanzania. *From https://stock.adobe.com/. File #: 45336885 by Peter Hermes Furian.*

The nature of this chapter—which charts highly contemporary developments in the field and uses the work of SV as a window into the ongoing social marginalization faced by people with albinism—is such that there is no substantive body of peer-reviewed literature to complement our analysis. This dearth of published material necessitates some use of in-house sources such as unpublished reports and surveys. It is hoped that by taking stock of recent developments and recommending ways forward, this chapter will contribute to and irrevocably extend the evolving conversation that has proliferated around albinism in Africa in recent years. Sharing our insights and learning from others takes us one step further in the fight to make persecution of people with albinism a distant memory.

II. TERMS AND CONTEXT

The relationship between albinism, disability, and human rights—across legal, medical, and humanitarian discourses—is not straightforward. Albinism is a rare, noncontagious, recessively inherited genetic condition. The disorder disrupts the body's production of melanin, reducing or eliminating pigmentation in the skin, eyes, and hair. This melanin deficiency causes complex visual impairment, altering retinal development and nerve connections to the eye. It also weakens natural defenses against sun damage, placing people with albinism at heightened risk of skin cancer, especially in hot countries. In albinism, visual impairment can be improved with vision devices and alterations to lighting or seating position. The threat of skin cancer is similarly manageable by limiting UV exposure, wearing long-sleeved clothing, and applying sunscreen. The eligibility of albinism as a marker of medical "disability" is therefore dubious. For many advocates, defining albinism this way is not only inaccurate but also disempowering.

But—certainly where albinism is concerned—"disability" is socially produced as much as it is medically determined. Expanding our definition of disability to encompass socially originating barriers to civil participation, we see the goalposts around albinism shift. (Our analysis here follows a human rights–based approach to disability, as espoused in much contemporary international human rights law, and reflected particularly in the 2006 UN Convention on the Rights of Persons with Disabilities (CRPD). CRPD emphasizes the interaction between the physical features or impairments of a person's body and the society in which they live, to the effect that disability is understood as an "evolving" effect of that interaction.) In Africa particularly, albinism is poorly understood: myriad myths and superstitions enshroud the condition and obscure its genetic basis. In a 6-year study conducted on Ukerewe Island in Tanzania's Lake Victoria, filmmaker Harry Freeland interviewed 62 members of an albinism association and recorded hundreds of hours of observational footage (later forming the basis of 2012 BBC documentary *In the Shadow of the Sun*). Participants reported perceptions by others that albinism was contagious and a curse from God; that persons with albinism were not human and did not die but simply disappeared; and that sex with a woman with albinism could cure infertility and AIDS. The body parts of people with albinism were reported to possess magical properties and confer fortune when used in witchcraft potions. These findings have since been reflected nationally and continentally in the upsurge of ritual violence against persons with albinism in Tanzania, and across Africa more widely, over the last decade. As of March 21, 2017, there have been 190 documented murders of people with albinism and a total of 515 attacks, across 27 African countries since 2006 (Under the Same Sun, 2017a). Tanzania presents a uniquely severe case—with 76 murders and 73 nonfatal incidents—although the center of gravity has recently shifted to Malawi, where 10 have lost their lives since the beginning of 2015 (Under the Same Sun, 2017, unpublished data). Children—who are comparatively easy to abduct and whose "innocence" is thought to increase the potency of witchcraft—comprise the majority of victims (UN News Centre, 2016). In Tanzania, this has precipitated a flawed government policy of relocating children with albinism to protectorate centers, where they are separated from their families and kept behind high walls for their safety.

Stigma of this severity—and not some immovable medical reality—incapacitates persons with albinism and renders them "disabled." Because people with albinism are mistakenly believed to be subhuman in many parts of Africa, services are not built to meet their needs. Marginalization impedes their access to health services, restricts the delivery of education, and isolates affected individuals from their families, communities, and caregivers. These observations are consistent with findings from a series of consultations implemented across Tanzania in 2016 by SV in partnership with the World Bank Group: a spectrum of participants with albinism—at local, district, and national levels—reported others' ignorance of their condition as a determining factor in their marginalization from mainstream opportunities and services (Standing Voice and World Bank Group, 2016, unpublished report; UN Independent Expert, 2016a). In Tanzania, for example, inadequate provision of sunscreen, skin screening, and preventative health education has caused significant delays in diagnosis and high rates of skin cancer, with one old localized study estimating only 10% of people with albinism survive beyond the age of 30 years (Luande et al., 1985). The SV skin cancer prevention program (SCPP) has provided more up-to-date statistics: in the program's first visit to Musoma Regional Hospital in November 2015, a variety of skin conditions associated with sun exposure were observed. Erythema was found in 92% of patients; solar elastosis and keratosis in 79% and 70%; cheilitis in 23%; and fully developed skin cancer in 23% (Standing Voice, 2016, unpublished report). This is stark documentation of the ubiquity of cancerous and precancerous presentation in populations of people with albinism to whom skin care has not historically been provided. Stigma causes these fatalities, transforming otherwise preventable health risks into a public health crisis.

Available data show similar marginalization in education. There has (to our knowledge) been no comprehensive or credible survey recording access to education among children with albinism nationwide in Tanzania: a task impaired by limited state capacity in data collection and by the methodological difficulty of finding and registering children with albinism, who have historically been hidden away and who are statistically twice as likely to live in rural locations (National Bureau of Statistics and Office of Chief Government Statistician, 2014). Important interventions have nonetheless been made to capture the experiences of children with albinism who are in institutional environments. In 2011, prominent NGO Under the Same Sun (UTSS) partnered with the United Nations Children's Fund (UNICEF) to deliver a joint report into the protectorate centers, where hundreds of children with albinism have been housed since the escalation of murders in 2008 (Under the Same Sun, 2011). The report—encompassing 9 centers of the 33 that are known—revealed endemic physical, sexual, and verbal abuse; high levels of stigma; and evidence of psychological disturbance in some children (Under the Same Sun, 2013). Inadequate security, poor sanitation, and gross overcrowding were ubiquitous. Centers were also found to lack adequate educational infrastructure, with chronic scarcity of resources and a shortage of teachers sufficiently trained to understand the visual impairment of children with albinism and mitigate its impact on their learning. Knowledge of the dermatological implications of albinism was similarly low, to the effect that staff were leaving children outside in the sun without protective clothing: cancer in various stages was widespread. These problems persist today, and what was once deployed as a temporary installation—an emergency step taken to protect children from danger—has become counterproductive as a permanent fixture. The majority of protectorates were initially established to house blind or deaf children or those with learning difficulties; they were never designed to accommodate the hundreds of children with albinism they do now. Kept in camps, these children are deprived of mainstream educational participation and denied a stake in society. Separated from their families and communities, they are further marginalized from public life, aggravating stigma rather than working to neutralize it.

The challenges facing children with albinism extend beyond protectorates. SV's unpublished interviews with beneficiaries have revealed comparative scarcity of enrollment and elevated risks of neglect, abuse, and ostracism for children who are accessing mainstream education. Together with Coventry University and Advantage Africa, SV is currently nearing completion of a 2-year research project funded by the International Foundation of Applied Disability Research; this has yielded fresh evidence of the vital role of teachers and the need for greater investment in albinism-specific sensitization and training for education professionals. Without suitable accommodation of their needs, most students with albinism cannot read from standard distances. Ensuing struggles with academic performance further ingrain stigma and contribute to common perceptions that children with albinism do not belong in mainstream schools. This impacts employability later in life: participants in SV's 2016 consultations with the World Bank Group defined education and employment as interrelated concerns and identified discriminatory recruitment procedures, low confidence, and poor educational access and attainment as obstacles to paid employment (Standing Voice and World Bank Group, 2016, unpublished report). Independent research carried out by SV on Ukerewe Island in 2016 demonstrated that people with albinism earned an average of $42 a month compared with the national average of $77 in Tanzania (Standing Voice, 2015, unpublished survey; World Bank Group, 2016). Women with albinism on the island were found to earn significantly less still. In the Tanzanian (and African) context, albinism can therefore be defined as a disability when the latter is invoked as a sociomedical category.

III. AN EVOLVING FIELD

The situation of people with albinism in Africa has prompted a number of responses across the continent (and world) over the last decade. Since the murders first broke in the press around 2006, albinism has accrued considerable currency in international media and climbed high on the agendas of lawmakers, journalists, academics, advocates, public health professionals, governments, philanthropists, and development actors.

Early on, this process was punctuated by milestone exposés in the media. Vicky Ntetema, former Bureau Chief for BBC Tanzania, went undercover in 2006, posing as a buyer to uncover witch doctors' involvement in the trade network around the limbs of people with albinism (Ntetema, 2008). As hers and others' reports surfaced, filmmaker Harry Freeland was already in the midst of shooting *In the Shadow of the Sun*, a landmark BBC documentary exposing the depth of discrimination against people with albinism across Tanzania at that time. Violence proliferated in the years that followed, with albinism associations and civil society actors scrambling to respond.

The last few years have seen unprecedented mobilization to confront this issue in Tanzania and across Africa. What began with flashes of publicity—and a relatively disorganized structural response—has given way to an increasingly coordinated momentum, transporting this issue to the highest echelons of international advocacy. More and more, albinism has been the theme of conferences, inserted into human rights curricula, and taken to the center of public discourse on disability rights in Africa. In 2013—under mounting pressure from a number of actors, most notably UTSS Founder and Chief Executive Officer (CEO) Peter Ash—the United Nations Human Rights Council adopted a resolution calling for the prevention of discrimination and violence against people with albinism (2013). The year after, the UN General Assembly declared June 13 as International Albinism Awareness Day, with effect from 2015. Also in 2015, Ms Ikponwosa Ero was installed by the United Nations Human Rights Council as an Independent Expert tasked with monitoring the human rights situation of persons with albinism, and the first Pan-African Albinism Conference—organized by UTSS—was held in Dar es Salaam, unifying albinism associations from across the continent and opening pathways of collaboration. These events catalyzed even greater advances in 2016, where SV, the World Bank Group, the UN Independent Expert, and a number of other partners coordinated *Action on Albinism in Africa*, a consultative forum hosting delegates from 29 African countries and an array of disciplinary fields. (Other contributors included the Government of Tanzania; the Tanzanian Prime Minister's Office; the UN Country Team in Tanzania; the Tanzania Commission of Human Rights and Good Governance; the Office of the High Commissioner for Human Rights; UTSS; the Embassies of Norway, Ireland, and the United States in Tanzania; and the Canadian High Commission in Tanzania.) From civil society organizations to international NGOs, academics and researchers to national human rights institutes, dermatologists and optometrists, mental health professionals, specialists in education and disability rights, and government, a spectrum of albinism experts came together to articulate challenges, define priorities, and recommend best practices for the first time. A roadmap of measures was established to prevent violence and discrimination, protect persons with albinism, and determine accountability (UN Independent Expert, 2016a). The forum concluded with the creation of a new UN Albinism Think Tank dedicated to streamlining these contributions and pursuing their implementation in policy. This body, to which SV was elected, met for the first time in Nairobi in November, 2016. That same month, an albinism conference and high-level stakeholder meeting was hosted by the Center for Human Rights at the University of Pretoria, South Africa. This has all unfolded against a backdrop of escalating media coverage and a series of country visits by the UN Independent Expert to review the enforcement of human rights mechanisms to protect persons with albinism across the continent. In Tanzania, SV's in-depth consultative research with the World Bank Group has captured the challenges and priorities of persons with albinism at local, district, and national levels and developed participants' capacity to track the impact of their own and each other's interventions (Standing Voice and World Bank Group, 2016, unpublished report). (This has laid foundations for the establishment of sectoral committees to monitor and evaluate stakeholder activity in health, education, security and safety, and economic empowerment and livelihood: outputs to be developed in further collaborations between SV and the World Bank Group.) The United Nations Educational, Scientific and Cultural Organization (UNESCO) has taken a similarly facilitatory role, obtaining commitments and forecasts from key organizations and collating these in a central resource (UNESCO, 2017, unpublished proposal).

It is difficult to overstate the significance of this shift. Though admirable, progress around this issue has historically tended to be localized and inadequately coordinated at the national (and continental) level. People with albinism and related stakeholders have suffered a functional disconnect, typically lacking resources and channels to communicate their needs and identify collective priorities. This has exacerbated risks of duplication and stifled strategy development. The last few years have seen a burgeoning Pan-African effort to bridge this gulf and unify stakeholders across the field, a shift reflected and reinforced by the publication of this book, as the first to house a variety of disciplinary voices active in this issue.

IV. LESSONS

For all stakeholders, this process has yielded unprecedented opportunities to understand the needs of people with albinism and map interventions accordingly. According to the UN Independent Expert (2016a), consultations have broadly revealed a need to demystify albinism by spreading awareness of the condition across multiple strata of society; to reduce the vulnerability of people with albinism to violence, through direct security measures and wider structural empowerment; to improve economic security for persons with albinism; and to increase their access to quality health and education services. The importance of including and platforming people with albinism at all stages of development processes has been broadly underscored (Standing Voice and World Bank Group, 2016, unpublished report).

Constructive collaboration with press has also been endorsed (Standing Voice and World Bank Group, 2016, unpublished minutes; Standing Voice, 2016, unpublished minutes). Media responses to the situation of Africans with albinism have historically gravitated toward (and continue to focus on) ritual violence and murder. Much journalism has tended to sensationalize or simplify, reducing the complexity of people with albinism, and their lives and multifarious challenges, into tidy, singular headlines (van Heerden, 2016; Charlton, 2016; Howden, 2008). This has helped keep albinism on the map but not always with the most constructive or accurate depiction: so contested has been the very ownership of albinism as a topic within public discourse. Consultations have nonetheless emphasized the redemptive utility of the media as a lobbying weapon and instrument of public sensitization. Equally, they have cautioned that journalists should be trained to report facts truthfully, and that they should negate myths, avoid graphic imagery, and counterbalance negative coverage with positive exposure of people with albinism as role models and success stories (Standing Voice and World Bank Group, 2016, unpublished minutes; Standing Voice, 2016, unpublished minutes). Reports should neither platform perpetrators nor enumerate profits purportedly attached to ritual crimes. The media should amplify the needs and priorities of people with albinism as revealed and defined by the advocacy movement outlined above. All dimensions of this issue should be explored: not just the murders—extreme and alarming atrocities inviting urgent attention, of course—but other expressions of discrimination too. The genetic basis of albinism; its implications for health in resource-poor settings; elevated risks of skin cancer; the experience of visual impairment at home, at school, in the workplace; the everyday grind of prejudice: the media should open up a vital space for reflection and create a conversation about these neglected problems.

Various human rights provisions for people with albinism have also been contested and clarified. Participants have vocalized a need for mechanisms to identify and monitor the human rights situation of people with albinism specifically, which should not be homogenized under the rubric of disability rights (UN Independent Expert, 2016a). Equally, the political weight of "disability" as a term—its particular currency as a buzzword, or bargaining chip, capable of unlocking institutional resources and funds—has too been acknowledged. What has been concluded is that the situation of persons with albinism is complex and multifaceted and variable under national and regional parameters; hence, as the UN Independent Expert has claimed, "human rights challenges associated with albinism are multi-layered and [can] be addressed by various human rights mechanisms" (2016b, p. 13). A widely articulated priority has been to ensure states' full and effective enforcement of existing legislation whose remit can justifiably be said to encompass the rights of persons with albinism. This marks a notable shift in emphasis from earlier conversations, which prioritized the installment of new legislative measures specifically addressing the condition. This has been reflected in a renewed emphasis, too, on the accountability of service deliverers and the particular standards already incumbent on health professionals intervening to support people with albinism (UN Independent Expert, 2016a).

Consultations have lent fresh impetus to the eradication of protectorates (UN Independent Expert, 2016a). *Action on Albinism in Africa* in particular provided a platform for participants to share experiences of the centers and reflect on their inadequacy in the pursuit of integration and full civil participation for people with albinism. In a charge led by former Deputy Minister Dr. Abdallah Possi—and following the application of pressure by UNESCO, UTSS, and others—the Tanzanian government has in the last 12 months taken promising steps to decongest the centers and stem the flow of new admissions (Possi, 2016, personal communication). Discourse within development has solidified in favor of their abolition, though popular media continues, frustratingly, to lag in this regard (Palacios, 2015). It is the widely expressed hope of many actors in this field that children with albinism will be relocated to high-quality inclusive private schools, where they will develop and grow as respected members of their communities. UTSS has done particularly admirable work in this area, so far sponsoring 400 students with albinism from kindergarten through to college (Under the Same Sun, 2017c).

V. BEST PRACTICES

In the wake of this Pan-African movement and its unprecedented unification of stakeholders, there is a critical opportunity to define and exchange best practices. A full inventory of these is of course not possible in what follows. Instead, more profitably in a contribution of this length, we use the experiences of SV to illuminate some of the principal challenges and triumphs that have shaped the evolving engagement of development actors with this issue. It is the closing ambition of this chapter to identify our own best practices, specifically in advocacy and health—how they are built, tested, embedded, and replicated—and share these, in the same collaborative spirit that has animated and catalyzed so many advances over the last decade.

The primary obstacle facing NGOs—widely reflected in SV's programmatic experiences and strongly corroborated throughout the stakeholder consultations outlined above—is sustainability of intervention. When international development actors confront this issue, they intercept a complex human rights predicament and a population whose marginalization is as ancient as it is severe. In response to a problem so sprawling and decentralized, marked by a mass of interconnected failings, we need socially integrated solutions that penetrate the fabric of African society and work within its bounds. Our strategies must be culturally and operationally assimilable to in-country processes and structures: long-term improvements to the lives of people with albinism will otherwise lie beyond reach.

Advocacy

For some years, there have been a number of efforts implemented in Africa to promote awareness of albinism. To date, these have involved advocacy campaigns and community sensitization to educate the public and stress the equality of all citizens before the law. But the execution of this advocacy has not always been effective or adequately sensitive to context. In Tanzania—where SV is operationally concentrated—"advocacy" around albinism has often meant the top-down distribution of educational literature or the simple dissemination of information: booklets, guides, and other instructive resources, not always available in Swahili or appropriately cognizant of an audience's age. Advocacy messages have too often been transmitted in discourses remote from the lifeworld of target groups and have moreover placed insufficient emphasis on the mutually strengthening relationship between community well-being and well-being for persons with albinism, which are two sides of the same coin. Increasingly apparent has been an appetite (and urgent need) for advocacy to become a larger and more interactive process, through which emotional responses to people with albinism are altered, and communities are engaged, enabled, and empowered to produce their own solutions to the challenges facing this group (Standing Voice and World Bank Group, 2016, unpublished minutes).

Significant strides have been taken in this direction. UTSS has an Advocacy and Public Awareness program that has raised awareness in villages and towns across Tanzania, using seminars, workshops, dialog, publications, and a variety of other media to demythologize albinism (Under the Same Sun, 2017b). The Mennonite Central Committee—in partnership with the Tanzania Albinism Society (TAS), Kanisa la Mennonite, and Albino Peacemakers—has developed advocacy curricula focused not only on imparting knowledge but also on facilitating changes in attitudes and behavior. UNESCO has recently completed an awareness campaign using community radios and digital media campaigns to promote positive perceptions and practices toward people with albinism (UNESCO, 2016, unpublished report). *In the Shadow of the Sun*—the 2012 BBC documentary filmed by Harry Freeland, SV's Executive Director—has been used as an advocacy weapon to intervene against stigma in locations where reported violence has been highest. One screening—held in the Chato District of Geita, as an emergency response to the local murder of a 1-year-old baby with albinism in February 2015—drew an audience of thousands.

A recent SV project funded by the Wellcome Trust has consolidated this shift toward interactive advocacy and pushed it further still. In 2016, SV partnered with Dr. Patricia Lund of Coventry University (United Kingdom), the Bagamoyo Creative Arts Institute, and the Tanzanian branch of the African Institute of Mathematical Sciences to pilot an innovative community engagement project in Tanzania. The project mobilized scientists and artists from across Africa and brought these actors together to debunk myths and showcase the science of albinism in a socially accessible way. Participants built a core narrative around the fictional experiences of one man with albinism, rejected at birth by his father and again in adulthood by the family of his girlfriend. Participants visited community locations and played out these scenes of abandonment through unannounced pop-up performances, interlaced with songs, dances, and spoken-word poetry designed to explore and clarify the genetic origins of albinism. Passersby were immersed as contributors in these unfolding dramatic scenarios, and their impromptu responses—anger, sympathy, indifference, shock—were vocalized and absorbed, sometimes redirecting the narrative. Audiences were able to interject—a mechanism providing sobering windows into community stigma—and broker their own solutions to the dilemmas of the protagonist: how he should behave, and whether he should be included or rejected by the family of his girlfriend.

This project revealed the transformative potential of interactive theater as a vehicle for community explorations of albinism. Audiences could use drama to access and inhabit multiple perspectives, examining the consequences of erroneous beliefs for persons with albinism and their families in a dynamic, immersive way. The moment of performance was able to germinate a profound transformation of norms and crystallize a radical type of advocacy where solutions are authored by communities and not imposed top-down. Increased sensitization around albinism was evaluated to be a core project outcome, with participants and bystanders responding positively to creative, arts-based strategies of intervention and articulating improvements in their understanding of albinism and appreciation for the rights of persons with albinism as a result. (The play's denouement was positive in almost all cases, with communities widely advocating for the protagonist to be accepted by his girlfriend's family.) The success of this pilot provides a compelling evidence base to scale and disseminate these activities through a series of regional tours across Tanzania. The Prime Minister's Office has observed and endorsed the project and formally requested its outputs be repurposed for a state-sponsored campaign.

Health

The skin cancer crisis facing people with albinism in Africa has precipitated a number of interventions from actors across the continent and beyond. As outlined at the start of this chapter, skin cancer presents the greatest numerical threat to this population and has continued to claim lives at an alarming rate and scale due to the structural marginalization of people with albinism from frontline services, preventative education, and community, familial, and state support. Social exclusion at all levels has produced the conditions for skin cancer to proliferate. This situation has been universally articulated—and exposed as Pan-African—throughout the consultative process that has emerged in this field over the last few years (UN Independent Expert, 2016b).

Established in 2013 in Tanzania, SV's SCPP has aspired to combat this crisis by building a scalable model of care that can be exported as best practice into neighboring African states.

The program currently operates biannual clinic cycles at 33 locations across 8 regions of Northern and Western Tanzania and has enrolled more than 2000 patients with albinism nationwide. At each clinic, patients receive dermatological screening; preventative education; liquid nitrogen cryotherapy; and specialist sunscreen. We also provide referral for advanced surgery if needed and increasingly look to deliver on-site minor surgeries in clinical locations where institutional capacity is adequate to support such procedures. Together with their families, peers, and wider community members, patients are trained to understand, manage, and advocate for their own needs. We also distribute Kilisun (SPF 50+), a sunscreen specifically made for people with albinism and produced in-country, eschewing past dependency on foreign imports, which were costly, unreliably sourced, and vulnerable to chemical impurities. Pioneered by Spanish pharmacist Mafalda Soto Valdés and manufactured in Moshi at the Kilimanjaro Sunscreen Production Unit in the Kilimanjaro Christian Medical Center, Kilisun uses a water-in-oil emulsion with a very low irritation risk and high performance in hot weather. Alongside África Director, BASF, Fridda Dorsch, UNICEF, and UTSS, SV has supported the production of this life-saving sunscreen and expanded its availability nationwide, distributing thousands of bottles annually through SCPP.

In Tanzania, skin cancer prevention services for people with albinism have historically either been sporadic, geographically restricted, or delivered by mobile teams of international specialists remote from Tanzanian communities, providers, and institutions. Enormously valuable foundations have nonetheless been laid by a variety of actors: the Regional Dermatology Training Center, or RDTC—the largest dermatology training facility in sub-Saharan Africa and a key partner of SV—has been training dermatologists from across the continent for decades and delivering outreach services for people with albinism since 1992. As a training facility, RDTC has, however, lacked the resources (and institutional rationale) to scale these services nationwide, so its outreach has historically been localized to nearby Arusha, Kilimanjaro, and Tanga. Inadequate investment in dermatology as a discipline has meanwhile afforded few platforms for RDTC graduates to practice after qualifying and has made access to patients with albinism even harder to achieve. As the central civil society organization run by and for people with albinism at the grassroots level in Tanzania, TAS has also remained largely untapped in previous interventions, which have overlooked its potential—with 12,000 members and 124 local branches across 25 regions—as an organ of publicity and facilitator of patient access and retention.

SV has sought to learn from these lessons. We have built SCPP precisely by mobilizing the strengths of existing actors but coordinating their efforts in a more sustainable way. We partner with RDTC to identify and recruit dormant networks of dermatologists across Tanzania, who are enrolled in the program to treat the specific dermatological needs of persons with albinism. These dermatologists are then assigned to nearby clinical locations, creating a decentralized network of frontline providers who reside and work in patients' own communities and are fully funded by the Tanzanian government. (Since 2015, UNICEF has formed a parallel partnership with RDTC to deliver clinics in the southern regions of Mbeya, Iringa, and Njombe.) SV has similarly invested in the capacity of TAS, who advertize upcoming clinics to their members and ensure maximum publicity at the grassroots level. SCPP increasingly provides platforms for TAS to fulfill its own organizational objectives by advocating for the welfare of its members and holding government authorities to account. Clinics are delivered in schools and hospitals to maximize the institutional entrenchment of the program and train teachers and health professionals to understand the importance of preventative care. This strengthens the institutional support networks available to people with albinism and leaves a trail of ambassadors to advocate for their rights in a spectrum of settings nationwide. (The SV Vision Program—discussed by Dr. R. Kammer in Chapter 8—similarly trains teachers to understand and cater for the visual impairment of children with albinism in schools.) The Tanzanian government has meanwhile shown great receptivity to and increasing investment in the expansion of SCPP. The program has witnessed promising strides toward state adoption, with many district governments now covering selected costs; providing free venue space; facilitating patient transport; and offering other health professionals as support staff. The government pays the salaries of all program dermatologists.

SCPP has found success by engaging Tanzanian individuals, families, communities, and institutions to pursue structurally sustainable improvements to dermatological health care for people with albinism. Its programmatic philosophy—working with all stakeholders to build collective capacity and mutual accountability—has given the first glimpses of a solution to this historical problem. In locations where SV's clinics have been running longest, reductions in the presentation of skin cancer have been as high as 85.7% (Standing Voice, 2015, unpublished report). The Ministry of Health has formally endorsed SCPP (2016, unpublished letter). The UN Independent Expert has recognized the program model as a form of best practice (2017, personal communication), and delegates from 29 African states have issued formal requests for the replication of SCPP in their respective countries (Standing Voice, 2015, unpublished survey).

VI. CONCLUSION

The murder and mutilation of people with albinism in Africa stems from ignorance and stigmatization: a deeper structural crisis in which people with albinism have been demonized, marginalized, and "disabled" by society.

Historically, people with albinism and related stakeholders have lacked the platforms and resources to communicate and identify collective priorities and goals. The last few years have seen an unprecedented shift toward coordination that has invigorated this field and extended the promise of sustainable collaboration on a national and continental scale. This process has created new opportunities to understand the challenges facing people with albinism and forge collective solutions to confront these together. It is in this spirit and in the wake of these developments that we have shared the lessons we have learned and our own best practices in advocacy and health: how we build, test, and embed models of intervention that can be scaled and replicated across Tanzania and beyond.

Moving forward, it is the responsibility of all actors to protect and nourish the collaborative space that has opened up around this issue. Sharing our needs, priorities, and challenges, we are better equipped to monitor and understand the impact of our own and others' interventions across a diversity of locations, institutions, and spheres of expertise. As an international NGO committed to stimulating in-country responses, SV will endeavor to scale its services by consolidating delivery in Tanzania and expanding internationally to two other African countries by 2020. We will develop our programs qualitatively through increased provision of training to frontline health and education providers, civil society actors, government representatives, and people with albinism, together with their families and communities. This will be coupled with continued advocacy to reshape local responses to albinism through interactive engagement and lobby those in power to take greater accountability. Together, we can create the platforms and tools for people with albinism—and those positioned to influence their welfare—to take ownership of development and catalyze change. As leaders in this evolving conversation, people with albinism can illuminate the path to a brighter tomorrow.

References

Barton, L., May 12, 2006. When Albino Monks Attack. The Guardian. [Online]. Available from: https://www.the-guardian.com/film/2006/may/12/1.

Charlton, C., February 18, 2016. The 'ghost People' of Tanzania: The Albino Community Who Live in Fear of Being Hunted Down and Hacked to Pieces for Their Body Parts Which Are Treasured by Witch Doctors. MailOnline. [Online]. Available from: http://www.dailymail.co.uk/news/article-3452630/The-ghost-people-Tanzania-albino-community-live-fear-hunted-hacked-pieces-body-parts-treasured-witch-doctors.html.

Convention on the Rights of Persons with Disabilities. 2006. [Accessed 22 March 2017]. [Online]. Available from: http://www.un.org/disabilities/documents/convention/convention_accessible_pdf.pdf.

Howden, D., December 8, 2008. Independent Appeal: Killing Spree Leaves Albinos Living in Fear. Independent. [Online]. Available from: http://www.independent.co.uk/news/world/africa/independent-appeal-killing-spree-leaves-albinos-living-in-fear-1056640.html.

Harry Freeland. Dir. In the Shadow of the Sun (film), 2012. Inroad Films and Century Films, UK.

KatyPerryVEVO. Katy Perry, E.T. (Official) ft. Kanye West. [Online]. Available from: https://www.youtube.com/watch?v=t5Sd5c4o9UM.

Luande, J., Henschke, C., Mohammed, N., 1985. The Tanzanian human albino skin: natural history. Cancer 55 (8), 1823–1828. [Online] Available from: http://onlinelibrary.wiley.com.

National Bureau of Statistics and Office of Chief Government Statistician, 2014. Basic Demographic and Socio-Economic Profile Statistical Tables: Tanzania Mainland. Government of the United Republic of Tanzania. [Online]. Available from: http://www.tanzania.go.tz/egov_uploads/documents/Descriptive_tables_Tanzania_Mainland_sw.pdf.

Ntetema, V., July 24, 2008. In Hiding for Exposing Tanzania Witch Doctors. BBC News. [Online]. Available from: http://news.bbc.co.uk/1/hi/world/africa/7523796.stm.

Palacios, A., May 13, 2015. Albinism in Tanzania: Safe Havens in Schools and Support Centres – in Pictures. Guardian. [Online]. Available from: https://www.theguardian.com/global-development-professionals-network/gallery/2015/may/13/albinism-in-tanzania-safe-havens-in-schools-and-support-centres.

UN News Centre, March 22, 2016. 'Witchcraft' Beliefs Trigger Attacks against People with Albinism, UN Expert Warns. UN News Centre. [Online]. Available from: http://www.un.org/apps/news/story.asp?NewsID=53514#.WNJPrGSLQ18.

Under the Same Sun, 2011. Situation Assessment of the Centres of Displaced Persons with Albinism in the Lake Zone and Tanga Regions: Findings from Under the Same Sun Survey. [No publisher]. [Online]. Available from: http://www.underthesamesun.com/content/resources.

Under the Same Sun, 2013. NGO Report of Under the Same Sun (UTSS): Children with Albinism: Violence & Displacement. The UN Committee on the Rights of the Child. [Online]. Available from: http://tbinternet.ohchr.org/Treaties/CRC/Shared%20Documents/TZA/INT_CRC_NGO_TZA_18032_E.pdf.

Under the Same Sun, 2017a. Reported Attacks of Persons with Albinism. [No publisher]. [Online]. Available from: http://www.underthesamesun.com.

Under the Same Sun, 2017b. Under the Same Sun: Advocacy & Public Awareness. [Online]. Available from: http://www.underthesamesun.com/content/advocacy-public-awareness.

Under the Same Sun, 2017c. Under the Same Sun: Education Support. [Online]. Available from: http://www.underthesamesun.com/content/education-support.

United Nations Human Rights Council, June 24, 2013. Resolution 23/13: Attacks and Discrimination Against Persons with Albinism. A/HRC/RES/23/13. [Online]. Available from: http://www.un.org/en/ga/search/view_doc.asp?symbol=A/HRC/RES/23/13.

United Nations Independent Expert on the enjoyment of human rights by persons with albinism, 2016a. Action on Albinism in Africa Consultative Forum: Outcome Report. [No publisher]. [Online]. Available from: http://www. ohchr.org/EN/Issues/Albinism/Pages/AlbinismInAfrica.aspx.

United Nations Independent Expert on the enjoyment of human rights by persons with albinism, January 18, 2016b. Report of the Independent Expert on the Enjoyment of Human Rights by Persons with Albinism. A/ HRC/31/63. [Online]. Available from: http://www.albinism.org/atf/cf/%7BF9C8AAE8-DE7A-4923-BC56-52ABED8FD74D%7D/Report%20EN.pdf.

van Heerden, D., June 7, 2016. Hunting for Humans: Malawian Albinos Murdered for Their Bones. CNN. [Online]. Available from: http://edition.cnn.com/2016/06/07/africa/africa-albino-hunted-bones-malawi/.

Webb, C., July 30, 2013. Albino Group Slams Comedy for Offensive Portrayal. The Sydney Morning Herald. [Online]. Available from: http://www.smh.com.au/entertainment/movies/albino-group-slams-comedy-for-offensive-portrayal-20130729-2quzr.html.

World Bank Group, 2016. World Bank Open Data: Tanzania. [Online]. Available from: http://data.worldbank.org/country/tanzania?view=chart.

Interventions: Preventive Management, Empowerment, Advocacy, and Support Services

Jennifer G.R. Kromberg

University of the Witwatersrand and National Health Laboratory Service, Johannesburg, South Africa

O U T L I N E

Albinism in Africa
http://dx.doi.org/10.1016/B978-0-12-813316-3.00013-1

271

I. INTRODUCTION

This chapter will highlight issues associated with the management of albinism. The first issue is that of prevention and the possible interventions necessary in optimizing quality of life. There are two aspects to prevention: the first option is prenatal diagnosis and selective abortion and this has been covered in detail in a previous chapter (see Chapter 11) and the second involves minimizing the impact and complications of having the condition on the affected individual. This latter aspect includes the detrimental effect of having physical problems, because of a sun-sensitive skin and poor eyesight, and the damaging psychosocial effects of being stigmatized and viewed as unacceptable in the community in which one lives.

Preventive management, therefore, involves interventions such as early (soon after the birth) genetic and health counseling for the parents of an affected child, as well as early stimulation for the child. As the child grows, skin lesions can be prevented with the use of sun barrier creams, visual defects can be managed and the effect on vision minimized, and educational issues can be ameliorated by the use of visual aids and by informing teachers of the child's needs or placing the child in an appropriate educational environment. These issues will be summarized briefly here as they have been covered in some detail in other chapters of this book.

Generally, parents, with affected children, and both children and adults with albinism can gain by becoming empowered, and their health providers can benefit by understanding the nature of empowerment. Advocacy is often necessary on behalf of people with albinism, so that they can access good health care, educational, and other community services and thereby achieve their potential. These issues will be discussed in this chapter.

Families with affected members will also require support services and the development and benefits of these and of support and advocacy groups will be covered. International organizations such as Inclusion International, the World Albinism Alliance (WAA), and general Genetic Support groups, such as Genetic Alliance South Africa, can also be of assistance. Furthermore, the value of the available community resources needs to be recognized, and they need to be accessed and used for the benefit of the affected child and the family.

II. PREVENTIVE MANAGEMENT AND INTERVENTIONS

To prevent the complications of having albinism from becoming a heavy burden on the affected person and the family, many of the challenges associated with the condition can be anticipated and managed appropriately. Good management starts in early childhood and lasts a lifetime.

Health and Genetic Counseling

For the parents who have just had their first child with albinism, appropriate and relevant counseling is essential. Crisis counseling is often necessary, as the mother may be in a state of high anxiety and confusion, and to prevent desertion (which has happened a few times in the writer's experience) of the infant and promote acceptance, help may be required urgently. Ideally, counseling can be provided in hospitals by an informed pediatric medical social worker, especially one with some experience in counseling parents with a newborn child with a birth defect. Preferably, also in the newborn period, genetic counseling (see Chapter 10) with a trained experienced genetic counselor or medical geneticist, who understands the features and the genetics of albinism and can also show empathy, listen to, and respond to the couple's emerging needs, should follow. At the same time, such counselors can provide appropriate health education to the parents, so that the child's needs can be understood and met as early as possible. The majority of mothers in a recent study, on mothers with newborn infants with various disorders (including albinism), felt that counseling should be provided soon after the birth of the child (Morris et al., 2015). Such counseling should be available on an ongoing basis.

Later, counseling for the unaffected siblings should also be considered. Although no relevant research appears to have been conducted on children with a sibling with albinism, it has been found that siblings of a child with a disability can show some psychological impairment (Breslau, 1982). Breslau (1982) compared 237 siblings of disabled children (aged 6–18 years) with 247 siblings in a random sample of families without an affected child. The results showed that younger male siblings, especially those near in age to the affected child, may show poor psychological functioning. Also, the later responses of siblings to early life experiences may be gender specific, with males showing some interpersonal aggression and females some feelings of depression and anxiety. Another finding was related to birth spacing, and aggressive behavior in boys less than 2 years younger than the affected child was significantly increased compared with boys more than 2 years younger than the affected child. This finding suggests that parents with an affected child should consider delaying further reproduction and planning the next child for more than 2 years later. It may not be possible to generalize these findings to albinism, but nevertheless, they give some indication of the issues that arise in families in this situation.

Finally, counseling is also recommended for both older children and adults with albinism (see Fig. 13.1), and this has been covered in detail elsewhere in this book (see Chapter 9).

Early Stimulation

Where possible, parents should be advised to place their child with albinism into a neurodevelopment clinic for monitoring and an early stimulation program. Such programs are usually run by a multidisciplinary team of health professionals including an occupational therapist, physiotherapist, speech therapist, medical consultant, and social worker, psychologist, educationalist, and geneticist. In this team setting, the parents of the child are active partners and work in close cooperation with the team. The affected children can be stimulated on several levels, so that their development does not lag too far behind their peer group. It is well recognized that infants with low vision, such as those with albinism, are generally slower to explore, to crawl, and walk than other children of the same age, but early stimulation at an appropriate clinic, including a home exercise program, can assist the child to reach

FIGURE 13.1   A genetic counselor discussing genetics and research with a person with albinism. *Photograph courtesy of Dr. J. Kromberg, University of the Witwatersrand, South Africa.*

these milestones sooner. Regular assessment by the therapists at the clinic can also mean that any other developmental problems, related or unrelated to the albinism, which the child may have, can be detected early and treated, if necessary.

The Strive Towards Achieving Results Together (START) program was developed in the 1990s in Johannesburg, initially to assist mothers in stimulating their preschool children with developmental delay (Solarsh and Goodman, 1992). Later, it was expanded for use with preschool children with many different intellectual and physical disabilities (such as visual and hearing defects), including albinism, and for children who were environmentally deprived. The program was developed as a package, which could be used not only at hospital clinics but also by trained parents or caregivers at home and at residential institutions. It provided skills to the parents who participated, so that they felt empowered to help their affected children and more capable of meeting their needs.

Dermatology Assessment and Skin Care

As the child develops and starts being exposed to the sun, the parents should be encouraged to book an appointment at a dermatology clinic, so that they can learn how to care for the child's skin and prevent sun damage. The child's skin can then be assessed and the parents can be counseled regarding the high risks of skin cancer and the lifelong preventive measures that are essential from a young age. A nongovernmental organization (NGO) working in Tanzania has shown that providing regular skin care assessments, treatments, and sunblock to adults and children with albinism, over an 18 month period, can reduce skin cancer cases by 18% (Beale, 2015).

It is well recognized that not only children with albinism but also fair-haired children of every ethnic group, especially those living in tropical or subtropical areas of the world, can benefit from being protected from too much sun exposure and sun burn, in childhood.

(Because of the high rate of skin cancer in Australia, the Education department there has an instructive jingle for primary schools: "slip on your shirt, slap on your hat and slop on the cream" every day!). Parents of children with albinism also need to learn where they can access a regular supply of the necessary sun barrier cream for their child and how they can motivate him or her not only to use the cream but also to wear suitable clothing at all times and avoid sun exposure as much as possible (see Chapter 6).

Visual Assessment and Visual Aids

Once the child is older, visual assessment is required from a low-vision expert. Ideally, this should be done in the first year of life (starting as early as 6 months of age). The importance of early intervention has been stressed by ophthalmologists and optometrists (Raliavhegwa et al., 2001), especially in children with strabismus because the possible development of amblyopia can then be prevented (see Chapter 7). Also, visual acuity can be determined and appropriate visual aids provided, so that the child can use his available vision to its fullest potential and therefore benefit from early visual cues and later from schooling (see Chapter 8). In Johannesburg, for example, patients are referred to St. John's Eye hospital for a full visual assessment.

Schooling and Education

The choice of schooling for the child with albinism is important and, in South Africa and elsewhere in Africa, the choice is often between a school for the visually impaired and a local community school. The right choice might ease the burden of having the condition, and if the child adapts well and progresses at the selected school, it could prevent maladjustment and future psychological problems (see Chapter 9).

Later, the choice of tertiary training and career is also important (to prevent social and economic deprivation and gain quality of life), and the services of the Disability Office at tertiary institutions (such as the University of the Witwatersrand in Johannesburg, where many different services for visually impaired students are available) should be used. Regarding employment, indoor jobs should be preferentially selected and those which involve daily sun exposure and outdoor work should be avoided where possible.

A study carried out in Malawi has shown that sun sensitivity and visual impairment have prevented full participation of children with albinism in local mainstream schools (Lund et al., 2015). Also, recent abductions and mutilations have caused some parents to keep their children away from school and even sometimes to hide them. The findings showed that stigma, prejudice, and rejection (particularly paternal) occurred, while an involved supportive male figure led to more security and positive feelings about the future in the children. However, financing the needs of a child with albinism was very difficult (when income was derived from fishing and subsistence farming). Preventive management requires health education, protection of the child from an early age, appropriate advice to the educators, and the distribution of sunscreen and visual aids, if the good health of affected children is to be maintained and their schooling successfully completed.

Facing the Cultural and Psychosocial Issues

The major issues of stigmatization and discrimination associated with albinism have to be faced by the parents from the neonatal period onward and throughout their lives. They need to learn to counteract the misinformation that is widely believed and stated by many people and provide them with the accurate facts relating to albinism. Affected children need to grow up in a secure family that provides them with a stable, enabling, and empowering environment. They need to know who they are, why they have albinism, and how to handle negative attitudes. A supportive informed background will help to prevent them from succumbing to the inevitable name-calling and belittling attitudes that they will encounter in the community. However, Sindile Makha (2013), a 41-year-old woman with albinism, states that she has faced her issues and that it saddens her that albinism is seen as a problem, as this does not reflect in her life; she is a university graduate and an up-and-coming politician. She challenges others not to let albinism define or limit them; when challenges present they have to be conquered, and she does not see her albinism as a negative or a detraction.

Furthermore, a study (Butler-Jones, 2013) has shown that people with albinism in Tanzania who have appropriate education and community support show resilience, which leads to an improved condition, improved behavior in seeking medical help, and improved survival. The author recommended that the community assists people with albinism to feel a sense of belonging and that the work of agencies in the community that mobilize, educate, and empower affected people should be maintained and supported.

III. EMPOWERMENT

The parents of a child with albinism and the affected child or adult can benefit from being empowered. Through the process of empowerment, one gains confidence in one's own abilities and builds up enabling knowledge. The process involves accrediting oneself to influence social conditions and take action to improve life situations. "Empowerment is a purposeful process, which shifts the perception of powerless individuals, groups and communities, and enables them to assume greater capacity and to gain access to control over resources which affect their lives" (Rickards and Atmore, 1992, p. 27). Empowerment is necessary because of an existing imbalance, which may disable and alienate people with disabilities, and, as a result, their basic rights may be denied. Furthermore, the perceptions of self and those of others may contribute to the feelings of powerlessness. Generally, feelings of empowerment develop over time and require the changing of perceptions, the acquiring of new skills, and better understanding of the condition.

Although there appears to be no available literature on albinism and empowerment, the general statements given above and those that follow, in the light of the present author's many years of experience, appear to be applicable in the case of albinism.

Gutierrez (1990) suggests that there are three basic levels of empowerment: (1) the micro level, which is at the individual or personal level; (2) the macro level, which is at the level of society and collective action; and (3) the interface between the micro and macro levels, at which discussion arises as to how individual empowerment can influence group empowerment and how an empowered group can improve the functioning of its individual members.

In this section, the focus is mainly on empowerment at the micro level. According to Rappoport (1985) the concept of empowerment, at the personal level, suggests a sense of control over one's life involving cognition, motivation, and personality. It expresses itself at the level of emotions and ideas about self-worth and about being able to make changes in the community around us. The empowered individual at the interface has a sense of personal control, which can be combined with the ability to influence the behavior of others and enhance the existing strengths in other individuals and/or communities.

de Bruyn (1992) has emphasized that it is primarily within the family that people develop beliefs about themselves that give feelings of being worthwhile, of being self-confident, and of being personally powerful (and empowered). Why is the family so important in the issue of empowerment? A quote from Skynner (1979) eloquently addresses this question:

> The institution of the family stands in a peculiarly central and crucial position. It faces inward to the individual, outward to society, preparing each member to take his place in the wider social group by helping him internalize its values and traditions as part of himself. From the first cry at birth to the last words at death, the family surrounds us and finds a place for all ages, roles and relationships for both sexes. Our needs for physical, emotional and intellectual exchange, and for nurturance, control, communication and genital sexuality, can all exist side by side and find satisfaction in harmonious relationship to one another. It exists to make itself unnecessary, to release its members into the wider community as separate autonomous beings, only to recreate there images of itself anew. It has enormous creative potential, including that of life itself…

Four crucial psychological changes are required to empower unempowered people, including those with albinism, and move them from a state of inaction to one of action (Gutierrez, 1990):

1. Increasing self-efficacy: self-efficacy is the belief in the ability to produce and organize events in one's own life (Bandura, 1982). It includes strengthening ego functioning and developing a feeling of personal power, mastery, initiative, and the ability to act.
2. Developing group consciousness: this involves the development of an awareness of how the experiences of individuals and groups are affected by political structures. A critical perspective on society then develops, which results in recognizing that individual, group, or community problems can stem from a lack of power. A sense of shared fate is created so that individuals in the group can focus their energies on the causes of problems, rather than on changing their internal subjective states.
3. Reducing self-blame: this is related to consciousness raising. By assigning problems, where appropriate, to the existing power arrangements in society, individuals can be freed from blaming themselves for their negative situation. Self-blame can be associated with feelings of depression and immobilization. Reducing self-blame helps individuals to change focus and feel less deficient and more able to alter their situation.
4. Assuming personal responsibility for change: this counteracts some of the potential negative effects of reducing self-blame. Individuals who feel no responsibility for their problems may not make the effort to find solutions, unless they take some responsibility for future change. By actively attempting to identify solutions, individuals are more likely to make an effort to improve their lives.

However, the psychological changes that bring about empowerment are not necessarily single achievements, but they represent a continuing process of growth and change. Each aspect of the process may happen simultaneously, enhancing one another, but change can also occur throughout the life cycle. The process of empowerment is a way of interacting with the world, a way people with albinism can consider, and a way many of those leading the Albinism Societies across Africa have adopted.

Where social workers and counselors are available to people with albinism, they could provide them with empowering techniques through their helping relationships. Such interventions may be at the individual, family, or group level, and according to Gutierrez (1990) they include the following:

1. Accepting the client's (or the counselor's) definition of the problem and communicating the belief that the client is capable of understanding the situation.
2. Identifying the client's current level of functioning and sources of individual or interpersonal power and building on the existing strengths.
3. Analyzing the client's situation by firstly assessing how conditions of powerlessness are affecting it and secondly identifying sources of potential power such as forgotten skills, personal qualities that could increase social influence, helpful networks, and community organizations.
4. Teaching specific skills, for example, those required for problem-solving, community or organizational change, lifestyle issues (e.g., parenting, self-defense), and interpersonal relationships (e.g., assertiveness, self-advocacy).
5. Mobilizing resources and advocating for oneself or one's client, so that larger social structures provide what is necessary to empower individuals and/or client groups.

In the health setting, patients with albinism (and/or their parents) require empowerment to make sure that their health needs are met. Aujoulat et al. (2007) reviewed 55 articles on empowerment in patients with chronic diseases (including some genetic disorders). From their analysis of the findings from these articles, they showed that "the educational aspects of an empowerment-based approach were not disease specific" (p. 13). They also summed up their findings and arrived at a comprehensive definition of patient empowerment: "empowerment may be defined as a complex experience of personal change. It is guided by the principle of self-determination and may be facilitated by health care providers if they adopt a patient-centered approach of care which acknowledges the patients' experience, priorities and fears. In order to be empowering for the patient, therapeutic education activities need to be based on self-reflection, experimentation and negotiation so as to allow for the appropriation of medical knowledge and the reinforcement of psychosocial skills" (Aujoulat et al., 2007, p. 18).

Expanding on their previous work, Aujoulat et al. (2008) state that most studies on patient empowerment focus on two issues of the patient's experience of illness: managing regimens and relating to health-care providers. Other issues such as adapting to disrupted identities are generally overlooked. The outcome of empowerment is usually seen as the achievement of self-efficacy, mastery, and control. The authors describe patient empowerment as a process of personal transformation, which occurs through a double process of firstly "holding on" to previous roles and self-representations and learning to control the disease and its treatment, so as to differentiate oneself from the illness, and, secondly, "letting go," by accepting the necessity to relinquish some aspects of control, so as to integrate illness and illness-driven boundaries as being part of a reconciled self. If one draws a parallel with the experience of people with albinism, the stages toward reaching an empowered self might include, firstly, accepting the condition and managing the necessary regular and lifelong skin and eye treatments (as well as learning to cope with the inevitable stigmatization and discrimination) and, secondly, accepting that the condition limits some possibilities, such as working in the sun or in jobs that require excellent vision (these aspects of the condition cannot be controlled). As Aujoulat et al. (2008, p. 1228) argue "the process of relinquishing control is as central to empowerment as the process of gaining control."

Although referring to their experience with patients with diabetes, Funnell et al. (1991, p. 37) present a comprehensive definition of empowerment, which is relevant to patients with albinism in the health-care setting: "we have defined the process of empowerment as the discovery and development of one's inherent capacity to be responsible for one's own life. People are empowered when they have sufficient knowledge to make rational decisions, sufficient control and resources to implement their decisions and sufficient experience to evaluate the effectiveness of their decisions. Empowerment is more than an intervention or strategy to help people make behavior changes to adhere to a treatment plan. Fundamentally, patient empowerment is an outcome. Patients are empowered when they have knowledge, skills, attitudes and self-awareness necessary to influence their own behavior and that of others in order to improve the quality of their lives."

Later, Funnell and Anderson (2004) wrote a very instructive article with many practical ideas for how to empower patients, so that they self-manage their condition more effectively. Successful self-care behavior plans depend not only on managing the condition but also on tailoring the plan to fit the priorities of the patient in terms of goals, resources, culture, and lifestyle. Patients, such as people with albinism, need to set goals and make daily lifestyle decisions that are effective and fit their values, while considering the physiological and psychosocial factors that are associated with their condition. "This approach is based on three fundamental aspects of chronic illness care: choices, control and consequences" (Rubin et al., 2002, p. 29), which apply to the management of many genetic disorders, including albinism.

The authors (Funnell and Anderson, 2004) add that in the health-care setting, empowerment should be focused on the patient in a collaborative approach with the health care team, so that patients learn to use their innate ability to be responsible for their life and well-being and manage the realities of living with the condition. The empowered patient then reaps the benefits, such as an improved relationship with the health care professionals, more satisfaction with care, and better physical, psychosocial, and emotional outcomes. The health provider also gains professional satisfaction by observing the improved well-being and health of the patient.

Bravo et al. (2015) have suggested that a clear definition of patient empowerment is not available, and that the components of empowerment need to be clarified and its relationship to other constructs, such as health literacy, self-management, and shared decision-making, needs to be understood. Their research showed that these components included: "underpinning ethos, moderators, interventions, indicators and outcomes" (Bravo et al., 2015, p. 252), and they illustrated the interrelationships of these components with the three specified constructs.

The concept of patient empowerment, within the health care environment, assumes that empowered patients will manage their condition better and make rational decisions about treatments when necessary, so that morbidity is lessened and long-term health is the best it can be (McAllister, 2016). Research has shown that patients with genetic conditions using the clinical genetic services in the United Kingdom value the empowerment benefits they gain from a genetic counseling session. Analysis of the data collected in these studies showed that increases in patient empowerment scores after clinic attendance correlated significantly with both patient satisfaction and patient outcome scores (McAllister, 2016). These benefits from genetic counseling may involve not only physical benefits but also psychosocial benefits. In the case of albinism in Africa, such benefits might include communicating with someone who is familiar with the condition; feeling informed about the high risk of skin cancer and visual defects and how to manage them; and gaining in confidence due to a better understanding of the cause of albinism, the recurrence risks, and the misconceptions regarding cultural and mythical beliefs. In contrast to many severe genetic conditions, where patients may not be able to effect major health changes to their condition, empowered people with albinism can alter their behavior by taking the necessary preventative measures and, in the writer's experience, many do make these behavioral changes for their long-term benefit.

The empowerment concepts and strategies discussed here, hopefully, present a comprehensive perspective from which to analyze the situation of patients with genetic disorders, such as albinism, and determine how they can become empowered and interact with local power dynamics in their community (Heinecken and McCloskey, 1985). Understanding the fundamentals of empowerment and putting them into practice will help affected people to enjoy a better status and bargaining power and will improve their quality of life as individuals and participants in society.

IV. ADVOCACY

Advocacy may be defined simply as "active support as of a cause" (Holtzman, 1984).

The cause of albinism should be actively supported in Africa for several reasons. These include the following: it is a relatively rare disease and its need for recognition and its prevalence may not be appreciated; other causes are often competing, so the albinism cause must become more widely known and its appeal more effectively communicated; there is fear and misunderstanding about albinism and these need to be dispelled; and the limited knowledge about the condition in the public domain needs to be rectified. Advocates for albinism are required to work toward drawing attention to the condition, debunking the distortions and myths that surround it, reducing the prevailing ignorance, and participating in formulating policies that will advance the cause. For example, specifically, efforts should be made to reduce the influence of those traditional healers and witch doctors who appear to be perpetuating the damaging myths, stereotypes, and fear of disability (Mostert, 2016).

The concepts of empowerment and advocacy are linked in several ways because advocates for the disempowered can stimulate their empowerment. Advocates may be the parents of an affected child, adults (or occasionally children) with albinism, members or supporters in Albinism Societies, and/or health professionals such as those treating people with albinism, genetic counselors or clinical geneticists, journalists in the media, and/or lawyers. Clinical geneticists, for example, can act (and have acted) "as powerful advocates for families, especially those with academic and research reputations, because patients see them as well positioned to influence the progress of developments, winning research funding and then feeding research findings back to the families" (McAllister, 2016, p. 53).

Parents of affected children are possibly the most effective advocates for improved services, particularly health and educational services, because they are motivated by wanting the best for their child. However, people with albinism themselves can act as very effective advocates, not only by advancing education on their condition (for example, by telling their stories to the media) but also by approaching health services for better and more accessible service provision (and free sun barrier cream to prevent skin cancer). They can also lobby governments for funding for research to improve the understanding of the condition and legislators to form commissions to monitor, for example, employment opportunities for people with albinism and other disabilities. As a result of one legislator taking an interest in health and inherited diseases and with the help and expert assistance of five physicians in the field, the Hereditary Diseases Act was formulated and promulgated, in the United States, to "serve as a continuing public forum to consider genetic disorders" (Lapides, 1984, p. 181). A commission was then established and given the power to adopt rules and regulations to cover all programs for patients with genetic disorders and to ensure they had "voluntary participation, informed consent, confidentiality of information, reliability and accuracy of diagnostic procedures and availability of trained counselors who would insure nondirective counseling" (Lapides, 1984, p. 181). This is an example that albinism advocates could follow, while broadening their outlook and the cause they are supporting.

To promote advocacy of albinism and to provide interested people with the latest information, the First International workshop on Albinism in Africa was held in Douala, Cameroon, in July 2015. The 2-day meeting was promoted by Professor R. Aquaron (who has spent several decades investigating albinism in Cameroon) from France and his international committee. It was well attended, by 120 people, and many of the current issues associated with the condition were covered by eminent expert speakers, from various countries in Europe, the United Kingdom, Cameroon, South Africa, and elsewhere in Africa. Participants came from several African countries including Cameroon, Burundi, Benin, Ivory Coast, Nigeria, Niger, Mali, Malawi, Congo, South Africa, and Zambia. The workshop brought together many different people with a common interest in albinism, including basic researchers, clinicians, anthropologists, dermatologists, ophthalmologists, experts in human rights, university professors, members of associations who support albinism, albinism advocates, people with albinism, students, and members of the general public (Montoliu, 2015). Both scientific and social aspects of albinism were discussed. Raising positive awareness and advocating for people with albinism was a common aim for all organizations and people attending the workshop.

Several of the papers delivered at the workshop focused on advocacy. One session was entitled Effective Advocacy and the first speaker, John Chiti, a person with albinism and a singer, commented that music can be used as a tool for advocacy, for fund-raising, to gain an income, to provide a role model, as an emotional activity, and an effective means of communication. Another speaker, Bonface Massah, discussed the question of collaborating with researchers (Lund and Massah, 2015). He stated that the collection of data in seven African countries had provided governments with evidence regarding the prevalence of albinism and the extent of the current abuse. He used case studies to illustrate the beneficial impact academic studies have had on communities and the lives of families. He added that research should be linked to rights advocacy, local capacity should be developed, and material should be produced for this purpose.

Later Beale (2015) addressed the role of international organizations in supporting and promoting the human rights of people with albinism and advocating on their behalf. He emphasized that Africa has people with albinism who are advocates and their work is showing results. However, international organizations can help to fight the social marginalization of affected people. As an example, Standing Voice (SV), a group based in London, is working in Tanzania and focusing on health, education, and advocacy issues. It trains professionals, builds infrastructure, and implements frameworks for them (see Chapter 12). In addition, the organization develops arenas where advocates can expand the impact of their work, and, through empowering initiatives, it motivates individuals living or working alongside affected people to become advocates for positive change. SV promotes inclusion of people with albinism in every sphere of life; advocacy involves "information spreading outward" and affecting attitudes. SV identifies and focuses on local problems, trains and informs advocates, equips, guides, and updates them. Key stakeholders are people with albinism (and Albinism societies), dermatologists, hospital staff, health professionals, Ministers of Health, government officials, press clubs, and families. Advocacy offers opportunities for local research and limits dependency on external agencies.

Patient advocacy groups and parent support groups have been set up in many countries and for many genetic disorders since the 1960s (Hall, 2013). Because of the recent advances in human genetics, more information and more opportunities for management and treatment and potential therapies have become available. Furthermore, due to the advancement of the internet, social networking, and almost universal access to email, families with the same disorder have been able to link up and communicate with each other. The opportunity to learn from the experience of other families and other patient advocacy groups is enormous. However, knowledgeable health care professionals are required to ensure that the information being circulated is accurate and to provide advice where necessary. Groups focusing on the same genetic disorder can share diagnostic and medical information, practical advice about living with their condition, the natural history, and potential treatment, as well as supporting and participating in research and clinical trials. This has been the case with Albinism Societies all over the world.

Advocacy and support groups are usually started by one person, who has a personal interest in the condition (such as an affected child) and who seeks others in the same position. These families connect and share their information, experiences, and ideas. They may then formalize their group, hold regular meetings, elect a committee, consult a medical professional as a partner, eventually fundraise, rent an office, and employ staff. They will often develop advocacy plans, contact newspaper reporters to spread their news, to make their condition and the availability of the group more widely known and educate the public.

Such groups may have many different activities. Hall (2013) suggests that these include the following:

1. Information sharing within affected families, their relatives, and the public through conferences, meetings, newsletters, online, and social media.
2. Supporting and encouraging researchers to plan projects on their disorder (with some input from the group, where possible); inviting group members to volunteer as

subjects, actively participate, and even help to collect the data in some cases. Some groups will advocate for research regarding the natural history, which may not be well understood, and the application of new testing methods, if necessary, or therapies. Once such therapies become available the group may advocate for them and raise funds so that they become accessible to all affected individuals. They may even participate in the development of therapy interventions, which may extend beyond their own group.

3. Developing guidelines for the care and management of individuals with the disorder (in consultation with the medical professional involved).
4. Holding social events and support group meetings, where problems can be shared and emotional support offered. Also, the training of selected support parents in basic counseling skills can be undertaken (with professional help), so that they can provide a counseling service to new parents.
5. Creating a registry of affected individuals and/or a biobank of specimens and blood samples.
6. Using their internet skills to search the world literature to make sure their group is up to date with the latest relevant knowledge, medical information, and research findings on their disorder and is in touch with those investigating it. Families may also keep the doctors involved in their care informed about new developments and may contribute to new ideas about their condition.

Parent support and advocacy groups provide many resources, which are greatly appreciated by the affected families and which can be helpful to health professionals dealing with families with a newly diagnosed member.

In terms of advocacy for albinism on the international scene, Ikonwosa Ero, the Independent Expert on the Enjoyment of Human Rights by Persons with Albinism at the United Nations, recently said that there is no need to create new laws, instead disability rights advocates must use existing laws and enforce them to protect people with albinism. There should be a "call for action that ensures implementation of existing international standards" (Conference Report, 2017, Advancing the Rights of Persons with Albinism in Africa). For this purpose, she is suggesting the convening of a high-level meeting and the creation of regional action plans in Africa.

V. GENETIC ADVOCACY AND SUPPORT GROUP SERVICES

For all those involved, people with genetic disorders (such as albinism), their families, and health-care professionals, genetic advocacy groups play a vital part in dealing with the condition (Lin et al., 2003). Groups may start off as a regular informal gathering of few concerned families but may develop into large national and international associations. Initially, the aim might be to provide emotional support and assistance with managing the challenges of the disorder, but many organizations then undertake the roles of information provision, advocacy, and research (Terry and Boyd, 2002). Because advocacy has become a major function of many of these groups, the name "genetic advocacy groups" is now preferred over the previously used name "genetic support groups."

Setting Up Genetic Advocacy Groups

Genetic support and advocacy groups have been shown to be of benefit to their members over many years. In South Africa, for example, the first genetic advocacy group was set up, in the early 1970s, at the request of a couple who had a child with the recessively inherited and lethal Tay-Sachs disease. They wanted to meet other affected families and to establish awareness programs so that at-risk couples could prevent or avoid the tragedy of having an affected child. At their request, a meeting was held in 1973 in Johannesburg and concerned people from all over South Africa and Namibia attended. An organization was set up and, initially, named the Southern African Inherited Disorders Association (SAIDA). The aims of the Association were to (1) educate medical and paramedical professionals and the lay public about inherited disorders; (2) provide a fellowship for those affected with and by these disorders; and (3) support research into the causes, treatment, and prevention of such disorders.

The organization elected a committee, held regular meetings, developed branches in all the provinces of South Africa, and set about initiating activities. These included genetic support groups (the first three being for Tay-Sachs disease, Down syndrome, and cystic fibrosis); training support parents to offer a support service to new members; providing educational workshops and community awareness programs; circulating biannual informative newsletters; fund-raising to support (for example) international genetics experts to visit and update South African geneticists; starting toy libraries for disabled children and early stimulation programs; and supporting research. These activities continued for about 30 years (during which time, in 1994, the Albinism group was initiated and became very active), until fund-raising became more difficult and some activities went into abeyance. However, in 2015 the Association was resuscitated, a new committee was formed with new ideas, and new branding that resulted in the name being changed to Genetic Alliance, South Africa, to align with world trends.

During its time SAIDA initiated about 24 different genetic support groups and had many different Chairmen, including the Director of the Albinism group. Support groups found it beneficial to participate in the organization where they received stimulation and motivation from other groups, shared experiences and awareness programs, and had access to dedicated medical practitioners and genetic counselors for advice and guidance, when necessary.

Lessons From Inclusion International

Because people with albinism are often excluded from their communities, in African countries, as well as elsewhere in the world, it is worthwhile briefly examining the aims and objectives of Inclusion International. This organization focuses on people with disabilities, particularly those with intellectual disabilities, but the general principles it puts forward apply to people with albinism as well. The organization was founded in 1991 and the goals were:

- to strengthen the capacity of national members
- to promote human rights for affected people
- to build alliances
- to promote practical initiatives for sharing effective strategies

It was noted that the primary issues for parents were, firstly, education for all and, secondly, overcoming stigma. The basic assumptions were that the ideal scenario was for the affected child to go to the local school; live a normal life; and be treated and supported in the same way as his/her siblings were. To achieve these goals the law, such as international human rights laws and local laws regarding rights to education, should be used.

The organization promoted advocacy and proposed that individual parents should lobby target people and groups; develop support groups; build political support; and promote new initiatives by families. Strategies should involve Ministries of Health and Education, universities, teachers, local pioneers, parents, and professionals. They also supported the organization of meetings, seminars, and conferences to disseminate relevant information.

Inclusion International promotes positive perspectives and assumes teacher and student growth. Albinism Societies can and do learn from organizations such as these and can apply the same basic strategies in the management of their associations (for further information, see the website: Inclusion-international.org).

VI. SUPPORT GROUPS FOR ALBINISM

Albinism Societies Across Africa

It is informative to read through some of the website entries of the various Albinism Societies that have been set up in countries across Africa. These entries give insights into where organizations have been established, what services they are providing, the lobbying they are pursuing, the steps governments are taking, and the awareness, advocacy, and other proactive programs that are being initiated and implemented. Some of the countries, such as Cameroon, have more than one active society. In the following section a picture of some of the activities of a few of the societies will be presented.

The Albinism Societies in the Cameroon

The Albinism Societies of Cameroon are very active, and in July, 2015, they assisted in organizing and running the First International Workshop on Albinism in Africa held in Douala, Cameroon.

According to an entry on their website, the society states that across Africa people with albinism generally live with fear and rejection and many are outcasts. However, in Cameroon, many affected people say their lives are getting better, especially with assurances from the government that they are committed to improving the health and welfare of affected people. Attitudes to albinism vary in different parts of the country. In Buea, for example, at the foot of Mt Cameroon, people with albinism go about their daily tasks with no problem. But elsewhere in the country affected people are hunted and killed because witch doctors say medicine made from their tissues can bring wealth and good luck. Furthermore, sometimes, affected babies can be considered a sign of misfortune and buried alive or denied breast milk (www.albinism.org).

Jean Jaques Ndondonmou, President of the World Association for the Defense of the Interests and Solidarity of Albinos and Manager of the Public Contracts Awards Committee in Cameroon, states that things are improving, sensitization to the issues is increasing, there are more graduates with albinism, and fewer reports of the systematic elimination of affected babies.

Albinism Society, Kenya

The Albinism Society of Kenya was started in June 2006 as an NGO in conjunction with the local Disabled Persons Organization. Two recent newspaper articles illustrate the dynamic nature of the society and some of the current activities:

1. From the Citizen newspaper 15/6/16: "Conflicting perceptions of albinism."
   Members of the society demonstrated in Nairobi recently with posters asking "Who will protect us?" and "Albinism is not a cure for HIV."
   A survey by a local NGO states that Kenyans have conflicting perceptions on albinism: 8/10 say they are the same as everyone else, 7/10 say most of the time albinos need to be cared for, while 2/10 say people with albinism live with discomfort most of the time. Altogether, 1763 respondents from across Kenya contributed their opinions to the survey in April and May 2016 and the results were released to mark International Albinism Awareness Day in June 2016. The NGO states that the contradictory statements of the respondents may be due to lack of exposure to affected people because only 3/10 (28%) report having a person with albinism living in their community. Unfortunately, 1/10 Kenyans think that persons with albinism have supernatural powers; this is problematic because these types of beliefs can lead to discrimination and even violence against affected people. A further 16% say they don't know if affected people have these powers or not. Stereotypes and attitudes are further reinforced by 68% who believe that affected people need to be treated with more care than others.

2. From The Standard 13/6/16. "Albinism is just a condition. Accept us."
   Just over a decade ago we only knew the devastation albinism inflicted on those living with the condition. But this sad narrative has been changing in the last few years with High Court judge, Ms. Mumbi Ngugi and politician Mr. Isaac Mwaura MP (both people with albinism) blazing a trail in their respective fields. Today we have largely transformed what it means to live with albinism.
   More effective protective and preventive programs have been established, which now save lives. Awareness has soared and research surged. Social misconceptions have been reversed by awareness programs implemented by the National Council for Persons with Disabilities (NCPWD), the Albinism Society, Kenya, the Albinism Empowerment Network, and many other nonstate actors. The NCPWD is geared to changing negative social attitudes and has made possible access to free sunscreen lotions in all county hospitals. But considering that many affected people live in abject poverty, more needs to be done, lotions need to be delivered to rural health centers and free eye care and glasses are required. In September 2015 a man with albinism died after he was attacked and his ear and fingers were chopped off. One death is too much.
   As we gather as leaders we affirm our commitment to fight against the stereotypes associated with albinism. We stand with our people. We focus on building a country free of unnecessary barriers and full of a deeper understanding that we are one, irrespective of skin color. *For further information:* www.albinismsocietyofkenya.org.

Albinism Society of South Africa

The Albinism Society of South Africa (ASSA) is currently very active. The Director is Ms. Nomasonto Mazibuko (see Chapter 14), who is a founding member of the society and has been the National Director for the past 6 years. The group was founded in 1994 and now has established branches in all the provinces of South Africa, except for the Western Cape, where another smaller and more recently started society operates. ASSA has about 1000 members in Gauteng province alone. It offers regular support groups and counseling sessions for parents, as well as awareness and advocacy programs, among its many activities. Ms. Mazibuko recognizes that there has been progress in the Gauteng province, partly because of the activities of ASSA and its partnership with the provincial Department of Social Development (which provides funding for some of the ASSA projects).

The month of September is recognized as Albinism month in the country, and a big, well-sponsored, and well-supported parade was held on September 24, 2016, in Johannesburg, as part of the local ongoing awareness program. Radio interviews and newspaper articles also appeared.

A newspaper article entitled "Dispelling myths and encouraging albino role models. Public must be educated out of dangerous superstitions" appeared in The Star, the most widely read daily newspaper in Johannesburg, on Monday, September 19, 2016. It quoted Ms. Mazibuko as saying "these public awareness programs are imperative to combat discrimination and false information, which led to some of those living with albinism being killed for their body parts." She added that South Africa has a very advanced constitution, which states that discrimination is a human rights violation; everyone has a right to life, and the murdering and mutilating of people with albinism is completely unacceptable. "People must see that we are human" she emphasized.

The Society has some excellent role models at present, including a young lawyer. This lawyer, Thando, explained that support from her family helped her through her difficult childhood, when she had to endure taunting and discrimination from her classmates, although, she says, she herself never really acknowledged she was different (The Star, September 19, 2016). However, after qualifying as a lawyer and while working in a legal company, she took up the opportunity to do part-time modeling as she thought that people need to be confronted with albinism, so that they realize that affected people are normal people. Also, she wanted to give others with albinism the opportunity to identify with her and see her as an inspiring role model.

Another ambassador, on behalf of people with albinism, is a top athlete, Hilton Langenhoven, who participated in the 2016 Paralympic games in Rio de Janeiro and won a gold medal for long jump. A newspaper article on his Paralympic win appeared on the front page of The Star September 12, 2016, entitled "Langenhoven strikes gold – for every albino." He stated that "albinism is a day-to-day challenge wherever you are from, wherever you go. I'm glad to contribute toward that perception, that stigma about albinism in our country and in Africa as well. Albinos are known to be excluded from society and hidden in a box. I say we don't have to hide, we don't have to hide our disability. We can be powerful, we can motivate the nation." (For further information: www.albinism.org.za).

Under the Same Sun (Working in Tanzania)

Under the Same Sun (UTSS) is an international NGO, founded by Peter Ash (from Canada), which acts globally but, at present, has focused on Tanzania due to the present crisis there. From there, they are reaching out across Africa to stimulate a movement that roots out discrimination and plants seeds of empowerment for persons with albinism. UTSS is passionately committed to ending the often deadly victimization of persons with albinism and promoting their social inclusion. It also advocates for the well-being of affected people who in many parts of the world are misunderstood and marginalized.

They state that all persons are made in God's image and, as such, are worthy of love, respect, and, above all, dignity.

UTSS has three active programs in Tanzania:

1. The education program provides funding for selected children with albinism in high-quality boarding schools and tertiary institutions.
2. The advocacy and public awareness programs use the national and international media and bodies such as the African Union and the United Nations. They educate to diffuse the myths that surround the condition and publish data on the atrocities caused by belief in these myths. They run "Understanding Albinism" seminars across the country.
3. Low-vision clinics: regular clinics are offered to affected people and they are provided with spectacles and visual aids.

Recently the Canadian High Commission in Tanzania has provided funding to help launch a UTSS radio campaign project, which will further expand their advocacy and awareness programs (www.underthesamesun.com).

Albinism Foundation of East Africa

This NGO aims to promote the rights of people with albinism and the realization of their full potential. Its mission is to improve the lives of affected people in East Africa by initiating appropriate programs on health, education, survival, participation, and development through public awareness, advocacy, economic empowerment, and social inclusion (see www.albinismfoundationea.com).

In addition to the groups mentioned above, there are many other albinism support and advocacy groups across Africa, in countries such as Angola, Benin, Burkina Faso, Burundi, Central African Republic, Congo, Gambia, Ghana, Guinea, Ivory Coast, Liberia, Malawi, Mali, Mozambique, Namibia, Nigeria, Rwanda, Senegal, Sierra Leone, Tanzania, Uganda, Zambia, and Zimbabwe (this information was obtained from the WAA website, https://worldalbinism.org, but the list is likely to be incomplete).

Albinism Support Groups in Other Countries

The Albino Fellowship, United Kingdom and Ireland

The British Society, named the "Albino Fellowship," has been functioning for nearly 40 years. It was founded by Dr. WOG Taylor, a Scottish ophthalmologist, in 1979 (Hill, 1990), and it has a worldwide membership, from which affiliated organizations have been established in North America and Australia. The vision of the Fellowship, according to their website, is that: "All people with albinism have the opportunity to realize their full potential." Their mission is to "provide information about albinism; raise positive awareness of the condition; arrange opportunities for people to meet and for support; and encourage the sharing of experiences" (www.albinism.org.uk).

People with albinism living in the United Kingdom have benefited greatly from the practical advice, information and support received from the society, its services, and public awareness programs. Members of the medical profession and research scientists have also been provided with useful information. The Fellowship assisted in organizing the First International Colloquium on the Neurological Genetic Aspects of Albinism, which was held in 1989 at Oxford University, to celebrate the 10th anniversary of the Albino Fellowship (see Chapter 1). The papers presented at this colloquium were collected and published in the journal *Ophthalmic Paediatrics and Genetics*, 1990, volume 11 (No. 3).

National Organization for Albinism and Hypopigmentation, United States

The National Organization for Albinism and Hypopigmentation (NOAH) is a well-known support group founded in 1982 in the United States. It was appropriately named NOAH, by one of the founding members, the world expert on albinism, Dr. Carl Witkop, from the University of Minnesota, in Minneapolis, United States. It has been very active for many years, has held well-attended biannual congresses on albinism, and has a very informative website (www.albinism.org).

The organization has an admirable vision: "We envision a world where people with albinism are empowered to be fully functioning members of society, where barriers and stigma of difference no longer exist, and where people with albinism have a quality of life that is rewarding, dignified and fulfilling." The mission of NOAH is "to act as a conduit for accurate and authoritative information about all aspects of living with albinism and to provide a place where people with albinism and their families in the US and Canada can find acceptance, support and fellowship." Information is available for adults with albinism, parents, caregivers, educators, medical professionals, and students doing research. They hold a list of interested health professionals, recommended personally by members, which includes doctors, optometrists, ophthalmologists, and dermatologists. They also provide many different services, such as a school kit that includes information, resources for parents and teachers, and a kit for students involved in conversations and advocacy.

Other countries such as Australia, Brazil, China, Czech Republic, Denmark, Fiji, Finland, France, Germany, India, Italy, Japan, Malaysia, Nepal, New Zealand, the Netherlands, Norway, Pacific Islands, Russia, Samoa, Spain, Sweden, Taiwan, Togo, and Turkey have active organizations. The Spanish society, for example, was started in 2006 and now has many members, some of whom have migrated to Spain from North and Central African countries (www.albinismo.es). Furthermore, several Latin American countries have formed an organization for people with albinism, in South America.

Recently, in 2014, the European organizations set up a Federation of European Associations for people with albinism (Fotso, 2015). Together, they hold an annual European Day on Albinism and an annual workshop on the condition. This event is attended by many affected people and researchers and scientists involved in investigating albinism, and the latest information on the condition is presented and discussed by experts in the field.

International Albinism Support Groups

World Albinism Alliance

The World Albinism Alliance (WAA) was formed in 2011 after the plight of people with albinism, particularly those living in Africa was brought to the attention of the United Nations; the infringement of their human rights was noted and the importance of connecting affected people across the world was recognized.

The new world alliance aims to: "ensure that people with albinism are treated with dignity and respect according to the law, heritage and culture of their birthright consistent with the

United Nations Universal Declaration of Human Rights and its Convention on the Rights of Persons with Disabilities; to provide complete and accurate information relative to albinism to those with the condition, medical professionals, educators and the public; and to promote scientific research to more fully understand and manage albinism" (Constitution of the WAA, 23/4/2011, p. 2). The alliance has on its website a useful table of all the societies who have joined the alliance with their website addresses. According to the President, in 2017, it has 117 participating groups from 56 countries. WAA is involved with various worldwide activities including awareness programs, such as the International Albinism Day on 13 June each year (https://worldalbinism.org).

Information on finding genetic support/advocacy groups and resources for patients with albinism are available from the WAA website and for those with genetic disorders in general (including albinism) from Uhlmann (2009). Uhlmann (2009, pp. 105–106) also has two useful tables on how patients can find information about genetic disorders and on questions to consider when evaluating support group literature.

VII. HUMAN HEREDITY AND HEALTH

One of the newest enterprises on the African continent is that now named Human Heredity and Health (H3Africa). In June 2010 the US National Institute of Health, the Wellcome Trust, based in the United Kingdom, and the African Society of Human Genetics announced their plan to enhance the ability of local scientists to research chronic and infectious diseases in Africa (The H3Africa Consortium, 2014). H3Africa was formed with a 5-year budget and it is focused on capacity building as well as specific scientific goals. Research grants have been awarded directly to African institutions where principal investigators are based, so that African scientists can develop their own research agendas and projects. Also, the development of international collaborations is encouraged. The criteria for success in the 5-year project include the publication of articles with African lead authors, in high-impact journals; increased funding for research; establishment of biorepositories; and contributions to the reversal of the African "brain drain." This organization could be of assistance to African researchers who wish to focus their projects on albinism in Africa, in future.

Furthermore, public engagement in genetic and genomic issues is becoming increasingly important, in Africa and elsewhere, partly because of the recognition of the necessary process of benefit sharing (Chadwick, 2016). H3Africa is involved in addressing aspects of public engagement and informed consent in multiple populations in various social contexts and in both rural and urban areas.

VIII. SUMMARY AND CONCLUSION

People with albinism will continue to be born in Africa, where the mutated genes for the condition are so common, for the foreseeable future. For that reason the development of strategies to assist them to cope with their condition and achieve to the best of their potential is essential. Once a person has the condition, there is, at present, no available curative treatment, and health will deteriorate if precautions are not taken to manage the complications.

If the means of preventing these complications are understood and programs to implement appropriate assessments and treatment are put in place and utilized, affected persons need not suffer the indignity of having damaged skin and poor vision, leading to poor education, lack of employment, and, consequently, the likelihood of poverty. Initially, the family plays an essential role and they need to understand: what causes the worsening of the complications associated with the condition; how to prevent deterioration from occurring; how to access the best skin care, visual aids, and education for their child; how to manage discrimination; and then they need to teach their children how to care for themselves.

Even in the neonatal period the parents would benefit from receiving health and genetic counseling to help them understand the condition and how best to assist their child in coping with it. Then they should recognize the child's needs for stimulation and early education and arrange to have these needs met. The child will benefit from visual and skin assessments from an early age and the daily application of sunscreen for his/her whole life. Also, to prevent the development of skin cancer, hats and protective clothing should always be worn. To preempt possible educational problems, teachers need to be informed about the nature of albinism and the affected child's educational needs, and where necessary, educational services for the partially sighted need to be accessed. Recognition of the discrimination associated with the condition is essential, and strategies to deal with it need to be developed and put into practice.

Empowerment of the family and the individual with albinism is a priority, if they are to gain confidence in their own ability to take the necessary action to access (or demand) the necessary services, in order to prevent the complications associated with the condition developing. Such empowerment will lead to affected persons managing their health, psychosocial, and other problems more adequately, resulting in a better quality of life.

Advocates for albinism may have albinism themselves or may be health professionals dealing with affected patients and their families, community workers, or the staff of nongovernmental agencies. Effective advocates can promote the wide public acknowledgment of the rights of affected people and can create positive change in communities. Both advocacy and support groups are of great benefit to affected individuals and their families, providing them with the support, understanding, and insights of other affected people and informed professionals, as well as with opportunities to share experiences with others in the same situation and with access to many useful resources.

Albinism support groups have developed and are still growing in many African countries. They are offering essential educational information and a supportive environment to their members, undertaking many awareness programs in their communities and lobbying government departments to provide better services for affected people. Similarly, support groups have developed in other countries around the world to serve the needs of affected people and their families. In addition, WAA was set up in 2011 to bring the plight of people with albinism to the world stage.

References

Aujoulat, I., d'Hoore, W., Deccache, A., 2007. Patient empowerment in theory and practice. Polysemy or cacophony? Pat. Educ. Couns. 66, 13–20.

Aujoulat, I., Marcolongo, R., Bonadiman, L., Deccache, A., 2008. Reconsidering patient empowerment in chronic illness: a critique of models of self-efficacy and bodily control. Soc. Sci. Med. 66, 1228–1239.

Bandura, A., 1982. Self-efficacy mechanisms in human agency. Am. Psychol. 37, 122–147.

Beale, J., 2015. The role of international organizations in promoting the work of African advocates promoting the rights and well-being of people with albinism in Tanzania. In: 1st International Workshop on Oculocutaneous Albinism in sub-Saharan Africa, Douala, Cameroon, 24–25 July, Abstract book, p. 77.

Bravo, P., Edwards, A., Barr, P.J., Scholl, I., Elwyn, G., McAllister, M., 2015. Conceptualising patient empowerment: a mixed methods study. BMC Health Serv. Res. 15, 252–265.

Breslau, N., 1982. Siblings of disabled children: birth order and age-spacing effects. J. Abnorm. Psychol. 10 (1), 85–96.

Butler-Jones, C., 2013. The Impact of Resilience on Help-Seeking Behavior Among People with Albinism in Tanzania (Master's dissertation). Chicago School of Professional Psychology, Chicago, USA.

Chadwick, R., 2016. The expanding scope of gen-ethics. In: Kumar, D., Chadwick, R. (Eds.), Genomics and Society. Ethical, Legal, Cultural and Socioeconomic Implications. Elsevier Academic Press, London, pp. 69–81.

Conference Report, 2017. Advancing the Rights of Persons with Albinism in Africa, 9–10 November 2016., p. 8 Pretoria, South Africa.

de Bruyn, R., 1992. Empowerment within the family. In: Kromberg, J.G.R., Hansen, C.J., Hunter, F.L. (Eds.), Empowering Ourselves and Others. SACHED, Johannesburg, pp. 21–26.

Funnell, M.M., Anderson, R.M., Arnold, M.S., Barr, P.A., Donnelly, M., Johnson, P.D., Taylor-Moon, D., White, N.H., 1991. Empowerment: an idea whose time has come in diabetes education. Diabetes Educ. 17, 37–41.

Funnell, M.M., Anderson, R.M., 2004. Empowerment and self-management of diabetes. Clin. Diabetes 22, 123–127.

Fotso, C., 2015. The Association and Social Networks: vectors and integrators for albino people in society. In: 1st International Workshop on Oculocutaneous Albinism in sub-Saharan Africa. Douala, Cameroon. 24–25 July, Abstract book, p. 85.

Gutierrez, L.M., 1990. Working with women of colour: an empowerment perspective. Soc. Work 35, 149–153.

Hall, J.G., 2013. The role of patient advocacy/parent support groups. S. Afr. Med. J. 103 (12 Suppl. 1), 1020–1022.

Heinecken, J., McCloskey, J.C., 1985. Teaching power concepts. J. Nurs. Educ. 24, 40–42.

Hill, A.R., 1990. William Taylor and the albinism fellowship. Editorial Ophthalmic Paediatr. Genet. 11 (3), 155–157.

Holtzman, N.A., 1984. Advocacy issues. Introductory remarks. In: Weiss, J.O., Bernhardt, B.A., Paul, N.A. (Eds.), Genetic Disorders and Birth Defects in Families and Society: Toward Interdisciplinary Understanding. Birth Defects: Original Article Series, vol. 20. March of Dimes, New York, pp. 161–162.

Lapides, J.L., 1984. Advocacy issues. The legislature. In: Weiss, J.O., Bernhardt, B.A., Paul, N.W. (Eds.), Genetic Disorders and Birth Defects in Families and Society: Toward Interdisciplinary Understanding. Birth Defects: Original Article Series, vol. 20. March of Dimes, New York, pp. 181–183. 4.

Lin, A.E., Terry, S.F., Lerner, B., Anderson, R., Irons, M., 2003. Participation by clinical geneticists in genetic advocacy groups. Am. J. Med. Genet. 119A, 89–92.

Lund, P., Massah, B., Lynch, P., 2015. Education of children with albinism in Malawi: barriers and facilitators. In: 1st International Workshop on Oculocutaneous Albinism in sub-Saharan Africa, Douala, Cameroon, 24–25 July, Abstract book, p. 71.

Lund, P., Massah, B., 2015. Why collaborate with researchers? How do families with albinism benefit from academic studies? In: 1st International Workshop on Oculocutaneous Albinism in Subsaharan Africa, Douala, Cameroon, 24–25 July, Abstract book, p. 75.

Makha, S., 2013. Access to health services. In: Report of the National Conference of Persons with Albinism. 25–27 October, Ekhurhuleni, Gauteng, South Africa, p. 20.

McAllister, M., 2016. Genomics and patient empowerment. In: Kumar, D., Chadwick, R. (Eds.), Genomics and Society. Ethical, Legal, Cultural and Socioeconomic Implications. Elsevier Academic Press, London, pp. 39–68.

Montoliu, L., July 30, 2015. In: Meeting Report: 1st International Workshop on Oculocutaneous Albinism in sub-Saharan Africa. Blog of the European Society for Pigment Cell Research, pp. 1–6.

Morris, M., Glass, M., Wessels, T.-M., Kromberg, J.G.R., 2015. Mothers' experience of genetic counseling in Johannesburg, South Africa. J. Genet. Couns. 24, 158–168.

Mostert, M.P., 2016. Stigma as Barrier to the Implementation of the Convention on the Rights of Persons with Disabilities in Africa. African Disability Rights Yearbook. Pretoria University Law Press, Pretoria, pp. 3–24.

Raliavhegwa, M., Oduntan, A.O., Sheni, D.D.D., Lund, P.M., 2001. Visual performance of children with oculocutaneous albinism in South Africa. J. Med. Genet. 38 (Suppl. 1), S35.

Rappoport, J., 1985. The power of empowerment language. Soc. Policy 17, 15–21.

Rickards, J., Atmore, E., 1992. Empowerment through preschool educare of children, of teachers and of parents. In: Kromberg, J.G.R., Hansen, C.J., Hunter, F.L. (Eds.), Empowering Ourselves and Others. SACHED Trust, Johannesburg, pp. 27–35.

Rubin, R.R., Anderson, R.M., Funnell, M.M., 2002. Collaborative diabetes care. Pract. Diabetol. 21, 29–32.

Skynner, A.C.R., 1979. One Flesh: Separate Persons. Principles of Family and Marital Psychotherapy. Constable, London.

Solarsh, B., Goodman, M., 1992. The S.T.A.R.T. programme. In: Kromberg, J.G.R., Hansen, C.J., Hunter, F.L. (Eds.), Empowering Ourselves and Others. SACHED Trust, Johannesburg, pp. 46–51.

Terry, S.F., Boyd, C.D., 2002. Researching the biology of PXE: partnering in the process. Am. J. Med. Genet. 106, 177–184.

The H3Africa Consortium, 2014. Enabling the genomic revolution in Africa. Science 344 (6190), 1346–1348.

Uhlmann, W.R., 2009. Thinking it all through: case preparation and management. In: Uhlmann, W.R., Schuette, J.L., Yashar, B.M. (Eds.), A Guide to Genetic Counselling. John Wiley and Sons, Hoboken, New Jersey, pp. 93–131.

World Albinism Alliance, 2011. Constitution, 23 April. Website: https://www.worldalbinism.org.

A Personal Perspective: Living With Albinism

Nomasonto G. Mazibuko[1], Jennifer G.R. Kromberg[2]

[1]Albinism Society of South Africa, Johannesburg, South Africa; [2]University of the Witwatersrand and National Health Laboratory Service, Johannesburg, South Africa

I. INTRODUCTION

In the early 1970s, in Johannesburg, a research project was initiated on oculocutaneous albinism. At that time this genetic condition had been observed widely across South Africa and its neighboring countries and a few studies had been carried out, mainly by cancer specialists, working in Johannesburg and some of the rural areas in the country. However, many questions still remained unanswered, for example, how common was the condition, were there psychosocial side effects to being affected, how did the community react to having affected people in their midst, and were they stigmatized and discriminated against in any way? A team from the Department of Human Genetics at the South African Institute for

Medical Research (now the National Health Laboratory Service) set out to answer some of these and other questions.

Initially, the research projects were based in Soweto, an accessible satellite city attached to Johannesburg, where over a million black people lived, at that time. Health services had been set up for them at the nearby Baragwanath (now Chris Hani Baragwanath) hospital, one of the largest hospitals in the Southern hemisphere, and at local health clinics scattered over the suburbs of Soweto. Primary and high schools were built and staffed so that the children could be educated locally. Employment was mostly available in the Johannesburg city center and suburbs, and employees traveled there daily by train and taxi.

To pursue the research project Jennifer Kromberg, then a medical social worker, accompanied by a research assistant (initially Eva Molantoa, an anthropologist, and later Moipone E Zwane, a nurse) traveled to Soweto. They visited health clinics and schools to identify people with albinism and their families and request their participation in the research project. Most of these families were pleased to find someone knowledgeable with whom to discuss the condition and how to manage it, as well as to learn about its cause.

One of the families who was identified early on in the project was the Ngwenya family. This was a remarkable family, with ten children, five of whom had albinism. The second youngest was Nomasonto Grace Ngwenya, and it is her life story and perspective that is the topic of this chapter. She and Jennifer Kromberg have had a long association and friendship, and based on this shared history, as well as on several interviews in 2016 and 2017, and one in 2017 with her second daughter, Senamile, Nomasonto's story has been written and appears below.

II. MEETING NOMASONTO GRACE: BIRTH AND CHILDHOOD

Nomasonto (given this name because she was born on a Sunday) Grace Ngwenya was born on May 14, 1950, the ninth child of her parents, in the Johannesburg suburb of Sophiatown. This area was occupied mostly by black people (many of whom owned their own houses) and by those of mixed ethnicity, at the time. Her parents (who were both Zulu but were not related) were not surprised that she had albinism, as they had three other older affected children. Her father was Zulu born in Ingagane near Newcastle, in the present day Kwazulu Natal (KZN) province of South Africa. However, when he was young his family moved to Johannesburg, in search of better opportunities. Her mother was a leap year baby, born on February 29, 1916, of Zulu parents, on the Robinson Deep gold mine near Johannesburg (where her father worked as a miner). The family then moved to live in Western Native Township and later to Sophiatown.

Nomasonto spent her first few years in a big house with several bedrooms, a large kitchen and pantry, and a cottage in the back garden, rented by tenants. They also had a large garden, where her mother grew apricot trees, a grape vine, mealies (maize), and cabbages. When Nomasonto was still a child the family were forcibly removed from this house in Sophiatown, because of the separate development plans of the apartheid government of the day, and sent to live in a house in the peri-urban area of Meadowlands, adjacent to the newly developing Soweto. Her parents lived there for the rest of their lives. Nomasonto's daughter Senamile comments that her grandmother was a wonderful woman, and her house was always warm and welcoming and full of relatives and friends.

Nomasonto's father had a job as a driver for Freedman's Abattoirs, which involved identifying cattle for sale in the rural areas, then loading them onto a large cattle truck, and driving it back to the abattoir. He remained in this job until his death, in his 50s, in a road accident in 1960, near Thaba Zimbi in the North Eastern part of South Africa. Nomasonto was 10 years old at the time of his death and she remembers him fondly, saying that she had a good singing voice and her father used to ask her to come and sing to him and his friends. Nomasonto's mother was a housewife for most of her life; however, when there were only the last three children left at home her husband said to her "you only have three children to care for now so you can go to work." She found a job at a branch of the OK Bazaars, a large department store chain, and worked there for 4 years. She then developed hypertension and retired. She died of a heart attack, aged 79 years in Meadowlands, and was greatly missed by her own and her extended family.

Of Nomasonto's nine siblings all survived to adulthood except the seventh child, a boy named Mandla, who had albinism; he died of diphtheria at the age of 2 or 3 years of age. There were three girls and two boys with albinism (including Mandla) and three girls and two boys who were unaffected (see family pedigree in Fig. 14.1). Her mother used to cope with the large family by pairing the children, an older one had to take responsibility for a younger one, and in this way the younger ones were always looked after and supervised. Her relationship with her mother was good and she reports that her mother was always fair, giving all the children equal care and opportunities.

III. SCHOOLING AND HIGHER EDUCATION

Nomasonto Grace started preschool at the nursery school attached to the local Anglican Church. The church, and particularly some of the priests, was very active in the antiapartheid movement in those days. She had a wonderful teacher who she still remembers, called "teacher Francis," and this teacher recalls that she was a lively child, who could sing and dance and loved to participate in the traditional dance and drama classes. These early experiences led Nomasonto to a strong, lifelong faith and dedication to the Anglican Church.

Initially, Nomasonto went to school at Thobeka Lower Primary school in Meadowlands and had to walk about 8 km to get to school every day. A couple of years later she moved to Emzimvubu Primary School, which was much nearer home. At school, she enjoyed the school feeding scheme but also often took two cents and a tin mug with her, to get an extra peanut butter sandwich and a mug of milk. These were supplied by a nongovernmental organization called the African Children's Feeding Scheme. She does not recall any particular stigmatization, discrimination, or abuse, due to her albinism, during those years (see Fig. 14.2). However, she states that she came from a large family (she being the fourth affected child in the family), which was well known in the area, and the other schoolchildren were used to seeing affected children.

She completed her schooling at Meadowlands High school (where she met her future husband). She did well there and decided to become a teacher, as did three of her sisters (as she says, there were only two careers for women to choose from, in those days, nursing and teaching). Her mother found sponsorship (through her employer and the Anglican Church) for Nomasonto to attend Eshowe Teachers' Training College in KZN. The first day she arrived

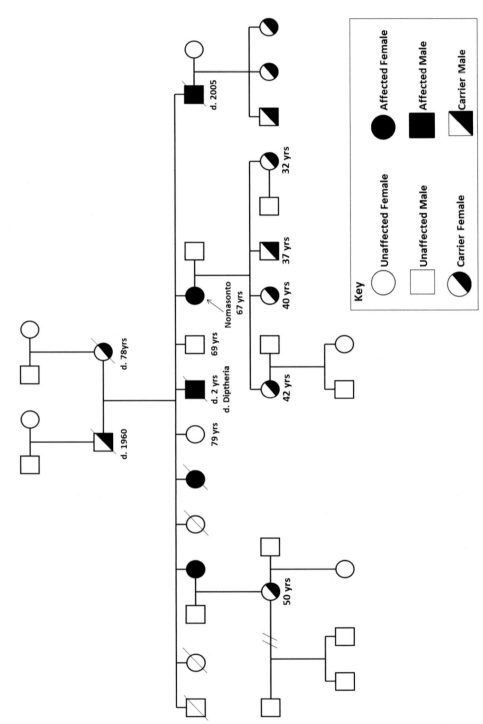

FIGURE 14.1   Family pedigree. *Image courtesy Dr. J. Kromberg, University of the Witwatersrand, South Africa.*

FIGURE 14.2   Nomasonto in her high-school uniform.

there the matron of the college residence was surprised to see her and told her she was at the wrong place; the matron had never met a person with albinism before. She was taken to the college principal who sorted out the matter, explaining to the matron what albinism was. Sometimes Nomasonto found things difficult, and she frequently had to explain her condition to others, but because of her outgoing and friendly personality she enjoyed her college experiences. She trained as a primary school teacher, successfully completing the required 2-year course. As she left the college with her certificate, her younger brother Anthony (Tony) arrived to undertake his training.

IV. WORKING LIFE

She started her first teaching job in April 1973 at Thembalethu Lower Primary School in Meadowlands. There she worked with her older sister, Joyce, who was the school principal. Nearly 20 years later, two deputy principal jobs came up at the school and she applied for one of them. Although, initially, she was offered the job (as she had the most experience and best curriculum vitae compared with the other applicants), she was then rejected by the school as unsuitable and, as she understands it, was unfairly discriminated against because of her albinism. Furthermore, her sister was accused of nepotism and both were unfairly fired. This obvious discrimination against two of her children upset her mother greatly and she died of a heart attack a few years later.

Driving has always been a problem for people with albinism; however, Nomasonto felt she had fair vision and so she decided to apply for a driver's license. At the licensing station the officials told her that she should not be driving; however, they agreed to test her and, after testing, granted her a license. Nevertheless, Nomasonto decided not to drive. She says that she felt she was a "safety hazard" on the road, and that the bright lights of the oncoming cars disturbed her. "Subconsciously" she knew she could not cope and recognized that she should not be driving. She is aware that in some other countries people with albinism have been declared legally blind.

In 1993 Nomasonto applied for a place on the Project Management Diploma course, at the University of the Witwatersrand. This course was offered when the African National Congress (ANC) had won the first free and fair election in South Africa and were taking over the government and trying to upskill the people of the country. She was accepted on the course, did very well, and was one of the few students who qualified. Also, Mr. P. Kahn, the lecturer on the course, recommended her for a project grant, and when it was awarded to her, Nomasonto was able to start the Albinism Society of Soweto, later to become the Albinism Society of South Africa (ASSA). Furthermore, he supported her application to US Aid for funding to travel to the annual National Organization for Albinism and Hypopigmentation (NOAH) Congress in the United States. She received this travel grant and went to Chicago for the meeting. This was her first exposure to people with albinism from all over the world and it led to her realization that albinism was not unique to South Africa. As she says "it was the first time I saw people from other countries with albinism!" Since that time she has been to three more NOAH meetings in the United States (at various cities across the country).

In 1994 Nomasonto and her sister Joyce both found jobs in the Provincial Department of Education, at the time that the ANC came to power in South Africa. Nomasonto worked in the examination and results department and had eight people working under her, by the time she retired in 2010. Joyce was employed in the early childhood development unit until she retired in 2005 (she died of breast cancer in April 2016).

Recently, the South African Gender Commission approached Nomasonto to be a Commissioner on their Board. She accepted this appointment and her work has taken her to meetings all over the country, brought her into contact with many women's rights abuses, and also with the issues, and specific abuses, faced by women with albinism. She is now a strong advocate not only for the rights of people with albinism but also for the rights of all women. She is particularly concerned that women with albinism, and all disabled women, be treated with respect and dignity, as, in fact, should all women.

Because of her verbal abilities, natural confidence, and willingness to speak up in public, Nomasonto was asked to represent the Gender Commission in March 2016 and 2017 (as the champion of gender issues), at the 60th and 61st Annual Global Conferences on the topic, at the United Nations in New York. This year (2017) she was promoting and spearheading the demand for equal pay for equal work value, for disabled as well as all women.

Nomasonto, following in her mother's footsteps, has always had a strong religious faith and long-term affiliation to the Anglican Church. The church embraced the family many years ago and assisted them in various ways, starting in her mother's time. Nomasonto has been active in the Mothers' Union and was the member responsible for enrolling new members for some time. A video entitled "People of the Way," which included Nomasonto's family story, was commissioned by Bishop Brian Germond and widely screened, as an educational exercise, in Anglican churches and elsewhere.

V. MARRIAGE AND CHILDREN

In 1975 Nomasonto married Benedict Mazibuko, a Zulu man who was not related to her and not affected with albinism. She had known him since high school, where he was a classmate, and said he was a kind and considerate person, who respected her and her rights. They had shared many experiences, went to the same youth club, socialized together, and knew each other well. She stated that he had seen her grow up and he had thought out his decision to propose marriage to her very carefully. They have both worked for most of their lives together, he as an electrician in his own business and she as a teacher. They have now been married for 41 years, living in the same house, since 1977, in Soweto, and have had a long and happy life. According to their second child, Senamile, her father always affectionately refers to her mother as "sthandwa sami" meaning "my love." Altogether, they have four unaffected children and, of all her siblings, Nomasonto has the most children.

The oldest child is a girl, Nompumelelo (Mpumi), born in 1975, who, after finishing school, qualified with a hospitality diploma. She then went to work at the City Lodge Group of hotels, and later worked for a bank, in the car loans department, for 3 years. She is now Nomasonto's personal assistant and runs the ASSA office in the Johannesburg city center. Mpumi has two daughters, aged 11 and 6 years, and lives in her own house. The second child is another girl, Senamile (meaning happiness), born in 1977. She has a degree in Marketing, is an entrepreneur and the Director of her own business, she is unmarried, and has her own home. The third child is a boy called Mandla, born in 1980. He has a qualification from Vega Advertising College and is the Art Director of his own business. The last child is a girl, Noluthando, born in 1985, who is a qualified teacher (University of South Africa) and sports scientist with a degree from the Tshwane University of Technology. She is teaching at a private primary school and is married to an accountant. Nomasonto is very pleased that all her children, as is the case with most of her siblings, have tertiary qualifications and are independent.

According to Senamile, Nomasonto was a very loving, warm, and protective mother, but she was sometimes somewhat overprotective, limiting contact with others and making social interaction difficult. She always had plenty of energy, was versatile and assertive, and pushed her way through problems, where necessary. She insisted that all her children wore hats all the time and used zinc ointment regularly on their skin (and, as a child, Senamile was nervous and worried that she might slowly turn white and be like her mother). Nomasonto sent them all to private schools, so that they had a good education; she gave them a solid grounding and was very disciplined, but always encouraging. While growing up, and even now, Senamile prewarns her friends who visit the house educating and informing them of her mother's condition. This is to prevent the discomfort, discrimination, and typical stereotype behavior often directed at people with albinism. As Senamile says "we have met lots of people who react inappropriately to people with albinism." Furthermore, she finds she is continuously educating acquaintances about albinism. She spends a lot of her time at home with her mom and dad. To Senamile, her mother is "more than just a mum, she is a good friend too."

VI. TALKING ABOUT GENETICS

Nomasonto has discussed the genetics of albinism with her children, family, nieces, and nephews. She says that she would love to have a grandchild with albinism because she would like someone in her nuclear family to look like her. "People say the second granddaughter looks like me, but she does not have albinism so it is not really true!… It is a nice feeling to have someone in the family that really looks like you." She has told her children to tell their prospective partners that their mother has albinism, as she does not want any possible partner not to be alerted to that fact and to be taken by surprise when they meet her or they have offspring with albinism.

Regarding being a carrier for an albinism gene mutation, Senamile says it does bother her. There is still discrimination and stigmatization in society that makes life difficult for people with albinism and, therefore, she would be worried about having an affected child. She added that some of her long-term friends, although they know about Nomasonto, have hidden the fact that one of their relatives also has albinism; they still seem to be ashamed and embarrassed about admitting to such a family history. Therefore, if she was at high risk and had a high-risk pregnancy Senamile stated that she would want prenatal diagnostic testing, but if the infant in the uterus was found to be affected she would not request termination (partly because of her strong religious beliefs). Senamile added "such a child with albinism would be better in our family than anywhere else, we have the genetic history and we would be more receptive than a family who knows nothing about albinism."

Because carrier testing is now available for albinism, Nomasonto will be passing this information on to her children and their partners. She will also be informing ASSA members at one of their next monthly meetings. Nomasonto commented that many tests are available for Down syndrome these days, even tests during pregnancy, and she felt research on albinism was lagging behind. This observation led to a discussion on prenatal diagnosis for albinism and she mentioned that she is totally against termination of pregnancy for albinism in the fetus. She added "every unborn child with albinism deserves a fair chance in life. People with albinism are of normal intelligence, generally well accepted, and capable of doing anything." Nevertheless, she stated that women should know what options they have and be given choices.

VII. STARTING THE ALBINISM GROUP

Albinism Society of Soweto: In 1994 Nomasonto founded the first Albinism group in her family home in Meadowlands. She adds, initially "we ran it out of our own pockets," but "my family and the members of our church were very supportive." In the early days Nomasonto named the group the Albinism Society of Soweto and she had it registered with the Department of Social Development as a nonprofit organization. She appointed a Board, consisting of interested professionals, which included a lawyer, a doctor, a teacher, a psychologist, and others. Motsamai Makume, Joel Dikgole, Dr. Peter Ngatane, Dr. C Rataimane, and Joyce Buku were some of the members. Sometime later she was offered office space by Anchor Life, an insurance company, in their building in Plein Street, central Johannesburg. This offer was the result of a radio program run by a well-known radio personality, Shado Twala, who had asked her listeners for pledges to assist in the support of the albinism group.

Albinism Society of South Africa (ASSA): After the group was formally registered in 1995 it grew fast and it was decided that it should be expanded beyond the boundaries of Soweto. It then spread nationally and became known as the ASSA. In 1998, funding was coming in and a manager was required to run the Society. At that time, Nomasonto's younger brother Tony, who also had albinism and was a trained teacher, was unemployed and he was appointed as group manager. He was studying for his MBA, then changed to focus on marketing and later gained his marketing diploma. A new board of seven members, four of whom had albinism and three were unaffected, was then appointed to run the society. Tony managed the organization for about 7 years and during that time he was also the Chairman of the Southern African Inherited Disorders Association, which consisted of about 25 genetic support groups, for a couple of years. One of his projects was an annual Essay Competition, which was open to all high-school students and ran for a few years, raising awareness of albinism at this level. Tony developed epilepsy at the age of 40 years and then diabetes as he grew older and he died of complications from these conditions, aged 58 years, in 2010.

Leadership activities: Nomasonto took over the leadership of ASSA after Tony's death. She has been actively involved since her retirement from full-time work in 2010 and is presently the Director. Her personal assistant and office manager is her oldest daughter Mpumi. The Society is very busy dealing with calls from the public and tackling the discriminatory issues faced by many people with albinism. She described and discussed the latest case, in which a woman with albinism was murdered and buried in a shallow grave in Ecabazini, KZN. In another case a mother, living in Northern KZN near the Mozambique border, tried to sell her three affected children for R100,000.00 (about US$8000) each to a traditional healer from Mozambique. The healer could not pay the required amount, became angry (because the price was so high), and went to the police to report the case. The woman was arrested and the court case was heard in Mangozi, KZN, in July 2016. The same year ASSA also organized a march outside the University of the Witwatersrand, Johannesburg, in support of the drive to stop the killings of people with albinism by traditional healers, who wanted body parts to make "muti" (medicine). Furthermore, in February 2017, the President of the Republic of South Africa, Jacob Zuma invited Nomasonto Mazibuko, as a special guest, to Parliament for the State of the Nation address. He acknowledged her commitment to advocacy for people with albinism and to the UN Declaration and he condemned the killing of innocent people with the condition.

Recently, a meeting with traditional healers in Soweto was arranged by Nomasonto for discussions regarding the use of body parts from people with albinism for medicine and to explain to them that there is no truth in the belief that such medicine has any power to bring success or good luck at all. She added that all riches come from hard work, so why should the body tissues (such as the skin or hair) of people with albinism have any power? (She commented that greed has increased and this is the main problem!). She says she was well received by the healers and will be repeating the exercise in the province of KZN. She has also linked up with the government Department of Culture, Religion and Traditional Healers and has their support. Nomasonto has been discussing the widely held myth that affected people do not die like other people do, with the department officials and groups of healers. In addition, she has developed strong relations with the National Office of the South African Police Services, this is a strategic relationship to create awareness and ensure protection and considerate security for people with albinism.

Nomasonto has attended the congress of the UK Albinism Group in Glasgow, as well as various meetings in Africa, such as the Tanzanian meeting in June 2016. In 2013 the United Nations declared albinism to be a disability (mainly because of the associated visual impairment) and this drew worldwide attention to the condition. Nomasonto was very pleased about this development; she states that being classified in the Disability group brings a sense of belonging to affected people. She adds that it is no longer necessary "for people with albinism to dye their hair, in an attempt to be black," and, now, affected people should be able to gain self-confidence from their community and from their own homes. She also commented that the fashion, at present, is to have fair hair and so people with albinism have an asset from which they can benefit! The UN Declaration also demands support from the government, especially the Departments of Social Development and Health. The creating of facilities for eye and skin care for people with albinism is long overdue.

Many new ASSA programs have been initiated by Nomasonto. One is supported by the South African government Department of Women, Children and People with Disabilities and involves the distribution of dignity packs to all affected girls. Another has resulted from her visits to the offices of the newspapers and her requests for large-print newspapers. She says she had succeeded with City Press and a large-print newspaper was delivered to her house regularly, but, because of editorial management changes, this delivery has ceased and she now has to return to renew her request. She has also lobbied the Education Department to print large-print school textbooks. Furthermore, Bibles are now available in large print and she has one since she cannot comfortably read any print that is less than 18 pt. She also has an appointment with the Director of South African Statistics as she wants people with albinism to be counted in the next census (this was done for the first time recently, in Namibia and Tanzania), so that an official figure, for how many people with albinism live in South Africa, can be obtained.

ASSA now has branches in all the South African provinces except the Western Cape. The KZN and Gauteng branches are particularly active. September is "Albinism Month" and so focused activities are arranged all around the country. On September 24, 2016, there was a big occasion at Ghandi square in Johannesburg, supported by Nando's, a popular local restaurant. There was much publicity and an article on albinism appeared in a major daily newspaper, with a wide circulation. This article was entitled "Dispelling myths and encouraging albino role models" (The Star, Monday September 19, 2016, p. 5). Nomasonto was interviewed for the article, as was a young person with albinism, Thando Hopa (who has been mentored by Nomasonto for some time), who is now both a qualified lawyer and a popular model in the local fashion industry. Thando stated that when she was approached to do modeling by a well-known dress designer, "I wasn't really interested but I looked at my own childhood and I didn't have role models that looked like me that I could identify with, so I thought it was a good opportunity to become a model and inspire others like me." When she was a child, Thando reported, her parents were wonderful and always embraced and supported her, so much so that it was only on her first day at school when she started noticing she was different, "because children wouldn't sit next to me, and after some time came the name calling... I just had to continuously go through that dynamic of being absolutely loved and absolutely not accepted."

Another striking article appeared on the front page of the same popular newspaper earlier in the month. It was entitled "Langenhoven strikes gold – for every albino" (The Star, September 12, 2016, p. 1). Hilton Langenhoven is a person with albinism and an excellent athlete, who was a member of the South African team for the Paralympic games in Rio de

Janeiro in 2016. He won gold in the long jump on September 10, 2016. He said that through his winning of medals (he also won three gold medals in 2008 in Beijing in the 200m, long jump, and pentathlon), as a "vision-impaired individual who has done the nation proud… perception is changing regarding disabilities. It's absolutely fantastic if we can contribute to that going forward."

At ASSA group meetings Nomasonto tries to sensitize people; she encourages them not to hide their affected children but to "put them in the public domain," and, for example, to ask other children with albinism to their child's birthday parties. Also, she feels that there is a need for further research projects so that knowledge on the condition can be increased. Funding must be found both to strengthen the Society's work and to support research projects (see Figs. 14.3 and 14.4).

Regarding research, Nomasonto is very supportive and states that it has benefited people with albinism greatly in the past. Particularly when antiactinic skin cream was provided to research subjects as was the case previously, over several decades. She will encourage her members to participate in any future projects that might be envisaged. A meeting will be held to discuss such projects, and the part she and the ASSA members can play in them, in due course.

Pan African Albinism Alliance: According to Nomasonto there was an attempt to start a Pan African Albinism Alliance in 2015, but this was not registered at that time. However, it is hoped that registration in Kenya will soon be successful and Nomasonto has been approached to be the first regional president. At present 16 countries are registered with the Alliance.

FIGURE 14.3   Nomasonto and Jennifer at a research meeting in 2017 with members of the Albinism Society. *Photograph courtesy Dr. J. Kromberg, University of the Witwatersrand, South Africa.*

FIGURE 14.4   Nomasonto's older sister Sylvia (79) at the research meeting in 2017. *Photograph courtesy Dr. J. Kromberg, University of the Witwatersrand, South Africa.*

The United Nations special representative on albinism, in Geneva, Ikponwosa Ero (originally from Nigeria), is trying to unite African societies, to promote networking, and to encourage them all to join the Alliance.

VIII. THE FUTURE

When one asks Nomasonto about the future she becomes very enthusiastic. She says she is amazed at the development of the South African Albinism Association, and that it has gained African and international status. She is also delighted that she can give hope and peace to new members, provide them with mentorship, and help them to feel that "albinism is a manageable condition." She observes that most of her members have developed new insights into their condition and can hold their heads up high.

Her dream for the future is to provide every affected person with free sunscreen and skin care, eye care and spectacles, and elementary toys for the stimulation of the young children. She would also like to establish an albinism clinic (or a "one-stop shop"), where families could attend regularly to obtain all their assessments and treatment in one place.

Regarding research, she would like to initiate a study on a group of children with albinism who could be followed over about 20 years in a long-term project. Their physical and social development could be tracked, so that the researchers could learn what happens during the growth of children with albinism, and develop interventions to manage any difficult experiences.

The future, therefore, holds many unknowns, but so many wonderful experiences have happened and so many new experiences can happen, that it should be both challenging and rewarding and she is looking forward to it with great enthusiasm.

IX. CONCLUSION

The life history of Nomasonto Mazibuko (nee Ngwenya) is informative and worth the telling. It is somewhat different from that of Alex Munyere (2004) who was born in Kenya into a family with no history of albinism, was somewhat neglected, and had a difficult childhood. However, they both are determined and capable individuals who made the best of their experiences and opportunities and became respected professionals leading independent and successful lives. Both are passionately committed to the cause of disability, albinism, and the development of an inclusive society.

Although being born into a protective and caring family, Nomasonto later had to face society on her own. Coping with schooling may have been fairly easy because of her family support, but later she had to go away to Teachers' Training College and deal, continuously, with the looks and questions of curious people. She was fortunate that she was assertive and had an outgoing personality, so she could manage these somewhat aggravating circumstances. She also had a functional network and found employment as a teacher, which stood her in good stead for many years. In addition, she had the active support of her church and her mother's connections.

Nomasonto made a sound marriage with a supportive and caring husband. Together, they produced and raised four healthy children, providing them with a solid education, constant encouragement, and a disciplined family home. They all benefited from tertiary education and became independent people.

After her teaching career came to an end, Nomasonto took a management diploma at the university and then, with some seed funding, she started the ASSA. This is a very dynamic and proactive organization, providing services to people with albinism, spreading awareness about the condition in the community, and tackling the cultural beliefs that are detrimental to the well-being of affected people. For example, Nomasonto has spoken at meetings of traditional healers about the fact that the body parts of people with albinism have no medicinal value at all.

The future holds exciting prospects for Nomasonto. She is both surprised and delighted at the continuing success of the Albinism Society, its worthwhile projects, and good national and international reputation. She is involved in the Pan African Albinism Alliance, which she sees as a dynamic development, with networking potential across the continent of Africa, and she feels that awareness is increasing and the field is opening up for further worthwhile research on the condition. She can now give hope and a positive outlook to families with a newborn child with albinism and to affected people themselves, throughout the country.

Reference

Munyere, A., 2004. Living with a disability that others do not understand. Br. J. Spec. Educ. 31 (1), 31–32.

Summary and Conclusion

Jennifer G.R. Kromberg[1], Prashiela Manga[2,3]

[1]University of the Witwatersrand and National Health Laboratory Service, Johannesburg, South Africa; [2]New York University School of Medicine, New York, NY, United States; [3]University of the Witwatersrand, Johannesburg, South Africa

I. INTRODUCTION

This book aimed at comprehensively covering topics relevant to albinism in Africa from the first-recorded reports of the condition several thousand years ago to the latest scientific developments. The field has developed from mere descriptions of unusual people seen in the community to the examination of minute sections of their DNA. Some topics have been covered in these pages at greater depths than others.

In this last chapter the key points raised on each topic and discussed at length in each chapter will be summarized briefly for the reader. Furthermore, we have identified potential gaps in our knowledge and propose possible future research, ranging across the psychosocial, medical, biological, and molecular fields. Finally, possible future developments in the field are covered.

II. GENERAL SUMMARY

The history of the condition is rich with descriptions of persons with albinism observed all over the world (Pearson et al., 1913). Probably one of the most famous is that obtained from the extra-Biblical book, in which Noah is described as having white skin, white hair, and pale eyes that shone like the rays of the sun (Sorsby, 1958). Such coloring would have been very unusual for people living in the Middle East and therefore all the more remarkable. The fact that his parents were first cousins adds support for the story because albinism is a recessive disorder.

European travelers in Africa (mainly in North Africa initially, but later in North, Central, and sub-Saharan Africa) returned with stories of the darkly pigmented people of Africa and the occasional one among them who was not pigmented but had fair skin, hair, and pale eyes (Pearson et al., 1913). Some explorers recorded the circumstances under which they found these unusual people and described the way that they were treated and the beliefs that surrounded the cause of their condition. Further reports described the fact that the skin was so sensitive that it was susceptible to skin cancer and that vision was poor. A Portuguese explorer then named people of this description as "Albini" (or white), and although local groups have their own names for the condition, this name has been retained through the centuries, as albino, and it is now used internationally, although the term "people with albinism" is generally preferred.

From reports received in the last few centuries it became clear that life was not easy for people with albinism. Not only did they have skin and eye problems but also their psychosocial circumstances were often difficult. Incidents of stigmatization, rejection, and even infanticide began to appear (Livingstone, 1857). Occasional examples of the special care provided for affected people by the kings of a few Central Africa ethnic groups were described. Sometimes they were believed to have special powers and chosen as traditional healers, but mostly they were considered to be handicapped or disabled and in need of care (Pearson et al., 1913).

Researchers, often not only anthropologists but also scientists, initiated a few small and large-scale studies on albinism in Africa in the 1800s and 1900s. The largest collection of data was compiled by Pearson et al. (1913) working in London. They also collected as many old and new reports as they could from all over the world and corresponded with medical doctors, missionaries, and others working in Africa and elsewhere, asking them to provide information on people with the condition in their area. This drive resulted in the publication, between 1911 and 1913, of a classic 12 volume monograph. Some of the reports went back to the time of Pliny and one of his reports on white-skinned people from Africa appears in Latin. The monograph included pedigrees, maps of where affected people had been observed, relevant descriptions from many of the explorers, and more detailed clinical reports from the medical doctors. Several reports recognized and described the different types of albinism in which the pigmentation varied; these clinical observations were later confirmed by researchers such as Loewenthal (1944).

Contemporaneously, scientists in England were investigating the cause of albinism and concluding that it was a recessive trait (Garrod, 1908); thus the mutant gene carried in single dose showed no effect but in double dose caused a lack of pigmentation throughout the body, so that the skin, hair, and eyes were affected. Later, it was discovered that if two people with albinism had children they were usually all affected, but occasionally they were all unaffected, thus it was concluded that mutations in more than one gene could cause albinism.

In the mid-20th century a dentist, who dedicated his life to studying albinism in the Americas (and carried out a couple of projects in Nigeria), Professor Carl Witkop, conducted many research projects on a wide range of aspects of the condition (Witkop et al., 1983). He assessed prevalence in various communities, tried to differentiate between the types, and provided detailed information on the Hermansky–Pudlak syndrome (albinism associated with a bleeding disorder). His team continued the work after he retired and produced much unique data on the clinical characteristics and molecular biology of albinism.

Still later, the role of melanin was investigated and it began to be suggested that not only did the disruption of melanin synthesis cause people with albinism to have depigmented skin, with no protection from the ultraviolet rays of the sun, but also it could affect the development of the optic tract in the embryo (Guillery, 1996). This resulted in the abnormal crossing over of the optic fibers to the left and right-hand sides of the brain, causing basic visual defects, myopia, and poor binocular and 3D vision.

Finally, in the 1980s and 1990s, molecular biologists studying DNA from people with albinism discovered, firstly, genetic mutations causing albinism in many of those with oculocutaneous albinism (OCA) type 1, followed by a mutation associated with the condition in most cases of OCA2, the common type found in Africa, and subsequently two pathogenic mutations causing OCA3. Since then many mutated pigment-related genes (about 21 to date, see Montoliu et al., 2013) have been found to be responsible for causing albinism and the associated syndromes in different populations. The field is therefore becoming very complex, however, in Africa particularly sub-Saharan Africa the commonest type is OCA2 and the commonest associated mutation is a 2.7 kb deletion on chromosome 15q11-12.

The clinical characteristics of African people with albinism have been described by researchers working in Nigeria, South Africa, Cameroon, Tanzania, Zimbabwe, and elsewhere. The skin is known to be susceptible to cancer, particularly squamous cell and basal cell carcinomas; however, this can generally be prevented by the use of antiactinic creams from a very young age and throughout life, by reducing sun exposure to a minimum, and by wearing long-sleeved protective cotton clothing and broad-brimmed sun hats. Visual defects include nystagmus, which improves with age, occasionally strabismus, poor visual acuity, which usually benefits from lensed spectacles, and photophobia for which dark glasses are recommended. Intelligence is known to be within the normal range. Fertility may be reduced not because of physiological causes, but due to the stigma attached to albinism and the limited opportunities for finding a suitable partner (Kromberg, 1985). Life expectancy might be reduced if the skin is not protected, if solar keratoses develop early in life, and if lesions are not properly treated or ignored and treated too late.

The prevalence of albinism has been studied in various countries. The rates in Africa range from 1 in 1000 in a few small isolates to 1 in 9000, and average at about 1 in 4000–5000 (Hong et al., 2006). This rate is about three times higher than that found in European countries where only about 1 in 16,000 people have albinism. However, in some isolates in Central America, in small communities, rates have reached as high as 1 in 200 (Woolf and Grant, 1962). Prevalence data from Asian countries are scarce, but an estimated minimum rate of about 1 in 18,000 is suggested in one area of China. Rates are influenced by the number of consanguineous unions occurring in a community (and cultural traditions determining selection of marriage partners), carrier frequency in the population, some as yet unconfirmed and/or unknown selective advantage of carriers (Kromberg, 1987), and, previously, by rates of infanticide.

Detailed prevalence studies are presently proceeding in several countries, including Namibia, Tanzania, and Zimbabwe. In the former two countries, the national census has included a question about having albinism and the information obtained is providing more accurate prevalence estimates. Also, the information collected some years ago in Zimbabwe is being reanalyzed and new findings on how common the condition is there are becoming available.

Molecular biologists are investigating the biology and biochemistry of pigmentation and the spectrum of mutations causing the condition in different parts of the world. In particular, the role of the proteins encoded by albinism-related genes is being studied and the ways in which they interact and affect biological functions are being identified. This will be particularly important in understanding the impact of mutations on phenotype, for example, if individuals carry mutations at more than one locus. Pigment-related genes have also been implicated in determining risk of actinic keratosis (Jacobs et al., 2015) and cutaneous squamous cell carcinoma (Asgari et al., 2016), independent of their role in skin pigmentation.

The dermatological aspects of albinism have been studied by many clinicians and medical specialists and scientists over the years (e.g., Luande et al., 1985). The serious damage caused by sun exposure on the unprotected skin of people with albinism has long been recognized. The skin damage results in development of keratoses and skin cancer if untreated. Squamous and basal cell carcinomas are common among people with albinism, but melanomas are rare. In some tropical areas of Africa, such as Nigeria, 100% of affected people had premalignant or malignant lesions by the time they were 20 years of age, resulting in a shortened life span (Okoro, 1975). These cancers usually developed on the exposed areas of the body and commonly on the face, neck, and head. Treatment was often not available or inaccessible. Furthermore, lack of education could lead to affected individuals and their families ignoring the signs of skin damage and delaying presentation at a clinic until the condition became very serious, when it was too late for treatment to be effective. Families and their affected members need to understand that treatment is most successful in the early stages of the skin damage. They also need to know that people with albinism should use sun barrier creams as soon as they start going out in the sun, as toddlers, and for the rest of their lives, to prevent the development of skin cancers.

People with albinism have visual disabilities and reduced visual acuity; these can vary greatly from one affected individual to another and even in affected siblings, however, all should have a visual assessment prior to going to school. Because melanin synthesis is required for the proper development of the embryonic optic tract and people with albinism cannot make sufficient melanin, their vision is affected. The primary defect is misrouting of the optic tract, which should run from each eye to both sides of the brain. Instead, in albinism, a disproportionate number of fibers are routed to each side. This abnormal decussation results in the lack of binocular vision and poor visual acuity. Low-vision rehabilitation is available (e.g., Kammer and Grant, 2014), and an informed optometrist should be consulted. Visual aids and spectacles can assist the student in the classroom and the adult in later life, although many with a more severe visual disability may never be able to see well enough to drive a motor car. An example of a low-vision rehabilitation program in Tanzania shows what can be achieved in this field of endeavor.

Psychosocial aspects of albinism have been systematically explored in a few studies. The birth of an infant with albinism has been shown to result in shock, distress, and withdrawal in many of the parents and in reduced maternal-infant interaction in the first few months of life (Kromberg et al., 1987). However, most mothers learn to accept the situation, adapt to the unexpected child, and care appropriately for his/her needs. Adjustment is difficult for the family and especially the affected person because of both the health problems and the associated stigma. The development of social identity and self-image may also be problematic in this environment. However, intelligence has been shown to be in the normal range and, with support from their family, a little help from their teachers, and reasonable educational opportunities, affected children should be able to reach their potential. Nevertheless, quality of life is affected in most cases because of community attitudes and ignorance, poor and inaccessible health services, and the African climate, which make good health difficult to maintain in affected people.

Cultural beliefs can be life threatening in areas where traditional healers have promoted the mistaken idea that tissue, bone, and skin from affected people can make powerful medicine, which, they say, can bring good luck to those who consume it. Another myth with significant negative impact is the widely held belief that people with albinism do not die naturally as others do (Kromberg, 1992). Both these myths need to be counteracted and debunked so that affected people can be accepted as human beings and can live their lives without unnecessary interference.

For people with albinism and their relatives to understand the genetics of the condition and the way in which it is inherited, genetic counseling is recommended (Kromberg and Jenkins, 1984). Such counseling can inform affected families of the nature of the disorder, the diagnosis and prognosis, the risks of recurrence, options for dealing with these risks, and how best to manage and cope with the medical and psychosocial complications of the condition. Risks vary from 1 in 2 to 1 in 3600 depending on the family history. Genetic counselors can listen to the family's story, assess their situation, provide the necessary information, and offer empathic support. Although not available in many African countries, the genetic counseling profession is growing slowly and uptake of the service is increasing in South Africa (Kromberg and Krause, 2013) as well as in Cameroon (Wonkam et al., 2011).

In Africa, at present, genetic testing for albinism is only available in South Africa. Such testing is used to confirm a diagnosis (although this is rare); identify the pathogenic mutation; test at-risk people to determine if they are carriers for albinism; and occasionally test an at-risk pregnancy (where both parents are carriers) to establish if the fetus is affected. This latter test is requested by some at-risk couples who seek termination of pregnancy if there is an affected fetus and by others who want to know the status of the fetus, to be prepared for the birth. However, it is important that prenatal testing be preceded by genetic counseling so that couples are aware of their choices and the implications and consequences of their request, thus allowing them to make informed decisions.

Poor education among the general public and a lack of respect from unaffected people lead to the many marginalization and human rights issues that individuals with albinism have to face. However, since albinism has been declared a disability by the World Health Organization (WHO) and the United Nations (in 2006), these issues are being recognized and tackled more realistically. Several nongovernmental organizations are now working in countries where the prevalence is highest and having some success (such as reducing the rates of skin cancer and the effects of the visual defects in Tanzania). They are providing efficient services to people with albinism and organizing effective public advocacy and information programs, thereby improving the level of knowledge in the community, reducing the stigmatization based on ignorance, and increasing the quality of life of affected people.

The main key to living a healthy life with albinism is preventing the complications of the condition and limiting the concomitant risk factors. For the child to develop normally, psychologically, the mothers (and fathers, if possible) should have counseling soon after the birth to ameliorate the poor maternal-infant interaction which might occur. They should also have genetic counseling so that they are fully informed about the condition. Then, to prevent possible delayed achievement of developmental milestones due to low vision, the child should have a stimulation program that can be undertaken by the informed mother in the home. Skin and visual assessments should be sought, so that the caregiver knows how to prevent sun damage and becomes aware of how to assist in the case of low vision. Health professionals working with the families should be aware that they often require empowerment strategies so that they advocate on their own behalf rather than relying on others. Support services, such as albinism support groups, can assist in this regard. Albinism associations, which have been active in Europe and America for many years (Hall, 2013), are developing all over Africa, to the benefit of affected people and their families. Also, international associations, such as the World Albinism Alliance, are increasing awareness programs internationally, so that the general public is becoming better informed on the condition.

A personal perspective on albinism has been written (in collaboration with JK) by the Director of the Albinism Society of South Africa, who has lived with her condition for the past 66 years. She describes how her family supported her, made sure she cared for her skin and avoided sun exposure, used the visual aids she needed, and received a good education. Then she outlines how she managed the mating game, found a good partner, established a continuing long and stable marriage, and had four unaffected children. Later, she set up the albinism society (securing a committee that included health professionals, teachers, and lawyers), provided support for needy and bewildered affected individuals, networked with the right influential people, and gained services for the society members. She now fights stigma and discrimination wherever she sees it, acts as an advocate on behalf of her members, runs successful public awareness programs, gives radio and newspaper interviews, encourages and participates in the development and organization of research programs, and communicates with like-minded people across Africa and internationally.

III. RESEARCH AND ALBINISM

Research on Rare Diseases and Implications for Research on Albinism

Research on albinism is given relatively low priority in many countries, especially in those where the prevalence of the condition is low. However, research on so-called rare diseases, among which albinism may be counted, is drawing more attention and warrants some discussion because it will impact on the possibilities for increasing research on albinism in Africa.

An article in the Lancet in 2008 tackled the question of "Why rare diseases are an important medical and social issue" (Schieppati et al., 2008). Rare diseases are those that occur in no more than 1 in 2000 people, but about 5000 disorders fit into this category (according to WHO) so that millions of people are affected (about 30 million in Europe alone). Together they are an important public health issue and a challenge to the health and medical professions. Most rare diseases are genetic disorders, which often impact physical and mental abilities as well as life expectancy, may be very disabling, and result in a poor quality of life for the affected individual. As has been shown to be the case with albinism (see Chapter 12), patients with rare diseases often have poor experiences regarding social and economic opportunities and health care. Furthermore, there is little information on their condition, limited availability of and accessibility to appropriate treatments, and difficulties in funding the necessary research. As yet, most rare diseases have no cure, so that the "unmet needs of patients with rare diseases offer new avenues for investment" (Schieppati et al., 2008, p. 2040). Due to the work of advocacy groups (such as the National Organization for Rare Disorders in the United States, see http://www.rarediseases.org), the important issues around these somewhat neglected rare diseases are being brought to the attention of the public and financial support for research programs have increased. Additionally, albinism is common among underrepresented minorities in the United States where a concerted effort is being made to address healthcare inequalities.

Although the "life costs" of living with albinism have not been calculated, there is some information available on the costs of other rare diseases (Mikami et al., 2016). Because of the advances in knowledge about genetic diseases and the growing awareness of such conditions, the number of affected patients and their caregivers is estimated to be much greater than previously expected. The European Union recognized these disorders as a public health issue three decades ago and has been financing relevant projects. Equal access to health care has been a major challenge for patients with rare diseases, as it is for people with albinism. However, this situation seems to be improving, partly because of the activities of advocacy groups, the accessibility of information on the internet, and the progress being made in medical genetics. Mikami et al. (2016) considered two cases (Huntington's disease and phenylketonuria) and applied a cost-benefit analysis to them. They then recognized that rare diseases not only affect the patient but also reduce the quality of life in the caregivers and significant others. Furthermore, they stated that having the same rare disorder as another patient does not mean that the same life costs apply because there is a great diversity between patients and within families. They concluded that their research shows that the cost of living life with a rare disorder can be high and they advocate that more interdisciplinary exchange and research is required.

Impact of Genetic Support Providers and Advocacy Groups on Research

A professional group that has the opportunity to stimulate research on albinism and other rare disorders is that composed of clinical geneticists and genetic counselors. These professionals provide genetic services and counseling and often initiate research into specific disorders. Many of them also have a role in local genetic support groups, which is usually a voluntary one, to help affected families and patients, but the health professional can often benefit too (Lin et al., 2003). Participating geneticists admit that they have a unique chance to meet many families with a disorder and broaden their outlook on various aspects of the condition. Also group meetings may present an opportunity for research, providing the professionals with practical access to research participants, who can often benefit immensely from research partnerships. However, a good relationship between the health professional and group requires transparency and mutual understanding (Terry and Boyd, 2001). Guidelines for health care professionals, who participate in Genetic Support Groups, based on their respect for patient confidentiality, have been formulated by the Genetic Alliance in the United States (Lin et al., 2003), and groups could benefit by using these and/or adapting them to meet their needs.

To promote research, individuals with genetic conditions, lay advocacy genetics groups, and researchers have been building partnerships (Terry and Boyd, 2001). Such partnerships can lead to productive interactions that are of benefit to the patients, the group, and the research investigator. The disorder advocacy group can play a central role as a coordinator in collaborative research endeavors focused on the biology, epidemiology, and genetics of the condition. Some of the elements of this specific partnership (which revolved around researching the biology of pseudoxanthoma elasticum, an inherited disorder affecting the skin, retina, and arteries) may be useful to other groups who wish to build lay/professional collaborations (Terry and Boyd, 2001).

Genetic disease advocacy groups can also participate in clinical research on their condition (Landy et al., 2012). A survey of 124 organizations in the United States focused on how such organizations participated and perceived their contributions to research programs. The results showed that they participated in recruitment (91%), collected data (75%), provided financial support (60%), and assisted with study design (56%). Most felt (68%) that their participation had contributed to an increase in research on their condition and that researchers should consult their local group to recruit participants (58%) and select research topics (56%). Although many advocacy groups are in their infancy in Africa, it is important that they are similarly included in clinical research efforts. In Africa, where colonial history has left a legacy of distrust, advocacy groups could foster alliances between researchers and affected individuals. They could also ensure that there is protection of the interests of the participants.

Researchers may assume that all the subjects participating in a research project are doing so for altruistic reasons, but in fact there are many interests that motivate participation in research enterprises (Merz et al., 2002). The commercialization of research findings has given rise to many ethical challenges. All participants in any research process should be given a voice when decisions regarding ownership, control of research results, and sharing of any benefits are to be made. It is advisable that researchers should negotiate these issues before a project is initiated.

One of the objectives of the World Albinism Alliance (in which many African organizations participate), regarding scientific research, is to coordinate "efforts to ensure that research opportunities to learn more about albinism and its management are fully explored internationally" (WAA, Constitution, 2011). Hopefully, this development will lead to the stimulation of more research projects and the publicizing of research findings widely and timeously.

Because genetic factors have been implicated in common health problems in developing nations (Kumar, 2016) and the WHO has acknowledged that even in developing countries the provision of genetic services is cost-effective (Kromberg and Krause, 2013), there is a huge potential for growth in the genetics education and research fields (Kumar, 2016). Recently, the Afro-UK Genetic Education Forum was set up in collaboration with the African Society of Human Genetics (www.afshg.org). Although it is necessary to prioritize the limited resources in developing countries and to focus on the common socioeconomic and health problems (such as those resulting from poverty, malnutrition, unsafe water, and infectious diseases), it is important to invest in genetics to prepare for future development (Kumar, 2016).

Molecular biology and genetic research has broadened our knowledge of cellular processes in health and disease, allowing new pathways to be targeted for development of diagnostic tools and therapeutic interventions, which will lead to the reorientation of the relationship between human biology and medicine in the future (Clarke, 2016). However, it will take some time for these advances to be of practical use in medicine and especially in rare disorders. Psychosocial research needs to keep in step with these new developments, so that the effects on the patients involved can be understood and managed and their quality of life appropriately improved.

IV. FUTURE RESEARCH TOPICS

Suggestions for research arise from many sources: people with albinism themselves and members of support groups, findings from research projects, which expose deficits in the understanding of the condition, researchers who explore the literature on the topic and find many unanswered questions, authors, such as those who contributed to this book, who raise unexplored issues, and workshops where issues are discussed and unanswerable questions arise. In general, it is widely acknowledged that more academic research is required on many different aspects of albinism. The few suggestions for future projects mentioned below will be divided into three groups, those involving either psychosocial issues, medical and clinical aspects, or molecular genetic studies.

Psychosocial Studies

Albinism in Children

Few studies have covered the aspects of albinism which are specific to affected children. From the psychosocial point of view, research has shown that maternal-infant interaction in the newborn period may be inadequate (Kromberg et al., 1987), but no long-term study has been carried out to show what the effects of poor neonatal bonding on the social and psychological development of the child with albinism might be. Such a study is necessary to inform adequate and relevant psychological counseling for such children. Children with albinism commonly face stigmatization, teasing and bullying due to their condition. However, no studies have investigated these phenomena, or how they affect the child's developing self-concept, or why some children then succumb to the taunting and become withdrawn, while others learn to cope and adapt.

Prenatal Diagnosis

Presently there are few reports of prenatal diagnosis for albinism, and these are from countries other than African countries (apart from a very small South African study by Kromberg et al., 2015). Furthermore, no research has been conducted to determine the opinions of African people at high risk, whether they would request such testing during pregnancy, and how they would act after receiving results. If these services are to be offered in future, such data are necessary.

Quality of Life

Quality of life, in the case of people living with albinism, has not been studied much in Africa (apart from one study carried out in Malawi, Braathen and Ingstad, 2006). If health professionals are to understand the condition and its impact on the affected individual, the quality of their life and the problems they face in daily living must be documented. How the patients themselves view the physical, psychosocial, emotional, social, cognitive, cultural, sexual, and spiritual aspects of their lives needs to be investigated. The findings will assist healthcare professionals to promote realistic adaptation to the condition and to facilitate an improvement in the general well-being and quality of life in affected individuals.

Development of Self-Concept and Social Identity

These psychosocial concepts have not been studied in people with albinism at any depth. Because they are dynamic concepts that can be affected by the environment and interactions with others, they need to be studied, probably by psychologists, in people with albinism living throughout Africa. The diversity of cultures and issues specific to particular African regions and ethnic groups will necessitate multiple studies. The results of such studies would provide valuable insights into the psychology of affected people and how informed psychological counseling could meet their specific needs.

Human Rights and Albinism

Some observations have been made on this topic, but research is required to clarify the actual nature and extent of the human rights abuses faced by people with albinism. Questions arise as to whether it is the right to take part in cultural life, to liberty, and/or to decent education, employment opportunities, good health, life and security, adequate food, and/or even the right to scientific research participation, which is abused in the case of people with albinism, or some or all of these? Further questions arise regarding the nature of the violence against people with albinism and the reasons why the false belief has taken hold that their body parts have magical powers; the expertise and research work of cultural anthropologists are required to investigate this problem and find solutions to putting a stop to this lethal practice. The findings from new research on these topics could provide the relevant insights necessary to direct appropriate action and to inform advocacy programs.

Genetic Counseling Services

The few available genetic counseling services are not used as much as they could be by families with a member with albinism, who could benefit greatly by receiving such counseling. The reasons for this poor uptake need to be investigated. Possibly, referrals are not being properly motivated or made, or the necessity and benefits of such counseling are not being understood by the patients and their healthcare providers (including the traditional healers that they consult), and/or the nurses who are involved in treating affected people have an inadequate understanding of the genetics of albinism. How parents behave and who they consult, after having an infant with albinism, also needs to be investigated, so that the appropriate people can be targeted for relevant education. All these issues need to be clarified, so that action can be taken to remedy the situation.

Medical and Clinical Research Studies

Dermatology Studies

A few studies have been carried out, some years ago, on patients attending skin clinics in various African countries (e.g., Luande et al., 1985). However, it would be worthwhile repeating these studies with larger samples across all age groups and types of albinism and with modern technology, assessment methods (including measuring levels of skin and hair pigmentation and documenting eye color in individuals with different types of albinism), and treatment taken into account. The findings from such investigations would give the dermatologists a better basis for explaining the implications of the diagnosis and the prognosis to the patient.

Visual Defects

What are the eye problems specific to the types of albinism found in Africa and how do they differ from one type to another? How do the interactions between the various albinism-related genes modify the impact of mutations on visual function? Can a therapeutic program be developed that can correct visual defects postnatally, or is a prenatal treatment the only effective option. Studies in mice using nitisinone to promote increased pigment production suggest that postnatal treatment may be effective, but further research is required to study the effects in humans and efficacy in various forms of albinism (Onojafe et al., 2011). These questions need investigating if more appropriate and detailed answers are to be provided to inquiring parents of an affected child.

Birth to Twenty Study

The Director of the South African Albinism Society has suggested that a Birth to Twenty study should be carried out on a series of children with albinism, similar to the longitudinal study on a cohort of children undertaken by Drs. Cameron and Yach at the University of the Witwatersrand (described by Richter et al., 2007). They should be enrolled in the study at birth and investigated on both physical and psychosocial levels at regular intervals until they turn 20 years of age. No such long-term study has been conducted in Africa, so that no information exists on how the children develop and cope over the first 20 years of their lives. Such a study would be very instructive not only to the medical and health personnel and social scientists performing it but also to the members of the Albinism Society who need more information to better understand their condition.

Epidemiological Research: Prevalence of Albinism

There is little available reliable data on the prevalence of albinism and the epidemiological aspects of the condition in many African countries. These data are required if targeted public health services are to be planned and implemented appropriately. The approach of including a question in a national population census may prove to be highly informative.

Development of Therapies

Several questions need to be addressed before a successful drug can be developed. Firstly, can a therapy that simply increases pigmentation in the skin and eyes significantly improve visual acuity without the need for prenatal treatment? Studies in OCA1B would suggest the answer is yes (Onojafe et al., 2011). Additional challenges for drug development include: the need to target multiple organs would require a systemic therapy that increases the risk for side effects; pharmacogenomics studies would be required to assess the population-specific impacts of the drug; and the prevalence of albinism in poor nations and communities would require a cost-effective treatment that can be easily provided in a rural setting where refrigeration and access to skilled healthcare workers may be limited. One factor that increases the chances that a drug will be developed is that several forms of albinism result from a defect in tyrosinase protein maturation and delivery to the melanosome; therefore one agent may be developed to treat several forms of albinism.

The possibility of treatments for genetic disorders by direct intervention on the genome is closer to becoming a reality with the development of novel methodologies such as clustered regularly interspaced short palindromic repeat, which can facilitate specific targeting and correction of genetic defects. The clinical pipeline for development of such a therapy, particularly when it would most likely require intervention at the fetal stage, may be long and challenging. Ethical issues and use of these therapies in third world countries are additional challenges that will need to be addressed.

Molecular Biology Research

Molecular Biology of Pigmentation

Although significant progress has been made with respect to our understanding of skin pigmentation, many questions remain unanswered. For example, what roles do the various albinism-related proteins, such as OCA2 and TYRP1, play in the melanocytes? To date the only albinism-related protein (with the exception of the syndromic forms of albinism) with a well-defined function is tyrosinase; however, regulation of the enzyme (expression, levels in the cell, activity) is highly complex and not completely delineated. For example, a polymorphism in the gene encoding Interferon Regulatory Factor 4 was found to modulate tyrosinase expression (Praetorius et al., 2013), while tyrosinase protein levels in the cell are affected by cholesterol levels (Hall et al., 2004).

In addition to the skin and hair, melanocytes are found in several additional sites in the body; most recently "melanocyte-like" cells have been found to impact risk of atrial arrhythmia (Levin et al., 2009). How mutations in pigment genes affect these cells may identify additional comorbidities in people with albinism.

Answers to these questions will allow us to develop a clear picture of all the factors that impact pigmentation and melanocyte function and contribute to the pathogenesis of albinism.

Oculocutaneous Albinism 2 and Ephelides/Dendritic Freckles

Research has shown that in Africa there are two main subtypes of OCA2: the type (OCA2a) in which pigmented freckles (ephelides) never develop and the type (OCA2ae) in which such freckles develop in sun exposed areas of the skin. Rates of skin cancer are significantly higher in patients with the former type than in those with the latter type (Kromberg et al., 1989). The difference between these two types at the molecular level has not yet been fully explained. In one recent study preliminary findings suggested that when the 2.7 kb deletion was present in homozygous state (84 patients) and when there was one 2.7 kb mutation and one nonsense mutation (e.g., R165X, 3 patients), these dendritic freckles were absent. However, when the 2.7 kb mutation is absent and there are other causative mutations, freckles were observed (Aquaron and Brilliant, 2015). These findings contradict those reported in South Africa where 47 individuals with OCA2 were studied (Stevens et al., 1995). Six individuals with freckles were found to be homozygous for the 2.7 kb deletion and 14 were heterozygous for the deletion. Further research is being planned to clarify this matter.

Identification of Pathogenic Mutations and Improved Genetic Testing for Albinism in Africa

Results of genetic testing in South Africa show that about 78% of patients with OCA2 are homozygous for the common 2.7 kb mutation (Stevens et al., 1995). When the remaining group of patients (63, of whom 40 were heterozygous for the common mutation) were screened only a further 9 mutations were found and no other common mutations were identified (Kerr, 2015; Kerr et al., 2000). Further research is required to elucidate the mutations in about one-fifth of those with albinism in South Africa (see Fig. 15.1). This endeavor will be important if efficient and comprehensive prenatal diagnostics are to be offered.

Oculocutaneous Albinism 2 and Genotype–Phenotype Correlations

No comprehensive genotype–phenotype correlations have been performed in Africa and such studies are required if a more informed prognosis is to be given to parents of affected children or to the affected persons themselves. Such studies are difficult, however, because a large number of genetic mutations and mutation modifiers are involved (Arveiler, 2015).

V. FUTURE DEVELOPMENTS IN THE FIELD

The future looks good for people with albinism, according to the Director of the Albinism Society of South Africa; she says that the fact that Albinism has become recognized as a disability gives affected people a place in the world, which they did not have previously. Furthermore, the tremendous development of Albinism societies all over Africa, the assertiveness of their leaders, the information, services, and support they have accessed for their members, and the new insights into the condition and advocacy programs they have developed, all hold promise for a better future for affected people. The international support from bodies such as the United Nations and WHO is also encouraging. The publicizing of the human crisis, stimulated by the demand for the body parts of people with albinism by traditional African healers for the making of medicines, has led to an international outcry.

FIGURE 15.1 Sputum collection for DNA extraction. *Photograph courtesy of Dr. J. Kromberg, University of the Witwatersrand, South Africa.*

Articles, such as one entitled "Albinism in Africa: a medical and social emergency" (Brilliant, 2015), have brought albinism into the public arena. These activities have resulted in international drives by nongovernmental organizations to assist in setting up sustainable health services and advocacy programs for people with albinism in several African countries (see Chapter 12).

The WHO has developed a policy of "Health for All" and proposed that the principles of this policy should be integrated into national, regional, and local health policies across all countries. However, there is, apparently, no method of assessing whether these policies are effectively integrated into country health systems and how these systems address human rights issues and social inclusion of people with disabilities. One research group investigated the core concepts of human rights and inclusion of vulnerable groups in the health policies (specifically the disability and rehabilitation policies) of four African countries (Mannan et al., 2013). The core concepts that are particularly relevant to people with albinism include: nondiscrimination; participation; protection from harm; integration; family support; prevention; capacity building; and quality of care. The findings showed that the four countries varied in their application of the core concepts and inclusion of vulnerable groups. The rankings of the countries were: Namibia high, Malawi and Sudan low, and South Africa low, for their rehabilitation policy. If vulnerable groups, such as people with albinism, are going to benefit from these policies in future, the "Health for All" document must be thoroughly integrated into the health policies of all the countries in Africa and effectively put into practice in the local health services at grass roots level.

The new era will bring "precision, or personalized, medicine" (Middleton Price et al., 2016) into the public domain. People with albinism will benefit when the implications of having a particular mutation associated with their albinism will mean a better insight into their prognosis and treatment. Precision medicine for rare diseases will use "knowledge of the underlying mutations and resulting protein defects to design individual therapies that aim to correct the specific problem of the patient" (Middleton-Price et al., 2016, p. 12). When the individual patient's genome is known, their susceptibility to illness can be explained, biomedical discovery can progress, and ways of achieving recovery can be identified. These developments will throw new light onto OCA in ways we cannot yet anticipate.

VI. CONCLUSION

In conclusion, this book has covered the condition of albinism, with particular focus on Africa, as broadly as possible from the viewpoint of the social scientist, medical and clinical researcher, specialist ophthalmologist, dermatologist and optometrist, genetic counselor, psychologist, molecular geneticist, nongovernmental organization workers, and a person with albinism herself. All these contributors have participated in presenting a variety of the issues associated with albinism from their own experience, as well as from reports presented in the literature, so that the reader can develop a wide-ranging view of the condition. However, it is recognized that there "is a great deal of individuality, personal history and accident in the ways in which a genetic disorder is experienced" (Richards, 1999, p. 3), and the diversity of responses from the individual, the family, and the community is great. So, although we will not have been able to cover the field completely comprehensively, it is hoped that the material presented here will be of benefit to the readers and to their future interactions with people with albinism, so that their quality of life is significantly impacted for the better.

References

Aquaron, R., Brilliant, M., 2015. Tentative genotype-phenotype correlations in Cameroonian and Tanzanian OCA2 patients. In: 1st International Workshop on Oculocutaneous Albinism in Sub-Saharan Africa. 24–25 July. Douala, Cameroon, Abstract book, p. 57.

Arveiler, B., 2015. Molecular genetics of albinism. In: 1st International Workshop on Oculocutaneous Albinism in Sub-Saharan Africa. 24–25 July. Douala, Cameroon, Abstract book, p. 27.

Asgari, M.M., Wang, W., Ioannidis, N.M., Itnyre, J., Hoffmann, T., Jorgenson, E., Whittemore, A.S., 2016. Identification of susceptibility loci for cutaneous squamous cell carcinoma. J. Investig. Dermatol. 136, 930–937.

Braathen, S.H., Ingstad, I., 2006. Albinism in Malawi: knowledge and beliefs from an African setting. Disabil. Soc. 21 (6), 599–611.

Brilliant, M., 2015. Albinism in Africa: a medical and social emergency. Int. Health 7, 223–225.

Clarke, A., 2016. Genetics, genomics, and society: challenges and choices. In: Kumar, D., Chadwick, R. (Eds.), Genomics and Society. Ethical, Legal, Cultural and Socioeconomic Implications. Elsevier Academic Press, London, pp. 21–37.

Garrod, A.E., 1908. Inborn errors of metabolism. III. Albinism. Lancet 2, 73–79.

Guillery, R., 1996. Why do albinos and other hypopigmented mutants lack normal binocular vision, and what else is abnormal in their central visual pathway. Eye 10, 217–221.

Hall, J.G., 2013. The role of patient advocacy/parent support groups. S. Afr. Med. J. 103 (12 Suppl. 1), 1020–1022.

Hall, A.M., Krishnamoorthy, L., Orlow, S.J., 2004. 25-hydroxycholesterol acts in the golgi compartment to induce degradation of tyrosinase. Pigment Cell Res. 17, 396–406.

Hong, E.S., Zeeb, H., Repacholi, M.H., 2006. Albinism in Africa as a public health issue. BMC Public Health 6, 212–219.

Jacobs, L.C., Liu, F., Pardo, L.M., Hofman, A., Uitterlinden, A.G., Kayser, M., Nijsten, T., 2015. IRF4, MC1R and TYR genes are risk factors for actinic keratosis independent of skin color. Hum. Mol. Genet. 24, 3296–3303.

Kammer, R., Grant, R., 2014. Albinism and Tanzania: development of a national low vision program. Visibility 8 (2), 2–9.

Kerr, R., 2015. Genetic investigations of the OCA2 gene in the aetiology of oculocutaneous albinism in sub-Saharan Africa. In: 1st International Workshop on Oculocutaneous Albinism in Sub-Saharan Africa. 24–25 July. Douala, Cameroon, Abstract book, p. 33.

Kerr, R., Stevens, G., Manga, P., Salm, S., John, P., Haw, T., Ramsay, M., 2000. Identification of P Gene mutations in individuals with oculocutaneous albinism in sub-Saharan Africa. Hum. Mutat. 15, 166–172.

Kromberg, J.G.R., 1985. A Genetic and Psychosocial Study of Albinism in Southern Africa (Ph.D. thesis). University of the Witwatersrand, Johannesburg, South Africa.

Kromberg, J.G.R., 1987. Albinism in Southern Africa: why so common in blacks? S. Afr. J. Sci. 83, 68.

Kromberg, J.G.R., 1992. Albinism in the South African Negro. IV. Attitudes and the death myth. Birth Defects Orig. Artic. Ser. 28 (1), 159–166.

Kromberg, J.G.R., Jenkins, T., 1984. Albinism in the South African Negro. III. Genetic counseling issues. J. Biosoc. Sci. 16, 99–108.

Kromberg, J.G.R., Krause, A., 2013. Human genetics in Johannesburg, South Africa: past, present and future. S. Afr. Med. J. 103 (12 Suppl. 1), 957–961.

Kromberg, J.G.R., Zwane, E., Jenkins, T., 1987. The response of black mothers to the birth of an albino child. Am. J. Dis. Child. 141, 911–916.

Kromberg, J.G.R., Castle, D., Zwane, E., Jenkins, T., 1989. Albinism and skin cancer in Southern Africa. Clin. Genet. 36, 43–52.

Kromberg, J.G.R., Essop, F.E., Rosendorff, J., August 2015. Prenatal Diagnosis for Oculocutaneous Albinism. Southern African Society for Human Genetics 16th Congress, 16–19 August. Pretoria, Abstract book, p. 28.

Kumar, D., 2016. Socioeconomic outcomes of genomics in the developing world. In: Kumar, D., Chadwick, R. (Eds.), Genomics and Society. Ethical, Legal, Cultural and Socioeconomic Implications. Elsevier Academic Press, London, pp. 239–258.

Landy, D.C., Brinich, M.A., Colten, M.E., Horn, E.J., Terry, S.F., Sharp, R.R., 2012. How disease advocacy organizations participate in clinical research: a survey of genetic organizations. Genet. Med. 14 (2), 223–228.

Levin, M.D., Lu, M.M., Petrenko, N.B., Hawkins, B.J., Gupta, T.H., Lang, D., Buckley, P.T., Jochems, J., Liu, F., Spurney, C.F., Yuan, L.J., Jacobson, J.T., Brown, C.B., Huang, L., Beermann, F., Margulies, K.B., Madesh, M., Eberwine, J.H., Epstein, J.A., Patel, V.V., 2009. Melanocyte-like cells in the heart and pulmonary veins contribute to atrial arrhythmia triggers. J. Clin. Investig. 119, 3420–3436.

Lin, A.E., Terry, S.F., Lerner, B., Anderson, R., Irons, M., 2003. Participation by clinical geneticists in genetic advocacy groups. Am. J. Med. Genet. 119A, 89–92.

Livingstone, D., 1857. Missionary Travels. John Murray, London.

Loewenthal, L.J.A., 1944. Partial albinism and nystagmus in Negroes. Arch. Derm. Syph. 50, 300–301.

Luande, J., Henschke, C., Mohammed, N., 1985. The Tanzanian human albino skin. Natural history. Cancer 55 (8), 1823–1828.

Mannan, H., McVeigh, J., Amin, M., MacLachlan, M., Swartz, L., Munthali, A., Van Roy, G., 2012. Core concepts of human rights and inclusion of vulnerable groups in the Disability and Rehabilitation Policies of Malawi, Namibia, Sudan and South Africa. Jnl. Disabil. Policy Studies 21, 1–15.

Merz, J.F., Magnus, D., Cho, M.K., Caplan, A.L., 2002. Protecting subjects' interests in genetics research. Am. J. Hum. Genet. 70 (4), 965–971.

Middleton-Price, H., Read, A.P., Donnai, D., 2016. Genetics in medicne.3. Precision medicine. In: Galton Institute Occasional Papers Third Series, No 5 London.

Mikami, K., Kent, A., Haddow, G., 2016. The "life costs" of living with rare genetic diseases. In: Kumar, D., Chadwick, R. (Eds.), Genomics and Society. Ethical, Legal, Cultural and Socioeconomic Implications. Elsevier Academic Press, London, pp. 193–206.

Montoliu, L., Gronskov, K., Wei, A.-I., Martinez-Garcia, M., Fernandez, A., Arveiler, B., Morice-Pickard, F., Riazuddin, S., Suzuki, T., Ahmed, Z.M., Rosenberg, T., Li, W., 2013. Increasing the complexity: new genes and new types of albinism. Pigment Cell Melanoma Res. 27 (1), 11–18.

Okoro, A.N., 1975. Albinism in Nigeria. A clinical and social study. Br. J. Dermatol. 92, 485–492.

Onojafe, I.F., Adams, D.R., Simeonov, D.R., Zhang, J., Chan, C.C., Bernardini, I.M., Sergeev, Y.V., Dolinska, M.B., Alur, R.P., Brilliant, M.H., Gahl, W.A., Brooks, B.P., 2011. Nitisinone improves eye and skin pigmentation defects in a mouse model of oculocutaneous albinism. J. Clin. Investig. 121, 3914–3923.

Pearson, K., Nettleship, E., Usher, C.H., 1913. A monograph on albinism in man. Research Memoirs Biometric Series VII, Dulau, London: Drapers Co.

Praetorius, C., Grill, C., Stacey, S.N., Metcalf, A.M., Gorkin, D.U., Robinson, K.C., Van Otterloo, E., Kim, R.S., Bergsteinsdottir, K., Ogmundsdottir, M.H., Magnusdottir, E., Mishra, P.J., Davis, S.R., Guo, T., Zaidi, M.R., Helgason, A.S., Sigurdsson, M.I., Meltzer, P.S., Merlino, G., Petit, V., Larue, L., Loftus, S.K., Adams, D.R., Sobhiafshar, U., Emre, N.C., Pavan, W.J., Cornell, R., Smith, A.G., Mccallion, A.S., Fisher, D.E., Stefansson, K., Sturm, R.A., Steingrimsson, E., 2013. A polymorphism in IRF4 affects human pigmentation through a tyrosinase-dependent MITF/TFAP2A pathway. Cell 155, 1022–1033.

Richards, M., 1999. Daily life and the new genetics: some personal stories. In: Marteau, T., Richards, M. (Eds.), The Troubled Helix. Cambridge University Press, Cambridge, pp. 3–59.

Richter, L., Norris, S., Pettifor, J., Yach, D., Cameron, N., 2007. Cohort profile: Mandela's children: the 1990 birth to twenty study in South Africa. Int. J. Epidemiol. 36 (3), 504–511.

Schieppati, A., Henter, J., Daina, E., Aperia, A., 2008. Why rare diseases are an important medical and social issue. Lancet 371, 2039–2040.

Sorsby, A., 1958. Noah – an albino. Br. Med. J. 2, 1587–1589.

Stevens, G., van Beukering, J., Jenkins, T., Ramsay, M., 1995. An intragenic deletion of the P gene is the common mutation causing tyrosinase-positive oculocutaneous albinism in southern African Negroids, in southern African Negroids. Am. J. Hum. Genet. 56 (3), 586–591.

Terry, S.F., Boyd, C.D., 2001. Researching the biology of PXE: partnering in the process. Am. J. Med. Genet. 106, 177–184.

Witkop, C.J., Quevedo, W.C., Fitzpatrick, T.P., 1983. Albinism and other disorders of pigment metabolism. In: Stanbury, J.B., Wyngaarden, J.B., Frederickson, D.S., Goldstein, J.L., Browne, M.S. (Eds.), The Metabolic Basis of Inherited Disease. McGraw Hill, New York, pp. 301–346.

Wonkam, A., Tekendo, C.N., Sama, D.J., Zambo, H., Dahoun, S., Bena, F., Morris, M.A., 2011. Initiation of a medical genetics service in sub-Saharan Africa: experience of prenatal diagnosis in Cameroon. Eur. J. Med. Genet. 54, e399–e404.

Woolf, C.M., Grant, R.B., 1962. Albinism among the Hopi Indians in Arizona. Am. J. Hum. Genet. 14, 391–400.

World Albinism Alliance, April 23, 2011. Constitution. Item 5.8., p. 3.

Glossary

The following list (compiled and edited by J. Kromberg and P. Manga) includes definitions, phrases, ministatements, acronyms, and abbreviations collated from a number of sources including contributions from authors of the chapters in this book. The publisher and editors do not claim any ownership and deny breach of any copyrights issues arising from inclusion in this glossary. A more complete list of genetic terms can be found in King R.C., Stansfield W.D., Mulligan P.K., 2006. A Dictionary of Genetics, Oxford University Press, Oxford.

ALBINISM-RELATED WEBSITES

Albinism support groups:
Albino Charity Organization: http://albinocharity.org/learn.php
Albinism Fellowship: http://www.albinism.org.uk/
The Albino Foundation: http://albinofoundation.org/
The Albinism Society of South Africa: http://www.albinism.org.za/
Asante Mariamu: http://www.asante-mariamu.org/about-us/our-partners/
National Organization for Albinism and Hypopigmentation: http://www.albinism.org
Salif Keita Global Foundation: www.salifkeita.us.
Standing Voice: http://www.standingvoice.org/
Under The Same Sun: http://www.underthesamesun.com/
Vision For Tomorrow Foundation: https://www.visionfortomorrow.org/
World Albinism Alliance: https://worldalbinism.org/

ACRONYMS

AD Autosomal dominant
arrayCGH Array comparative genome hybridization
AFEA Albinism Foundation of East Africa
AfSHG African Society of Human Genetics
AFZ Albino Foundation of Zambia
AIDS Acquired immune deficiency syndrome
AIMS African Institute for Mathematical Sciences
AK/AKS Actinic keratoses
ALBA Association des Personnes Albinos en Espagne
ANC African National Congress
AR Autosomal Recessive
ASHG American Society of Human Genetics
ASMODISA Association Mondiale pour la Defense des Interets et la Solidarite des Albinos.
ASSA Albinism Society of South Africa
BASF Baden aniline and soda factory
BBC British Broadcasting Corporation
BCC Basal cell carcinoma
BIP Binding immunoglobulin protein

cAMP Cyclic adenosine monophosphate

CEO Chief executive officer

CHRAGG The Commission of Human Rights and Good Governance

CHS Chediak–Higashi syndrome

cm Centimeters

CM Cutaneous melanoma

CRPD Convention of the Rights of People with Disabilities

CS Contrast sensitivity

CVS Chorionic villus sampling

DOPA L-3,4-dihydroxyphenylalanine

DPO Disabled Peoples Organization

EEL Evans Electroselenium Limited

ER Endoplasmic reticulum

GA-SA Genetic Alliance of South Africa

GDP Gross domestic product

GS Griscelli syndrome

H3Africa Human Heredity and Health, Africa

HIV Human immunovirus

HPS Hermansky–Pudlak Syndrome

HRQOL Health-related quality of life

HSP70 Heat shock protein

INGO International nongovernmental organization

IT Information technology

kb Kilobases

KZN Kwazulu Natal

MATP Membrane-associated transporter protein

MC1R Melanocortin-1 receptor

MIEP Malawi Integrated Education Program

MITF Microphthalmia transcription factor

MM Malignant melanoma

MRI Magnetic resonance imaging

MSH Melanocyte-stimulating hormone

NCWPD National Council for Persons with Disabilities

NEI-VFQ National Eye Institute Visual Functioning Questionnaire

NGO Nongovernmental organization

NGS Next-generation sequencing

NIH National Institute of Health, United States

NOAH National Organization for Albinism and Hypopigmentation, United States

NSGC National Society of Genetic Counselors

OA Ocular albinism

OCA Oculocutaneous albinism

OCA2A People with OCA2 without ephelides (freckles)

OCA2AE People with OCA2 with ephelides

OCA2B People with brown albinism

OCA3 People with rufous albinism

OCT Optical coherence tomography

OMIM Online Mendelian Inheritance in Man

PGD Preimplantation genetic diagnosis

PKU Phenylketouria

POMC Proopiomelanocortin

PWA Persons/people with albinism

PWS Prader–Willi syndrome

QOL Quality of life

RDTC Regional Dermatology Training Center

ROS Reactive oxygen species
RPE Retinal pigment epithelium
SA South Africa
SAIDA Southern African Inherited Disorders Association
SCA Sickle cell anemia
SCC Squamous cell carcinoma
SCPP Skin cancer prevention program
SNP Single nucleotide polymorphisms
SPF Sun protection factor
START Strive towards achieving results together
SV Standing Voice
TAAM The Albino Association of Malawi
TAS Tanzania Albinism Society
TV Television
TYR The tyrosinase gene
TYRP1 Tyrosinase-related protein-1
UK United Kingdom
UN United Nations
UNESCO United Nations Educational, Scientific and Cultural Organization
UNICEF United Nations Children's Educational Fund
UNISA University of South Africa
US/USA United States/United States of America
UTSS Under The Same Sun
UV Ultraviolet
VEP Visual evoked potential
VIPs The Vitiligo impact patient scale
WAA World albinism alliance
WHO World Health Organization

TERMS AND PHRASES

Actinic keratoses (AKS) These are common lesions of the skin considered to be the earliest stage in skin cancer development. Further changes in cell growth can turn AKS into a type of skin cancer known as squamous cell carcinoma.
Advocacy Active support of a cause.
Allele An alternative form of a gene at the same chromosomal locus.
Allelic heterogeneity Different alleles for one gene.
Amniocentesis The procedure of collecting a sample of amniotic fluid from the pregnant uterus.
Autosomal Determined by a gene on one of the chromosomes other than the sex chromosome.
Carrier A person who carries an allele for a recessive disease (*see heterozygote*) without the disease phenotype but can pass it on to the next generation; it is also used to denote a female carrying the mutation on one of the two X-chromosomes for an X-linked recessive disorder.
Carrier testing Carried out to determine whether an individual carries one copy of an altered gene for a particular recessive or X-linked disorder.
Chorionic villus sampling The sampling of tissue from the chorionic membrane of the embryo.
Chromosome Subcellular structures that contain and convey the genetic material of an organism.
Clinical genetics services Specialized service that offers diagnosis of genetic conditions, genetic risk information including information about specific genetic conditions and their inheritance and risks to unaffected and unborn family members, genetic testing, and supportive counseling to help the family make decisions and cope better with the genetic condition in their family. The service is offered to all members of a family in which a genetic condition may be present, not just those who have the condition.
Coding DNA (sequence) The portion of a gene that is transcribed into mRNA.

Codon A three-base sequence of DNA or RNA that specifies a single amino acid (the building blocks of proteins).

Complex diseases Diseases such as vitiligo, which are characterized by risk to relatives of an affected individual that is greater than the incidence of the disorder in the population. Multiple factors contribute to determining whether an individual will develop the disorder, including genetic and environmental factors.

Congenital Any trait, condition, or disorder that exists from birth.

Consanguinity Marriage between two individuals having common ancestral parents, commonly between first cousins; an approved practice in some communities who share social, cultural, and religious beliefs. In genetic terms, two such individuals could be heterozygous by descent for an allele expressed as coefficient of relationship, and any offspring could be therefore homozygous by descent for the same allele expressed as coefficient of inbreeding.

Deletion Loss of genetic material (may be applied to a chromosome or a gene).

Disease etiology Any factor or series of related events directly or indirectly causing a disease. For example, the genomics revolution has improved our understanding of disease determinants and provided a deeper understanding of molecular mechanisms and biological processes, including genetic and environmental factors such as sun and chemical exposures.

Disease management A continuous, coordinated healthcare process that seeks to manage and improve the health status of a patient over the entire course of a disease. In the case of albinism, this would include participation of healthcare workers who support parents of an affected child at birth and educate them about the needs of the child, as well as genetic counselors, social workers, vision specialist and dermatologist. The term may also apply to a patient population. Disease management services include disease prevention efforts and patient management.

DNA (deoxyribonucleic acid) The chemical that comprises the genetic material of all cellular organisms.

DNA sequencing Technologies through which the order of base pairs in a DNA molecule can be determined; the Sanger method is commonly used but now replaced by the next-generation sequencing (NGS) techniques.

Dominant An allele (or the trait encoded by that allele), which produces its characteristic phenotype G when present in the heterozygous form.

Empowerment Empowerment is a purposeful process that shifts the perceptions of powerless individuals, groups, and communities and enables them to assume greater capacity and to access control over resources that affect their lives.

Environmental factors May include sun exposure, chemical, dietary factors, infectious agents, physical, and social factors.

Enzyme A protein that acts as a biological catalyst that controls the rate of a biochemical reaction within a cell. In the case of skin pigmentation, the key enzyme is tyrosinase, which catalyzes the conversion of L-tyrosine to L-DOPA and L-DOPA to DOPAquinone.

Ephelides Brown pigmented ragged freckles about 1 cm in diameter.

Epidemiology The study of disease incidence and prevalence.

Epigenetic A term describing nonmutational phenomena, such as methylation and histone modification that modify the expression of a gene; epigenomics is also used in the same context but refers to several genomic regions having similar roles or functions.

Exon The sections of a gene that code for its functional product. Eukaryotic genes may contain many exons interspersed with noncoding introns. An *exon* is represented in the mature mRNA product—the portions of an mRNA molecule that is left after all *introns* are spliced out, which serves as a template for protein synthesis.

Family history An essential tool in clinical genetics. Interpreting the family history can be complicated by many factors, including small families, incomplete or erroneous family histories, consanguinity, variable penetrance, and the current lack of real understanding of the multiple genes involved in polygenic (complex) diseases.

Founder effect Changes in allelic frequencies that occur when a small group is separated from a large population and establishes in a new location.

Frame-shift mutation The addition (duplication) or loss (deletion) of a number of DNA bases that are not a multiple of three, thus causing a shift in the reading frame of the gene. This shift leads to a change in the reading frame of all parts of a gene that are downstream from the mutation leading to a premature stop codon and thus to a truncated protein product.

Gene The fundamental unit of heredity; in molecular terms, a gene comprises a length of DNA that encodes a functional product, which may be a polypeptide (a whole or constituent part of a protein or an enzyme) or a ribonucleic acid. It includes regions that precede and follow the coding region as well as introns and exons. The exact boundaries of a gene are often ill-defined because many promoter and enhancer regions dispersed over many kilobases may influence transcription.

Gene therapy A therapeutic medical procedure that involves either replacing/manipulating or supplementing non-functional genes with healthy genes. Gene therapy can be targeted to somatic (body) or germ (egg and sperm) cells. In somatic gene therapy the recipient's genome is changed, but the change is not passed along to the next generation. In **germline gene therapy**, the parent's egg or sperm cells are changed with the goal of passing on the changes to their offspring.

Genetics Refers to the study of heredity, gene, and genetic material. In contrast to **genomics**, the genetics is traditionally related to lower-throughput, smaller scale emphasis on single genes, rather than on studying structure, organization, and function of the whole genome.

Genetic counseling Genetic counseling has been defined as integrating interpretation of family and medical histories to assess the chance of disease occurrence or recurrence, education about inheritance, testing, management, prevention, resources, and research and counseling to promote informed choices and adaptation to the risk or condition.

Genetic screening Testing a population group to identify a subset of individuals at high risk for having or transmitting a specific genetic disorder.

Genetic test An analysis performed on human DNA, RNA, genes, and/or chromosomes to detect heritable or acquired genotypes. A genetic test also is the analysis of human proteins and certain metabolites, which are predominantly used to detect heritable or acquired genotypes, mutations, or phenotypes.

Genetic testing Strictly refers to testing for a specific chromosomal abnormality or a DNA (nuclear or mitochondrial) mutation already known to exist in a family member. This includes diagnostic testing (postnatal or prenatal), presymptomatic or predictive genetic testing, or for establishing the carrier status. The individual concerned should have been offered full information on all aspects of the genetic test through the process of "nonjudgemental and nondirective" genetic counseling. Most laboratories require a formal fully informed signed consent before carrying out the test. Genetic testing commonly involves DNA/RNA-based tests for single-gene variants, complex genotypes, acquired mutations, and measures of gene expression. Epidemiologic studies are needed to establish clinical validity of each method to establish sensitivity, specificity, and predictive value.

Genome The entire genetic material of a cell or organism.

Genomics The study of the structure and function of the whole genome of an organism. The term is commonly used to refer large-scale, high-throughput molecular analyses of multiple genes, gene products, or regions of genetic material (DNA and RNA). The term also includes the comparative aspect of genomes of various species, their evolution, and how they relate to each other.

Genotype The genetic constitution of an organism; commonly used in reference to a specific disease or trait.

Hemizygous The presence of a gene in only single copy, e.g., not only X-linked genes in males but also autosomal genes where one copy is deleted.

Heritability The proportion of variance of a characteristic due to genetic rather than environmental factors.

Heterogeneity (genetic) The occurrence of a single phenotype due to mutation of more than one gene (usually implies more than one genetic locus).

Heterozygote Refers to a particular allele of a gene at a defined chromosome locus. A heterozygote has a different allelic form of the gene at each of the two homologous chromosomes.

Heterozygosity The presence of different alleles of a gene in one individual or in a population—a measure of genetic diversity.

Homozygote Refers to same allelic form of a gene on each of the two homologous chromosomes.

Interventions Formalized efforts by healthcare providers to promote health, relieve disease symptoms or promote behavior to either improve mental or physical health, discourage behavior that puts health at risk, or reframe the beliefs and attitudes of those with health conditions or risks. Can be surgical, pharmaceutical, educational, or psychological.

Karyotype The chromosome constitution as displayed by a microscopic preparation of dividing chromosomes photographed and arranged in homologous pairs.

Linkage, genetic The occurrence of two genetic loci close enough on the same chromosome to interfere with independent assortment at cell division.

Locus The specific site on a chromosome at which a particular gene or other DNA landmark is located.

Loss-of-function mutation A mutation that decreases the production or function (or both) of the gene product.

Mendelian Following the patterns of inheritance proposed originally by Gregor Mendel. Inherited in a pattern clearly consistent with Mendel's laws: the law of segregation and the law of independent assortment.

Mendelian genetics Classical genetics focuses on monogenic genes with high penetrance. The Mendelian genetics is a true paradigm and is used in discussing the mode of inheritance.

Molecular genetic testing Molecular genetic testing for use in patient diagnosis, management, and genetic counseling; this is increasingly used in presymptomatic (predictive) genetic testing of "at-risk" family members using a previously known disease-causing mutation in the family.

Molecular genetic screening Screening a section of the population known to be at a higher risk to be heterozygous for one of the mutations in the gene for a common autosomal recessive disease, for example, screening for cystic fibrosis in the North European populations and beta thalassemia in the Mediterranean and Middle East population groups.

Morbidity The incidence of illness (number of people ill) in a given population (e.g., of patients). Often used as a healthcare outcome measure.

Mortality A measure of the number of deaths in a given time or the proportion of deaths in a given population (e.g., of patients). Often used "numbers per 1000" as a healthcare outcome measure.

Mutation A heritable alteration in the DNA sequence. The change from the normal to an altered form of a particular gene.

Muti Traditional medicine

Natural selection The process whereby some of the inherited genetic variation within a population will affect the ability of individuals to survive to reproduce (*fitness*).

Newborn screening Performed in newborns in state public health programs to detect certain genetic diseases for which early diagnosis and treatment are available.

Nonsense mutation Substitution of a single DNA base that leads in a stop codon, thus leading to the truncation of a protein.

Nucleotide A subunit of the DNA or RNA molecule. A nucleotide is a base molecule (adenine, cytosine, guanine, and thymine in the case of DNA) linked to a sugar molecule (deoxyribose or ribose) and phosphate groups.

Nystagmus The eyes move horizontally in a more or less rhythmical manner from side to side or in a rotary manner from the original point of fixation.

Ocular albinism A form of albinism causing depigmentation of the eyes only.

Oculocutaneous albinism A form of albinism causing depigmentation of the eyes and skin.

OMIM Acronym for McKusick's Online Mendelian Inheritance in Man, a regularly updated electronic catalog of inherited human disorders and phenotypic traits accessible on NCBI network. Each entry is designated by a number (*OMIM number*).

p The short arm of a chromosome.

Pedigree A diagrammatic representation of a family tree, showing health problems experienced by different members.

Penetrance The proportion of individuals with a particular genetic constitution who show its effect.

Pharmacogenomics A broad term, now increasingly applied to the identification of genes along with all regulatory sequences that could lead to new drug discovery, drug development, and assessing individual's variation in the efficacy or toxicity to a drug.

Phenotype The clinical and/or any other manifestation or expression, such as a biochemical immunological alteration, of a specific gene or genes, environmental factors, or both. The visible expression of the action of a particular gene; the clinical picture resulting from a genetic disorder.

Photophobia Abnormal intolerance of, or sensitivity, to light.

Polygenic/multifactorial Determined by multiple genes and usually also by nongenetic factors.

Polymerase chain reaction A molecular biology technique developed in the mid-1980s through which specific DNA segments may be amplified selectively.

Polymorphism The stable existence of two or more variant allelic forms of a gene within a particular population or among different populations.

Predictive testing Determines the probability that a healthy individual with or without a family history of a certain disease might develop that disease.

Preimplantation genetic diagnosis Used following in vitro fertilization to diagnose a genetic disease or condition in a preimplantation embryo.

Prenatal diagnosis Used to diagnose a genetic disease or condition in a developing fetus.

Prenatal testing Testing for a genetic conditions in a fetus or embryo before it is born.

Proband The affected individual through whom a family with a genetic disorder is ascertained (also called propositus).

Protein A protein is the biological effector molecule encoded by sequences of a gene. A protein molecule consists of one or more polypeptide chains of amino acid subunits. The functional action of a protein depends on its three-dimensional structure, which is determined by its amino acid composition.

Psychological distress Emotions such as anxiety, depression, and worry that are interfering with activities of daily living.

q The long arm of a chromosome.

Quality of life The general well-being of individuals and societies. Health-related quality of life describes quality of life in the health domain.

Recessive An allele that has no phenotypic effect in the heterozygous state; a defined phenotype could be expected in the homozygous or compound heterozygous state.

Recurrence risk The probability of an event recurring.

Risk communication An important aspect of genetic counseling, which involves pedigree analysis, interpretation of the inheritance pattern, genetic risk assessment, and explanation to the family member (or the family).

Risk perception The subjective judgment a person makes about the characteristics and severity of a specific risk.

Screening Carrying out a test or tests, examination(s), or procedure(s) to expose undetected abnormalities, unrecognized (incipient) diseases, or defects: examples are early diagnosis of cancer using mass X-ray mammography for breast cancer and cervical smears for cancer of the cervix.

Service evaluation In health care, an exercise designed to answer the question "what standard does this service achieve?"

Sensitivity (of a screening test) Extent (usually expressed as a percentage) to which a method gives results that are free from false negatives; the fewer the false negatives, the greater the sensitivity. Quantitatively, sensitivity is the proportion of truly diseased persons in the screened population who are identified as diseased by the screening test.

Single-nucleotide polymorphism (SNP) A common variant in the genome sequence; the human genome contains about 10 million SNPs.

Somatic All of the cells in the body, which are not gametes (germline).

Splicing A process, prior to transcription by mRNA, by which *introns* are removed and the exons adjoined.

Stem cell A cell that has the potential to differentiate into a variety of different cell types depending on the environmental stimuli it receives.

Stop codon A codon that leads to the termination of a protein rather than to the addition of an amino acid. The three stop codons are TGA, TAA, and TAG.

Strabismus Squinting

Sun protection factor A factor by which one can judge the effectiveness of a skin cream. People with albinism should use a broad spectrum cream of at least SPF 30 and apply it 30 min prior to exposure.

Syndrome A combination of clinical features forming a recognizable entity.

Ultrasound The use of sound waves to develop an image of the fetus in the uterus. Ultrasound is used to detect abnormalities in the fetus and to guide the needle in the procedure of amniocentesis and chorionic villus sampling.

Ultraviolet Ultraviolet rays of the sun are the most harmful rays. UV intensity is determined by the angle of the sun's rays (not the temperature). People with albinism should avoid outdoor activities between 10.00 a.m. and 4.00 p.m.

Index

Note: 'Page numbers followed by "f" indicate figures, "t" indicate tables.'

Printed in the United States
By Bookmasters